LIGHT-EMITTING DIODES
SECOND EDITION

Revised and fully updated, the Second Edition of this textbook offers a comprehensive explanation of the technology and physics of light-emitting diodes (LEDs) such as infrared, visible-spectrum, ultraviolet, and white LEDs made from III–V semiconductors. The elementary properties of LEDs such as electrical and optical characteristics are reviewed, followed by the analysis of advanced device structures.

With nine additional chapters, the treatment of LEDs has been vastly expanded, including new material on device packaging, reflectors, UV LEDs, III–V nitride materials, solid-state sources for illumination applications, and junction temperature. Radiative and non-radiative recombination dynamics, methods for improving light extraction, high-efficiency and high-power device designs, white-light emitters with wavelength-converting phosphor materials, optical reflectors, and spontaneous recombination in resonant-cavity structures, are discussed in detail. Fields related to solid-state lighting such as human vision, photometry, colorimetry, and color rendering are covered beyond the introductory level provided in the first edition. The applications of infrared and visible-spectrum LEDs in silica fiber, plastic fiber, and free-space communication are also discussed. Semiconductor material data, device design data, and analytic formulae governing LED operation are provided.

With exercises, solutions and illustrative examples, this textbook will be of interest to scientists and engineers working on LEDs, and to graduate students in electrical engineering, applied physics, and materials science.

Additional resources for this title are available online at www.cambridge.org/9780521865388.

E. FRED SCHUBERT received his Ph.D. degree with Honors in Electrical Engineering from University of Stuttgart in 1986 and is currently a Wellfleet Senior Constellation Professor of the Future Chips Constellation at Rensselaer Polytechnic Institute. He has made several pioneering contributions to the field of LEDs, including the first demonstration of the resonant-cavity light-emitting diode (RCLED). He has authored or co-authored more than 200 publications including *Doping in III–V Semiconductors* (Cambridge University Press, 1993, 0-521-01784-X) for which he was awarded the VDE Literature Prize. He is inventor or co-inventor of 28 US Patents and a Fellow of the IEEE, APS, OSA, and SPIE. He received the Senior Research Award of the Humboldt Foundation, the Discover Award for Technological Innovation, the RD 100 Award, and Boston University's Provost Innovation Fund Award.

Note: This book contains many figures in which color adds important information. For this reason, all figures are available in color on the Internet at the following websites: < http://www.cambridge.org/9780521865388> and < http://www.LightEmittingDiodes.org >.

LIGHT-EMITTING DIODES

SECOND EDITION

E. FRED SCHUBERT

*Rensselaer Polytechnic Institute,
Troy, New York*

CAMBRIDGE
UNIVERSITY PRESS

University Printing House, Cambridge CB2 8BS, United Kingdom

Cambridge University Press is part of the University of Cambridge.

It furthers the University's mission by disseminating knowledge in the pursuit of education, learning and research at the highest international levels of excellence.

www.cambridge.org
Information on this title: www.cambridge.org/9780521865388

First edition © E. Fred Schubert 2003
Second edition © E. Fred Schubert 2006

This publication is in copyright. Subject to statutory exception and to the provisions of relevant collective licensing agreements, no reproduction of any part may take place without the written permission of Cambridge University Press.

First edition published 2003
Second edition 2006
Second edition reprinted 2007
6th printing 2014

A catalogue record for this publication is available from the British Library

ISBN 978-0-521-86538-8 Hardback

Cambridge University Press has no responsibility for the persistence or accuracy of URLs for external or third-party internet websites referred to in this publication, and does not guarantee that any content on such websites is, or will remain, accurate or appropriate.

Contents

	Preface	page x
1	**History of light-emitting diodes**	**1**
	1.1 History of SiC LEDs	1
	1.2 History of GaAs and AlGaAs infrared and red LEDs	4
	1.3 History of GaAsP LEDs	8
	1.4 History of GaP and GaAsP LEDs doped with optically active impurities	9
	1.5 History of GaN metal–semiconductor emitters	15
	1.6 History of blue, green, and white LEDs based on GaInN p-n junctions	17
	1.7 History of AlGaInP visible-spectrum LEDs	19
	1.8 LEDs entering new fields of applications	21
	References	23
2	**Radiative and non-radiative recombination**	**27**
	2.1 Radiative electron–hole recombination	27
	2.2 Radiative recombination for low-level excitation	28
	2.3 Radiative recombination for high-level excitation	32
	2.4 Bimolecular rate equations for quantum well structures	33
	2.5 Luminescence decay	33
	2.6 Non-radiative recombination in the bulk	35
	2.7 Non-radiative recombination at surfaces	41
	2.8 Competition between radiative and non-radiative recombination	44
	References	46
3	**Theory of radiative recombination**	**48**
	3.1 Quantum mechanical model of recombination	48
	3.2 The van Roosbroeck–Shockley model	50
	3.3 Temperature and doping dependence of recombination	54
	3.4 The Einstein model	56
	References	57
4	**LED basics: Electrical properties**	**59**
	4.1 Diode current–voltage characteristic	59
	4.2 Deviations from ideal I–V characteristic	63
	4.3 Evaluation of diode parasitic resistances	67
	4.4 Emission energy	68
	4.5 Carrier distribution in p-n homojunctions	69
	4.6 Carrier distribution in p-n heterojunctions	70
	4.7 Effect of heterojunctions on device resistance	71
	4.8 Carrier loss in double heterostructures	75
	4.9 Carrier overflow in double heterostructures	78
	4.10 Electron-blocking layers	81
	4.11 Diode voltage	83
	References	84
5	**LED basics: Optical properties**	**86**
	5.1 Internal, extraction, external, and power efficiencies	86
	5.2 Emission spectrum	87

	5.3	The light escape cone	91
	5.4	Radiation pattern	93
	5.5	The lambertian emission pattern	94
	5.6	Epoxy encapsulants	97
	5.7	Temperature dependence of emission intensity	98
		References	100
6	**Junction and carrier temperatures**	**101**
	6.1	Carrier temperature and high-energy slope of spectrum	101
	6.2	Junction temperature and peak emission wavelength	103
	6.3	Theory of temperature dependence of diode forward voltage	104
	6.4	Measurement of junction temperature using forward voltage	108
	6.5	Constant-current and constant-voltage DC drive circuits	110
		References	112
7	**High internal efficiency designs**	**113**
	7.1	Double heterostructures	113
	7.2	Doping of active region	116
	7.3	p-n junction displacement	118
	7.4	Doping of the confinement regions	119
	7.5	Non-radiative recombination	122
	7.6	Lattice matching	123
		References	126
8	**Design of current flow**	**127**
	8.1	Current-spreading layer	127
	8.2	Theory of current spreading	133
	8.3	Current crowding in LEDs on insulating substrates	136
	8.4	Lateral injection schemes	140
	8.5	Current-blocking layers	142
		References	143
9	**High extraction efficiency structures**	**145**
	9.1	Absorption of below-bandgap light in semiconductors	145
	9.2	Double heterostructures	149
	9.3	Shaping of LED dies	150
	9.4	Textured semiconductor surfaces	154
	9.5	Cross-shaped contacts and other contact geometries	156
	9.6	Transparent substrate technology	157
	9.7	Anti-reflection optical coatings	159
	9.8	Flip-chip packaging	160
		References	161
10	**Reflectors**	**163**
	10.1	Metallic reflectors, reflective contacts, and transparent contacts	164
	10.2	Total internal reflectors	168
	10.3	Distributed Bragg reflectors	170
	10.4	Omnidirectional reflectors	181
	10.5	Specular and diffuse reflectors	184
		References	189

11	**Packaging**	**191**
11.1	Low-power and high-power packages	191
11.2	Protection against electrostatic discharge (ESD)	193
11.3	Thermal resistance of packages	195
11.4	Chemistry of encapsulants	196
11.5	Advanced encapsulant structures	198
	References	199

12	**Visible-spectrum LEDs**	**201**
12.1	The GaAsP, GaP, GaAsP:N, and GaP:N material systems	201
12.2	The AlGaAs/GaAs material system	206
12.3	The AlGaInP/GaAs material system	209
12.4	The GaInN material system	211
12.5	General characteristics of high-brightness LEDs	213
12.6	Optical characteristics of high-brightness LEDs	216
12.7	Electrical characteristics of high-brightness LEDs	218
	References	220

13	**The AlGaInN material system and ultraviolet emitters**	**222**
13.1	The UV spectral range	222
13.2	The AlGaInN bandgap	223
13.3	Polarization effects in III–V nitrides	224
13.4	Doping activation in III–V nitrides	226
13.5	Dislocations in III–V nitrides	227
13.6	UV devices emitting at wavelengths longer than 360 nm	231
13.7	UV devices emitting at wavelengths shorter than 360 nm	233
	References	236

14	**Spontaneous emission from resonant cavities**	**239**
14.1	Modification of spontaneous emission	239
14.2	Fabry–Perot resonators	241
14.3	Optical mode density in a one-dimensional resonator	244
14.4	Spectral emission enhancement	248
14.5	Integrated emission enhancement	249
14.6	Experimental emission enhancement and angular dependence	251
	References	253

15	**Resonant-cavity light-emitting diodes**	**255**
15.1	Introduction and history	255
15.2	RCLED design rules	256
15.3	GaInAs/GaAs RCLEDs emitting at 930 nm	260
15.4	AlGaInP/GaAs RCLEDs emitting at 650 nm	265
15.5	Large-area photon recycling LEDs	268
15.6	Thresholdless lasers	270
15.7	Other RCLED devices	271
15.8	Other novel confined-photon emitters	272
	References	273

16	**Human eye sensitivity and photometric qualities**	**275**
16.1	Light receptors of the human eye	275

16.2	Basic radiometric and photometric units	277
16.3	Eye sensitivity function	280
16.4	Colors of near-monochromatic emitters	283
16.5	Luminous efficacy and luminous efficiency	284
16.6	Brightness and linearity of human vision	286
16.7	Circadian rhythm and circadian sensitivity	287
	References	289
	Appendix 16.1 Photopic eye sensitivity function	290
	Appendix 16.2 Scotopic eye sensitivity function	291

17 Colorimetry — 292

17.1	Color-matching functions and chromaticity diagram	292
17.2	Color purity	300
17.3	LEDs in the chromaticity diagram	301
17.4	Relationship between chromaticity and color	302
	References	302
	Appendix 17.1 Color-matching functions (CIE 1931)	304
	Appendix 17.2 Color-matching functions (CIE 1978)	305

18 Planckian sources and color temperature — 306

18.1	The solar spectrum	306
18.2	The planckian spectrum	307
18.3	Color temperature and correlated color temperature	309
	References	311
	Appendix 18.1 Planckian emitter	312

19 Color mixing and color rendering — 313

19.1	Additive color mixing	313
19.2	Color rendering	315
19.3	Color-rendering index for planckian-locus illumination sources	323
19.4	Color-rendering index for non-planckian-locus illumination sources	324
	References	327
	Appendix 19.1 Reflectivity of test-color samples	328
	Appendix 19.2 Reflectivity of test-color samples	330

20 White-light sources based on LEDs — 332

20.1	Generation of white light with LEDs	332
20.2	Generation of white light by dichromatic sources	333
20.3	Generation of white light by trichromatic sources	338
20.4	Temperature dependence of trichromatic LED-based white-light source	340
20.5	Generation of white light by tetrachromatic and pentachromatic sources	344
	References	344

21 White-light sources based on wavelength converters — 346

21.1	Efficiency of wavelength-converter materials	347
21.2	Wavelength-converter materials	349
21.3	Phosphors	351
21.4	White LEDs based on phosphor converters	353
21.5	Spatial phosphor distributions	355
21.6	UV-pumped phosphor-based white LEDs	357

21.7	White LEDs based on semiconductor converters (PRS-LED)	358
21.8	Calculation of the power ratio of PRS-LED	359
21.9	Calculation of the luminous efficiency of PRS-LED	361
21.10	Spectrum of PRS-LED	363
21.11	White LEDs based on dye converters	364
	References	364

22 Optical communication — 367
22.1	Types of optical fibers	367
22.2	Attenuation in silica and plastic optical fibers	369
22.3	Modal dispersion in fibers	371
22.4	Material dispersion in fibers	372
22.5	Numerical aperture of fibers	374
22.6	Coupling with lenses	376
22.7	Free-space optical communication	379
	References	381

23 Communication LEDs — 382
23.1	LEDs for free-space communication	382
23.2	LEDs for fiber-optic communication	382
23.3	Surface-emitting Burrus-type communication LEDs emitting at 870 nm	383
23.4	Surface-emitting communication LEDs emitting at 1300 nm	384
23.5	Communication LEDs emitting at 650 nm	386
23.6	Edge-emitting superluminescent diodes (SLDs)	388
	References	391

24 LED modulation characteristics — 393
24.1	Rise and fall times, 3 dB frequency, and bandwidth in linear circuit theory	393
24.2	Rise and fall time in the limit of large diode capacitance	395
24.3	Rise and fall time in the limit of small diode capacitance	396
24.4	Voltage dependence of the rise and fall times	397
24.5	Carrier sweep-out of the active region	399
24.6	Current shaping	400
24.7	3 dB frequency	401
24.8	Eye diagram	401
24.9	Carrier lifetime and 3 dB frequency	402
	References	403

Appendix 1	Frequently used symbols	404
Appendix 2	Physical constants	408
Appendix 3	Room temperature properties of III–V arsenides	409
Appendix 4	Room temperature properties of III–V nitrides	410
Appendix 5	Room temperature properties of III–V phosphides	411
Appendix 6	Room temperature properties of Si and Ge	412
Appendix 7	Periodic system of elements (basic version)	413
Appendix 8	Periodic system of elements (detailed version)	414

Index — 415

Preface

During the last four decades, technical progress in the field of light-emitting diodes (LEDs) has been breathtaking. State-of-the art LEDs are small, rugged, reliable, bright, and efficient. At this time, the success story of LEDs still is in full progress. Great technological advances are continuously being made and, as a result, LEDs play an increasingly important role in a myriad of applications. In contrast to many other light sources, LEDs have the potential of converting electricity to light with near-unit efficiency.

LEDs were discovered by accident in 1907 and the first paper on LEDs was published in the same year. LEDs became forgotten only to be re-discovered in the 1920s and again in the 1950s. In the 1960s, three research groups, one working at General Electric Corporation, one at MIT Lincoln Laboratories, and one at IBM Corporation, pursued the demonstration of the semiconductor laser. The first viable LEDs were by-products in this pursuit. LEDs have become devices in their own right and today possibly are the most versatile light sources available to humankind.

The first edition of this book was published in 2003. The second edition of the book is expanded by the discussion of additional technical areas related to LEDs including optical reflectors, the assessment of LED junction temperature, packaging, UV emitters, and LEDs used for general lighting applications. No different than the first edition, the second edition is dedicated to the technology and physics of LEDs. It reviews the electrical and optical fundamentals of LEDs, materials issues, as well as advanced device structures. Recent developments, particularly in the field of III–V nitrides, are also discussed. The book mostly discusses LEDs made from III–V semiconductors. However, much of the science and technology discussed is relevant to other solid-state light emitters such as group-IV, II–VI, and organic emitters. Several application areas of LEDs are discussed in detail, including illumination and communication applications.

Many colleagues and collaborators have provided information not readily available and have given valuable suggestions on the first and second editions of this book. In particular, I am deeply grateful to Enrico Bellotti (Boston University), Jaehee Cho (Samsung Advanced Institute of Technology), George Craford (LumiLeds Corp.), Thomas Gessmann (RPI), Nick Holonyak Jr. (University of Illinois), Jong Kyu Kim (RPI), Mike Krames (LumiLeds Corp.), Shawn Lin (RPI), Ralph Logan (retired, formerly with AT&T Bell Laboratories), Fred Long (Rutgers University), Paul Maruska (Crystal Photonics Corp.), Gerd Mueller (LumiLeds Corp.), Shuji Nakamura (University of California, Santa Barbara), N. Narendran (RPI), Yoshihiro Ohno (National Institute of Standards and Technology), Jacques Pankove (Astralux Corp.), Yongjo Park (Samsung Advanced Institute of Technology), Manfred Pilkuhn (retired, University of Stuttgart, Germany), Hans Rupprecht (retired, formerly with IBM Corp.), Michael Shur (RPI), Cheolsoo Sone (Samsung Advanced Institute of Technology), Klaus Streubel (Osram Opto Semiconductors Corp., Germany), Li-Wei Tu (National Sun Yat-Sen University, Taiwan), Christian Wetzel (RPI), Jerry Woodall (Yale University), and Walter Yao (Advanced Micro Devices Corp.). I would also like to thank my current and former post-doctoral fellows and students for their many significant contributions to this book.

1

History of light-emitting diodes

1.1 History of SiC LEDs

Starting early in the twentieth century, light emission from a solid-state material, caused by an electrical power source, has been reported: a phenomenon termed *electroluminescence*. Because electroluminescence can occur at room temperature, it is fundamentally different from *incandescence* (or heat glow), which is the visible electromagnetic radiation emitted by a material heated to high temperatures, typically >750 °C.

In 1891 Eugene G. Acheson established a commercial process for a new manmade material, silicon carbide (SiC), that he termed "carborundum". The synthesis process was accomplished in an electrically heated high-temperature furnace in which glass (silicon dioxide, SiO_2) and coal (carbon, C) reacted to form SiC according to the chemical reaction (Filsinger and Bourrie, 1990; Jacobson et al., 1992)

$$SiO_2 \text{ (gas)} + C \text{ (solid)} \rightarrow SiO \text{ (gas)} + CO \text{ (gas)}$$

$$SiO \text{ (gas)} + 2C \text{ (solid)} \rightarrow SiC \text{ (solid)} + CO \text{ (gas)}.$$

Just like III–V semiconductors, SiC does not occur naturally. SiC, which has the same crystal symmetry as diamond, has a very high hardness. On the Mohs Hardness Scale, carborundum has a hardness of 9.0, pure SiC a hardness of 9.2–9.5, and diamond a hardness of 10.0. Because of its high hardness and because it could be synthesized in large quantities at low cost, carborundum was a material of choice for the abrasives industry.

In 1907, Henry Joseph Round (1881–1966) checked such SiC crystals for possible use as rectifying solid-state detectors, then called "crystal detectors". Such crystal detectors could be used for the demodulation of radio-frequency signals in early crystal-detector radios. Crystal detectors had been first demonstrated in 1906. Crystal–metal-point-contact structures were frequently tested during these times as a possible alternative to expensive and power-hungry vacuum-tube diodes, which were first demonstrated in 1904 (vacuum-tube diode or "Fleming valve").

Round noticed that light was emitted from a SiC crystallite as used for sandpaper abrasive. The first light-emitting diode (LED) had been born. At that time, the material properties were poorly controlled, and the emission process was not well understood. Nevertheless, he immediately reported his observations to the editors of the journal *Electrical World*. This publication is shown in Fig. 1.1 (Round, 1907).

A Note on Carborundum.

To the Editors of Electrical World:

SIRS:—During an investigation of the unsymmetrical passage of current through a contact of carborundum and other substances a curious phenomenon was noted. On applying a potential of 10 volts between two points on a crystal of carborundum, the crystal gave out a yellowish light. Only one or two specimens could be found which gave a bright glow on such a low voltage, but with 110 volts a large number could be found to glow. In some crystals only edges gave the light and others gave instead of a yellow light green, orange or blue. In all cases tested the glow appears to come from the negative pole. a bright blue-green spark appearing at the positive pole. In a single crystal, if contact is made near the center with the negative pole, and the positive pole is put in contact at any other place, only one section of the crystal will glow and that the same section wherever the positive pole is placed.

There seems to be some connection between the above effect and the e.m.f. produced by a junction of carborundum and another conductor when heated by a direct or alternating current; but the connection may be only secondary as an obvious explanation of the e.m.f. effect is the thermoelectric one. The writer would be glad of references to any published account of an investigation of this or any allied phenomena.

NEW YORK, N. Y.   H. J. ROUND.

Fig. 1.1. Publication reporting on a "curious phenomenon", namely the first observation of electroluminescence from a SiC (carborundum) light-emitting diode. The article indicates that the first LED was a Schottky diode rather than a p-n junction diode (after H. J. Round, *Electrical World* **49**, 309, 1907).

Round was a radio engineer and a prolific inventor who, by the end of his career, held 117 patents. His first light-emitting devices had rectifying current–voltage characteristics; that is, these first devices were light-emitting *diodes* or LEDs. The light was produced by touching the SiC crystal with metal electrodes so that a rectifying Schottky contact was formed. Schottky diodes are usually majority carrier devices. However, minority carriers can be created by either minority-carrier injection under strong forward-bias conditions, or avalanche multiplication under reverse-bias conditions.

The mechanism of light emission in a forward-biased Schottky diode is shown in Fig. 1.2, which displays the band diagram of a metal–semiconductor junction under (a) equilibrium, (b) moderate forward bias, and (c) strong forward bias conditions. The semiconductor is assumed to be of n-type conductivity. Under strong forward bias conditions, minority carriers are injected into the semiconductor by tunneling through the surface potential barrier. Light is emitted upon recombination of the minority carriers with the n-type majority carriers. The voltage required for minority carrier injection in Schottky diodes is larger than typical p-n junction LED voltages. Round (1907) reported operating voltages ranging between 10 and 110 V.

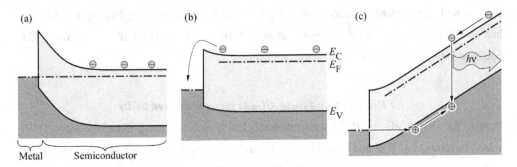

Fig. 1.2. Band diagram of a Schottky diode under (a) equilibrium conditions, (b) forward bias, and (c) strong forward bias. Under strong forward bias, minority carrier injection occurs, making possible near-bandgap light emission.

Light can also be generated in a Schottky diode under reverse-bias conditions through the avalanche effect in which high-energy carriers impact-ionize atoms of the semiconductor. In this process, holes are created in the valence band as well as electrons in the conduction band, which will eventually recombine thereby creating light. Additional light-generating processes in Schottky diodes under reverse-bias conditions have been reported by Eastman *et al.* (1964).

Lossev (1928) reported detailed investigations of the luminescence phenomenon observed with SiC metal–semiconductor rectifiers. The main use of these rectifiers was in solid-state demodulation radio-circuits that did not employ vacuum tubes. Lossev found that luminescence occurred in some diodes when biased in the reverse direction and in some diodes when biased in forward *and* reverse directions. The author was puzzled about the physical origin of the luminescence. He investigated whether light was generated by heat glow (incandescence) by testing the evaporation rate of a droplet of liquid benzene on the luminous sample surface. He found, however, that the benzene evaporated very slowly and correctly concluded that the luminescence was not caused by incandescence. He postulated that the process by which light was produced is "very similar to cold electronic discharge". The author also found that the light could be switched on and off very rapidly, making the device suitable for what he called a "light relay". The pre-1960 history of LEDs was further reviewed by Loebner (1976).

By the late 1960s, SiC films had been prepared by more careful processes (Violin et al., 1969), and p-n junction devices were fabricated, leading to blue light-emitting diodes. Electrical-to-optical power-conversion efficiencies were only 0.005% (Potter et al., 1969). In the ensuing decades, blue SiC LEDs were never substantially improved, because SiC has an indirect bandgap. Although many blue SiC LEDs were actually sold commercially in the early 1990s, they are no longer a viable product. In the end, the best SiC LEDs, emitting blue light at 470 nm, had an efficiency of only 0.03% (Edmond et al., 1993). SiC, the material of the very first LED, could no longer compete with III–V semiconductors.

1.2 History of GaAs and AlGaAs infrared and red LEDs

Prior to the 1950s, SiC and II–VI semiconductors had been well-known materials. Many II–VI semiconductors, e.g. ZnS and CdS, occur in nature. The very first LEDs had been made using SiC and there had been one publication by Destriau (1936) reporting LEDs made of zincblende (ZnS).

The era of III–V compound semiconductors started in the early 1950s when this class of materials was postulated and demonstrated by Heinrich Welker (1952, 1953). The class of III–V compounds had been an unknown substance prior to the 1950s and these compounds do not occur naturally. The novel manmade III–V compounds proved to be optically very active and thus instrumental to modern LED technology.

Bulk growth of the III–V compound GaAs commenced in 1954. In the mid 1950s, large single-crystal boules of GaAs were pulled from the melt. The sliced and polished wafers were used as substrates for the epitaxial growth of p-n junction diode structures, either by vapor-phase epitaxy (VPE) or liquid-phase epitaxy (LPE). Infrared (870–980 nm) LEDs and lasers based on GaAs were first reported in 1962 by groups working at RCA, GE, IBM, and MIT (Hall et al., 1962; Nathan et al., 1962; Pankove and Berkeyheiser, 1962; Pankove and Massoulie, 1962; Quist et al., 1962).

A sustained research effort on GaAs and AlGaAs/GaAs devices started in the early 1960s at the IBM Thomas J. Watson Research Center in Yorktown Heights, located about an hour's drive north of New York City. The IBM team consisted of well-known researchers such as Jerry Woodall, Hans Rupprecht, Manfred Pilkuhn, Marshall Nathan, and others.

Woodall (2000) recalls that his work centered on the bulk crystal growth of GaAs used to fabricate semi-insulating substrates for Ge device epitaxy, and n-type substrates to fabricate injection lasers via Zn diffusion. At that time, the GaAs-based injection laser had already been demonstrated at IBM, GE, and MIT Lincoln Laboratories. Rupprecht's interests were in impurity-diffusion theory and experiment along with experimental investigations into the newly discovered injection laser. Rupprecht was associated with a laser device physics group headed by Marshall Nathan, a co-inventor of the first injection laser (Nathan et al., 1962).

As Woodall developed a technique that lead to state-of-the-art horizontal Bridgman GaAs crystals, Rupprecht fabricated the materials into lasers and characterized them. This collaboration paid off immediately and continuous-wave (cw) operation of GaAs lasers at 77 K was attained (Rupprecht et al., 1963). They then learned of the liquid-phase epitaxy (LPE) technique pioneered by Herb Nelson at the RCA Laboratories in Princeton. The employment of LPE to grow GaAs lasers resulted in the achievement of 300 K lasers with lower threshold current densities than for Zn-diffused lasers. Stimulated by papers found in a literature search, Woodall set out to grow GaAs p-n junction diodes by using Si as an amphoteric dopant, i.e. Si atoms on Ga sites acting as donors and Si atoms on As sites acting as acceptors. This was an interesting idea, as hitherto LPE had been used to grow epilayers with only a single conductivity type.

The LPE conditions to form Si-doped p-n junctions were found very quickly. Si-doped GaAs p-n junctions were formed by cooling a Ga-As-Si melt from 900 to 850 °C to form Si donors and Si acceptors at the two temperatures, respectively. By examining the cross section of the chemically stained epitaxial layer, the lower layer, grown at 900 °C, was identified as being an n-type layer and the upper layer, grown at 850 °C, as a p-type layer. No loss in crystal quality was found in the regions of lower temperature growth. Furthermore, owing to band tailing effects caused by the highly doped, compensated region of the p-n junction, the LED emission occurred at 900–980 nm, far enough below the GaAs band edge (870 nm), so that the bulk GaAs substrate and the GaAs epilayer did not absorb much of the emitted light but acted as a transparent "window layer". LED external quantum efficiencies as high as 6% were attained, a major breakthrough in LED technology (Rupprecht et al., 1966). Rupprecht (2000) stated: "Our demonstration of the highly efficient GaAs LED is a typical example of a discovery made by serendipity." The quantum efficiency of the amphoterically doped GaAs LEDs was five times greater than that of GaAs p-n junctions fabricated by Zn diffusion. Si acceptor levels are deeper than Zn acceptor levels so that the emission from the compensated Si-doped active region occurs at longer wavelengths where GaAs is more transparent.

Being in the LED research business, the IBM group wondered if this doping effect could be extended to a crystal host with visible emission. There were two candidates, GaAsP and AlGaAs.

Whereas Rupprecht tried to do GaAsP epitaxy via LPE, Woodall set up an apparatus for AlGaAs. It was difficult to form good quality GaAsP epilayers by LPE due to the 3.6% lattice mismatch between GaP and GaAs. AlGaAs had problems of its own: "AlGaAs is lousy material" was the pervasive opinion at that time, because, as Woodall (2000) stated, "aluminum loves oxygen". This results in the incorporation of the "luminescence killer" oxygen in AlGaAs; in particular, in the vapor-phase epitaxy (VPE) process, but less so in the LPE process.

Without the support of IBM management, Rupprecht and Woodall "went underground" with their research, conducting the LPE AlGaAs epigrowth experiments after regular working hours and on the weekends. Woodall designed and built a "vertical dipping"-type LPE apparatus, using graphite and alumina melt containers. As an undergraduate student Woodall had majored in metallurgy at MIT and he remembered something about phase diagrams. He made an "intelligent guess" to select the Al concentrations for the LPE melts. He added Si to the melt for the first experiment, saturated the melt and then "dipped" the GaAs substrate while cooling the melt from about 925 to 850 °C. Finally, the substrate and epilayer were withdrawn from the melt, and the apparatus was returned to 300 K. Although no Si-doped p-n junction was observed, a 100 µm thick high-quality layer of AlGaAs had been grown with a bandgap in the red portion of the visible spectrum (Rupprecht *et al.*, 1967, 1968).

Visible-spectrum AlGaAs LEDs were also grown on GaP, a lattice mismatched but transparent substrate. Micrographs of the structure are shown in Fig. 1.3. When AlGaAs was grown on GaP substrates, the thermodynamics of LPE made the initially grown material Al-richer due to the Al distribution coefficient in the melt. As a result, the high-Al-content AlGaAs acts as a transparent window layer for the light emitted from the low-Al-content AlGaAs active region (Woodall *et al.*, 1972).

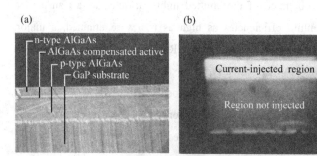

Fig. 1.3. (a) Cross section micrograph of AlGaAs LED grown on transparent GaP substrate. (b) Electroluminescence originating from current-injected region located under stripe-shaped contact viewed through transparent GaP substrate (after Woodall *et al.*, 1972).

Pilkuhn, also an "IBM'er" who had worked with Rupprecht on GaAsP LEDs and lasers (Pilkuhn and Rupprecht, 1965), had built a small battery-powered circuit with an LED emitting visible red light, which he showed to his colleagues and management at IBM (Pilkuhn, 2000). The reactions ranged from "nice but useless" to "great and useful". However, it was soon realized that the latter was true, i.e. that LEDs were extremely useful devices. The first application of the GaAsP LEDs was as indicator lights on circuit boards, where the LEDs indicated the status and proper function of the circuit board. LEDs were also used to show the status of the data processing unit of the classic IBM System 360 mainframe computer shown in Fig. 1.4.

Fig. 1.4. This classic 1964 mainframe computer IBM System 360 used high-voltage gas-discharge lamps to indicate the status of the arithmetic unit. In later models, the lamps were replaced by LEDs. The cabinet-sized 360 had a performance comparable to a current low-end laptop computer.

According to Rostky (1997), the first commercial GaAs LED was offered by the Texas Instruments Corporation in the early 1960s. The LED emitted infrared radiation near 870 nm. The manufacturing quantities of the product were low, probably caused by the high price for one LED, which reportedly was 130 US$.

The *resonant-cavity light-emitting diode* (RCLED) was first demonstrated in the AlGaAs/GaAs materials system (Schubert *et al.*, 1992, 1994). RCLEDs represented a new class of LEDs making use of spontaneous emission enhancement occurring in microscopic optical resonators or *microcavities*. Enhancement is greatest for wavelengths in resonance with the fundamental mode of the cavity. The emission enhancement is mediated by changes in the optical mode density within the cavity. RCLEDs have higher emission intensities along the optical axis of the cavity, which allows for higher coupling efficiencies to optical fibers.

At the present time, infrared GaAs/AlGaAs LEDs are widely used in video and audio remote controls and as sources for local-area communication networks. In addition, red AlGaAs/AlGaAs LEDs are used as high-brightness visible-spectrum LEDs with efficiencies higher than the

GaAsP/GaAs red LEDs but lower than the AlGaInP/GaAs red LEDs.

1.3 History of GaAsP LEDs

The beginning of visible-spectrum LEDs dates back to the year 1962 when Holonyak and Bevacqua (1962) reported on the emission of coherent visible light from GaAsP junctions in the first volume of *Applied Physics Letters*. Although the emission of coherent light was only observed at low temperatures, the devices worked as LEDs and emitted visible light even at room temperature. This publication marks the beginning of viable p-n junction LEDs emitting in the visible wavelength range.

Nick Holonyak Jr., who in 1962 worked at General Electric in Syracuse, New York, and who later joined the University of Illinois, had used vapor-phase epitaxy (VPE) of GaAsP on GaAs substrates. This technique is suited for large-volume growth of wafers in a research as well as a manufacturing environment. Holonyak (2000) recalled that when he first had made these LEDs, he had already envisioned many applications these new devices might have, including indicator lights, seven-segment numeric displays, and alphanumeric displays.

However, despite the early success of the Holonyak group, the goal of demonstrating a semiconductor laser, working at room temperature, remained elusive (Holonyak, 1963, 1964). It remained elusive for good reasons. The GaAsP material system grown on GaAs substrates has several problems which Holonyak and co-workers discovered.

Although excellent electrical junction characteristics were obtained (Holonyak *et al.*, 1963a), the optical properties degraded. When the phosphorus content in GaAsP was about 45–50%, a strong decrease in the LED radiative efficiency was found. These difficulties were attributed to the direct–indirect transition of the bandgap of GaAsP (Holonyak *et al.*, 1963b, 1966; Pilkuhn and Rupprecht, 1964, 1965). It was determined that the 300 K efficiency of GaAsP alloy devices dropped to less than 0.005% when the phosphorus concentration exceeded 44% (Maruska and Pankove, 1967).

The first commercial GaAsP LED was offered by the General Electric (GE) Corporation in the early 1960s. The LED emitted visible radiation in the red part of the spectrum. The manufactured quantities of the product were low, probably due to the high price, which was 260 US$ for a single LED. The product was offered in the Allied Radio catalog, a widely distributed catalog for amateur radio electronics (Rostky, 1997).

The Monsanto Corporation was the first commercial entity to start mass production of LEDs. In 1968, the company had set up a factory, produced low-cost GaAsP LEDs, and sold them to customers. The era of solid-state lamps had started. In the period 1968–1970, sales were skyrocketing, doubling every few months (Rostky, 1997). The Monsanto LEDs were based on GaAsP p-n junctions grown on GaAs substrates emitting in the visible red wavelength range (Herzog et al., 1969; Craford et al., 1972).

Monsanto developed a friendly collaboration with Hewlett-Packard (HP) Corporation, expecting that HP would make LEDs while Monsanto would provide the raw material – GaAsP. In the mid 1960s, Monsanto had sent one of its scientists from Saint Louis, Missouri, to Palo Alto, California, to help HP develop the LED business using Monsanto's GaAsP material. However, HP felt nervous about depending on a single source for the GaAsP material. The informal relationship ended and HP started growing its own GaAsP (Rostky, 1997).

For several years, from the late 1960s to the mid 1970s, the emerging market was in numeric LED displays, driven at first by calculators, then by wristwatches, following Hamilton Watch Corporation's introduction of the Pulsar digital watch in 1972. For a while, the early contenders, Monsanto and HP, took turns leaping into first place with a more advanced multiple-numeric or alphanumeric LED display (Rostky, 1997).

A key technical innovator and manager at Monsanto was M. George Craford, who has made numerous contributions to LEDs including the first demonstration of a yellow LED (Craford et al., 1972). It employed an N-doped GaAsP active region grown on a GaAs substrate. When Monsanto sold off its optoelectronics business in 1979, Craford joined HP and became the key person in the company's LED business. A profile of Craford, who for many years served as Chief Technical Officer, was published by Perry (1995). In 1999, HP spun off parts of its business (including the LED business) into Agilent Corporation which in turn co-founded Lumileds Lighting Corporation in 1999, as a joint venture with Philips Corporation. In 2005, Agilent sold its share of Lumileds to Philips.

It soon became clear that the large lattice mismatch between the GaAs substrate and the GaAsP epilayer resulted in a high density of dislocations (Wolfe et al., 1965; Nuese et al., 1966). As a result, the external efficiency of these LEDs was quite low, about 0.2% or less (Isihamatsu and Okuno, 1989). The importance of the growth conditions and thickness of a *buffer layer* was realized by Nuese et al. (1969) who pointed out that a thick graded GaAsP buffer layer yields improved brightness red LEDs. It is understood today that the thick graded buffer layer reduces the high dislocation density in the GaAsP epitaxial layer originating near the GaAsP-epilayer-to-GaAs-substrate boundary.

The direct–indirect transitions as well as the high dislocation density limit the brightness attainable with GaAsP LEDs. Today this material system is primarily used for low-cost, low-brightness red LEDs for indicator lamp applications.

1.4 History of GaP and GaAsP LEDs doped with optically active impurities

Ralph Logan's and his co-workers' pioneering work on GaP LEDs was done while working at AT&T Bell Laboratories in Murray Hill, New Jersey, in the early 1960s, where they developed a manufacturing process for GaP-based red and green LEDs. At that time, semiconductors had been employed to demonstrate both bipolar and field-effect transistors for switching and amplifying electrical currents. Engineers and scientists back then also began to realize that semiconductors would be perfectly suitable for light-emitting devices.

Logan (2000) recalls that his interest was stimulated by the first reports of GaP p-n junction LEDs by Allen *et al.* (1963) and Grimmeiss and Scholz (1964). Theses devices emitted red light at a useful efficiency so that the light could be clearly seen with the naked eye under ambient daylight conditions. The Grimmeiss–Scholz junctions had been reported to be made by alloying Sn, an n-type dopant, into p-type GaP.

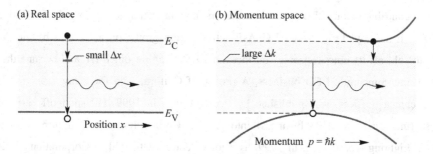

Fig. 1.5. (a) Real-space and (b) momentum-space optical transitions in GaP doped with an optically active impurity such as O or N, emitting in the red and green parts of the spectrum, respectively. GaP LEDs employ the *uncertainty principle* ($\Delta x \, \Delta p \geq h/2\pi$) which predicts that an electron wave function localized in real space is delocalized in momentum space, thereby making momentum-conserving (vertical) transitions possible.

GaP is an indirect-gap semiconductor that does not emit significant amounts of light due to the requirement of momentum conservation in optical transitions. Figure 1.5 shows the band diagram of GaP illustrating that the band extremum points occur at different values in momentum space. If GaP is doped with an **optically active isoelectronic impurity** such as N, strong optical transitions are obtained, as shown by Thomas *et al.* (1965), because the impurity levels are smeared out in momentum space. GaP doped with optically active impurities is a wonderful example of a practical device based on the **Heisenberg uncertainty principle**, which predicts that an impurity with a strongly localized wavefunction in position space (small Δx) will have a delocalized level in momentum space (large Δk), so that optical transitions can occur via the deep-level state.

The growth of GaP was accomplished by using platelets grown from a solution containing Ga and P. The platelets had lateral dimensions of 0.5 cm × 1 cm and were grown to a thickness of about 1 mm. This was the standard initial method of growing GaP and overcame the problem of dealing with the P overpressure required when growing GaP at high temperatures. No one at Bell Laboratories could immediately reproduce the remarkable results reported by Grimmeiss and Scholz. However, a big research effort in electroluminescence was launched at Bell Laboratories as a result.

In the solution growth of the GaP platelets, the dopants used were Zn and O (the latter from Ga_2O_3), but it was not generally realized that ordinary ambient air usually contains enough S to be a good n-dopant. The growth kinetics of the platelets had the intriguing result that the compensated melts generally produced an n-type layer on one platelet surface so that a p-n junction was formed under the GaP platelet surface. This was thought to explain Grimmeiss' results. Logan *et al.* (1967a) published these findings at once.

Logan's research group also reported the first demonstration of reproducible growth of efficient LEDs (Logan *et al.*, 1967b). These junctions were formed by growing an n-type GaP layer onto Zn-O-doped GaP wafers that were polished out of large solution-grown wafers with typical sizes of 2.5 × 2.5 cm^2. Logan *et al.* found that post-growth annealing in the range 400–725 °C could increase the LED efficiency by as much as an order of magnitude, yielding efficiencies exceeding 2%. The annealing was thought to diffuse the Zn to the O atoms, thereby increasing the density of isoelectronic Zn-O centers that mediated the electroluminescence.

At the end of the 1960s, ingots of GaP grown from melts at high temperature and pressure were becoming available, suitable for being cut into real substrates as we know them today. Green LEDs were formed with efficiencies as high as 0.6% by doping the GaP with N isoelectronic impurities (Logan *et al.*, 1968, 1971). The N was added in the form of GaN to the growth melts used to form the p-n junctions. While the external quantum efficiency of green LEDs is less than for the red LEDs, the sensitivity of the human eye to green light is more than 10 times higher than in the red, so the apparent brightness of the LEDs is comparable.

Other research laboratories such as IBM, RCA, and GE also looked into the possibility of making visible-spectrum LEDs that were more efficient than those made of GaAsP. Research on

GaP LEDs was pursued at IBM Corporation's Thomas J. Watson Research Center in Yorktown Heights in New York State. Manfred Pilkuhn and co-workers demonstrated an LPE-grown red GaP LED doped with Zn and O. The picture of a GaP LED with top and bottom contacts is shown in Fig. 1.6. The *IBM Research Journal* proudly reported that "brilliant red light" was emitted from the p-n junction. Note that in the 1960s, monochromatic colors were mostly generated by filtering incandescent light, so that the narrow-spectral-width LED light appeared to the observer as an impressively pure and "brilliant" color.

Fig. 1.6. GaP light-emitting diode grown by liquid-phase epitaxy emitting "brilliant red light" from the Zn- and O-doped p-n junction region (courtesy of Pilkuhn, 2000).

The active regions of Pilkuhn's GaP LEDs were co-doped with acceptors, e.g. Zn acceptors, and donors, e.g. Te, S, or Se donors, so that light was generated predominantly by donor–acceptor pair recombination processes. The energy of the light was below the bandgap of GaP. It was also found that co-doping of GaP with Zn and O resulted in a particularly large wavelength shift so that the emission occurred in the red wavelength range (Foster and Pilkuhn, 1965). Oxygen in GaP is neither donor nor acceptor but was identified as a deep level (see, for example, Pilkuhn, 1981).

Logan and co-workers, and their management team at AT&T Bell Laboratories immediately realized that there were many possible applications for LEDs. Indicator lamps were becoming useful in the telephone business. All such lights used at that time in the USA operated using 110 V. An example is the "Princess" telephone, which was intended to be used in bedrooms – the dial lit up when the phone was picked up from its cradle. The "Princess" was a prestigious telephone and the latest fad in the 1960s but had to be installed near a 110 V outlet. A service call to the local phone company was needed if the bulb burned out. If LEDs were to replace the

110 V light bulbs, the phone line could power the LEDs and a 110 V outlet would no longer be needed. In addition, GaP LEDs had an expected lifetime exceeding 50 years when used in telephones, much longer than 110 V light bulbs, so that this reliability promised substantial cost savings for the "Bell System" or simply "Ma Bell", as the phone company was called at that time.

More important was the multi-line "key telephone". This is the multi-line telephone used in large offices mostly by operators and secretaries where indicator lamps tell which line is being called and busy. To switch the telephone lines and the 110 V indicator lamps, a remote switch was used with dozens of wires to each phone. Installing and servicing these phones was very costly. In present-day telephones, the LED indicator lamps are powered over the phone line. A compatible circuit inside the phone handles the switching of the indicator lamps and phone lines. The savings in telephone manufacturing, installation, and service were impressive.

With the demonstration of the reproducible growth of efficient green N-doped GaP LEDs and red Zn-O-codoped LEDs, both of which were about equally bright and useful, the Bell Laboratories Development Department decided to manufacture the LEDs at its Reading, Pennsylvania, facility.

Telephone lines typically operate with a line voltage of approximately 40 V DC with currents of several milliamperes. The only effect of inserting an LED into this circuit is to reduce the drive voltage by approximately 2 V, a negligible effect, while the efficient LED makes a good indicator lamp. As a result, many phone models were equipped with an illuminated dial pad. Both red and green LED illumination was available, and green was the final choice made by telephone designers. Figure 1.7 shows a 1990 version of the AT&T Trimline telephone – still using GaP:N green LEDs for the dial pad illumination. Red and green LEDs were also incorporated in the multi-line "key telephones".

Should the reader ever be near Murray Hill, New Jersey, visiting the Bell Laboratories Museum, located at 600 Mountain Avenue, should be considered. Many technical artifacts including Logan et al.'s green GaP:N LED are displayed in the museum.

The Monsanto team applied N doping to GaAsP to attain emission in the red, orange, yellow, and green wavelength range (Groves et al., 1971; Craford et al., 1972; for a review see Duke and Holonyak, 1973). Many parameters, such as the emission and the absorption wavelength and the solubility of N in GaAsP and GaP were investigated. A useful growth method was vapor-phase epitaxy (VPE), because it allowed for N-doping in the vicinity of the p-n junction only. This resulted in less absorption of light in the layers adjoining the p-n junction and higher overall

LED efficiencies (Groves *et al.*, 1977, 1978a, 1978b). Today, GaP:N is the primary material for green emitters used in low-brightness applications such as indicator lights.

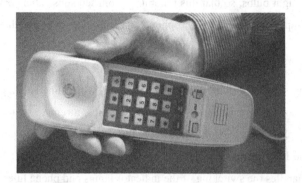

Fig. 1.7. AT&T telephone set ("Trimline" model) with the dial pad illuminated by two green N-doped GaP LEDs. The illuminated dial pad was one of the first applications of green GaP:N LEDs.

The first digital wristwatch with an LED display was released in 1972 by the Hamilton watch company. The watch became an instant furor and only its high price prevented it from becoming widely distributed. The digital Pulsar watch with an integrated calculator was released in 1975 and is shown in Fig. 1.8.

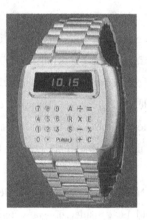

Fig. 1.8. Pulsar calculator watch with LED display released in 1975 by Hamilton Corporation. The first Pulsar LED watch (without calculator) was offered in 1972. It was based on GaInP LEDs (after Seiko, 2004).

Another early application of LEDs was the numeric display in pocket calculators. Figure 1.9 shows two programmable calculators of the mid 1970s, the Texas Instruments Corporation's SR-56 and the Hewlett-Packard Corporation's HP-67. Both used red GaAsP LEDs in the seven-segment numeric display. All calculators using LED displays shared a significant problem: the display could not be read under bright outdoor conditions because the light emitted by the LEDs was simply too dim; furthermore, the power consumption of LED displays was high so that the rechargeable battery running the calculator had to be recharged on a regular basis. Digital wrist watches using LEDs shared the same problem. Liquid crystal displays (LCDs), introduced at the end of the 1970s, had much lower power requirements. Consequently, LCDs totally replaced LED displays in calculators and watches by the beginning of the 1980s.

Fig. 1.9. Programmable pocket calculators Model SR-56 of the Texas Instruments Corporation and Model HP-67 of the Hewlett-Packard Corporation, both first manufactured in 1976. Seven-segment numeric characters composed of GaAsP LEDs were used in the display. The SR-56 came with a "huge" program memory of 100 steps. The HP-67 came with a magnetic card reader and had several freely programmable keys.

1.5 History of GaN metal–semiconductor emitters

In the late 1960s, the Radio Corporation of America (RCA) was one of the premier manufacturers of color televisions, using cathode ray tubes (CRT) with three electron guns to display images. At RCA's central research laboratory in Princeton, New Jersey, James Tietjen had become the director of the Materials Research Division, and he wanted to develop a flat-panel television display that could be hung on the wall like a painting. To create a full color image, the display must contain red, green, and blue pixels. Tietjen realized that red LEDs using GaAsP and green LEDs using GaP:N technology were already available. All that was needed for a flat TV based on LEDs was a bright blue LED.

In May 1968, Tietjen approached a young man in his group, Paul Maruska, and challenged him to find a method for growing single-crystal films of GaN, which Tietjen felt would yield blue LEDs. Maruska had been growing GaAsP red LEDs using the metal-halide vapor-phase epitaxy (MHVPE) approach. He gained much experience with the promises and perils of III–V compounds including phosphorus, a pyrophoric substance. On a day in 1968, phosphorus caused a garbage truck to catch fire on New Jersey's Route 1 in Princeton shortly after picking up some phosphorus-containing laboratory waste at the RCA Laboratories. The driver of the truck decided to immediately return the burning and smoking load to RCA and dump it on the front lawn of the research laboratories (Maruska, 2000).

When Maruska started working on GaN, he first went to the library at Princeton University and thoroughly studied copies of all the old papers on GaN from the 1930s and 1940s (Juza and Hahn, 1938). GaN had been prepared as a powder by reacting ammonia with liquid gallium metal at elevated temperatures. He chose sapphire as the substrate because it is a robust material that is not reactive with ammonia. Unfortunately, he misinterpreted the results of Lorenz and Binkowski (1962), who had reported the decomposition of GaN in vacuum at temperatures as low as 600 °C. All of his early GaN films were grown at temperatures below 600 °C to prevent decomposition, and hence were polycrystalline. Finally in March 1969, Maruska realized that in an ammonia environment, growth rather than decomposition would occur, and thus he raised the furnace temperature to 850 °C, the temperature typically used for GaAs growth. The sapphire appeared to be uncoated, because the GaN film was clear and had a specular surface. He rushed down to the RCA analytical center, and a Laue pattern revealed that the deposit was indeed the first single-crystal film of GaN (Maruska and Tietjen, 1969).

Maruska found that all of the GaN films were n-type without intentional doping. He sought to find a p-type dopant so that he could make a p-n junction. Zinc seemed to be an appropriate acceptor because it worked for GaAs and GaP. With heavy Zn concentrations, GaN films proved to be insulating. But they were never conducting p-type (Maruska, 2000).

During 1969, Jacques Pankove spent a sabbatical year at Berkeley University writing his classic textbook, *Optical Processes in Semiconductors*. When he returned to RCA Laboratories in January 1970, he immediately became interested and strongly involved in the new GaN films. Pankove *et al.* undertook a study of optical absorption and photoluminescence of thin-film GaN (Pankove *et al.*, 1970a, 1970b). The first example of electroluminescence from GaN was announced at RCA in the summer of 1971 (Pankove *et al.* 1971a). The sample consisted of an insulating Zn-doped layer which was contacted with two surface probes, and blue light centered at 475 nm was emitted. Pankove and co-workers then made a device consisting of an undoped n-type region, an insulating Zn-doped layer, and an indium surface contact (Pankove *et al.*, 1971b, 1972). This **metal–insulator–semiconductor** (MIS) **diode** was the first current-injected GaN light emitter, and it emitted green and blue light.

The RCA team speculated that magnesium might be a better choice of p-type dopant than zinc. They began growing Mg-doped GaN films using the MHVPE technique, and in July 1972, obtained blue and violet emission centered at 430 nm as shown in Fig. 1.10 (Maruska *et al.*, 1972). One of these Mg-doped blue light MIS emitters continues to emit light even today. Maruska *et al.* (1973) described these efforts in a paper entitled "Violet luminescence of Mg-doped GaN". Note that the GaN films, even though Mg doped, did not exhibit p-type conductivity, so that the luminescence in these films was probably mediated by minority carrier injection or impact ionization in the high-field insulating regions of the films. Pankove and the RCA team offered a model for the operation of these devices based on impact ionization and Fowler–Nordheim tunneling, because the characteristics were virtually independent of temperature (Pankove and Lampert, 1974; Maruska *et al.*, 1974a, 1974b). Of course, these devices were inefficient, and as a consequence, Tietjen, who had stimulated the work, now terminated it by ordering "stop this garbage" – words that Maruska (2000) still vividly remembers.

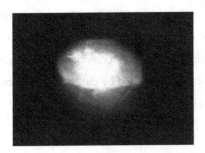

Fig. 1.10. Blue light emission found in 1972 caused by recombining electron–hole pairs created in a highly resistive GaN structure doped with Si and Mg (courtesy of Maruska, 2000).

1.6 History of blue, green, and white LEDs based on GaInN p-n junctions

After the research efforts of Pankove and co-workers had ended, work on GaN virtually ceased. In 1982 only a single paper was published on GaN. However, Isamu Akasaki and co-workers in Nagoya, Japan, refused to give up, and in 1989 they demonstrated the first true p-type doping and p-type conductivity in GaN. The stubborn Mg acceptors were activated by ***electron-beam irradiation*** (Amano *et al.*, 1989). It was later shown that a high-temperature ***post-growth anneal*** of Mg-doped GaN also activates Mg dopants in GaN (Nakamura *et al.*, 1994a). Superlattice doping (Schubert *et al.*, 1996) further enhances the activation efficiency of deep acceptors. These p-type doping breakthroughs opened the door to efficient p-n junction LEDs and laser diodes. Today, Mg-doping of GaN is the basis for all nitride-based LEDs and laser diodes.

Subsequent to the attainment of p-type doping, the first GaN p-n-homojunction LED was reported by Akasaki *et al.* (1992). The LED that emitted light in the ultraviolet (UV) and blue spectral range, was grown on a sapphire substrate. The result was presented at the "GaAs and Related Compounds" conference held in Karuizawa, Japan in 1992. The LED had an efficiency of approximately 1%. This was a surprisingly high value for the highly dislocated GaN material grown on the mismatched sapphire substrate. It was also the first demonstration that nitride LED efficiencies are not affected by dislocations in the same adverse manner as III–V arsenide and phosphide light emitters.

A name closely associated with GaN LEDs and lasers is that of the Nichia Chemical Industries Corporation, Japan. A team of researchers that included Shuji Nakamura and Takashi Mukai has made numerous contributions to the development of GaN growth, LEDs, and lasers. Their contributions included the demonstration of the first viable blue and green GaInN double-heterostructure LED (Nakamura *et al.*, 1993a, 1993b, 1994b) that achieved efficiencies of 10% (Nakamura *et al.*, 1995), and the demonstration of the first pulsed and cw GaInN/GaN current injection *blue laser* operating at room temperature (Nakamura *et al.*, 1996). Initially, a particular design, the two-flow organometallic vapor-phase epitaxy (OMVPE) growth-system design was used (Nakamura *et al.*, 1991). However, the use of two-flow OMVPE at Nichia Corporation has been discontinued (Mukai, 2005). Detailed accounts of the team's contributions were given by Nakamura and Fasol (1997) in the book *The Blue Laser Diode* and by the Nichia Corporation in the booklet *Remarkable Technology* (Nichia, 2004).

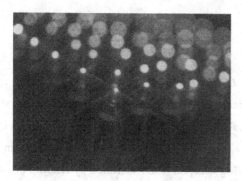

Fig. 1.11. Array of GaInN/GaN blue LEDs manufactured by the Nichia Corporation (after Nakamura and Fasol, 1997).

Blue LEDs made by the Nichia Corporation are shown in Fig. 1.11. A common application of high-brightness GaInN green LEDs is traffic signals as shown in Fig. 1.12. The earlier mentioned GaP:N green LEDs are not suited for this application due to their much lower brightness.

In 1990, when Nakamura entered the field of GaN devices while working for the Nichia Corporation, he was a 36-year-old engineer without a Ph.D., not a single publication, and no conference contribution (Nakamura and Fasol, 1997). At the end of the 1990s, he had become a Professor at the University of California in Santa Barbara and a consultant for the Cree Lighting Corporation, a fierce competitor of Nichia. In 2001, he strongly criticized the Nichia Corporation and Japanese society. In the book entitled *Breakthrough With Anger*, Nakamura (2001) stated, "There is something wrong with this country. Industry and universities are terribly sick."

Fig. 1.12. Green traffic signals are one of the ubiquitous applications of GaInN/GaN green LEDs.

The GaInN material system is also suited for **white LEDs**. There are different approaches to white LEDs, including white LEDs based on phosphor wavelength converters (see, for example, Nakamura and Fasol, 1997) and on semiconductor wavelength converters (Guo *et al.*, 1999). Much progress is expected in the area of white LEDs, since they have the potential to deliver a substantially higher luminous efficiency compared with conventional incandescent and fluorescent light sources. Whereas conventional light sources have typical (demonstrated) luminous efficiencies of 15–100 lm/W, white LEDs have the potential for luminous efficiencies exceeding 300 lm/W.

1.7 History of AlGaInP visible-spectrum LEDs

The AlGaInP material system is suited for high-brightness emission in the red (625 nm), orange (610 nm) and yellow (590 nm) spectral range and today is the dominant material system for high-brightness emitters in that wavelength range. Figure 1.13 shows some of the common signage applications of red and yellow AlGaInP LEDs.

The AlGaInP material system was first developed in Japan for visible-spectrum lasers (Kobayashi *et al.*, 1985; Ohba *et al.*, 1986; Ikeda *et al.*, 1986; Itaya *et al.*, 1990). Efforts started with AlGaInP/GaInP double-heterostructure lasers using $Ga_{0.5}In_{0.5}P$ as the active material, which is lattice matched to GaAs substrates. The bandgap energy of lattice-matched GaInP is approximately 1.9 eV (650 nm), making the material suitable for visible lasers emitting in the red. These lasers are used, for example, in laser pointers and in digital video disc (DVD) players.

Fig. 1.13. Examples of red and amber AlGaInP/GaAs LEDs used in signage applications.

The addition of Al to the GaInP active region allows one to attain shorter emission wavelengths including the orange and yellow spectral region. However, $(Al_xGa_{1-x})_{0.5}In_{0.5}P$ becomes an indirect semiconductor at Al compositions of $x \approx 0.53$, so that the radiative efficiency strongly decreases at wavelengths near and, in particular, below 600 nm. Consequently, AlGaInP is not suited for high-efficiency emission at wavelengths below 570 nm.

Subsequent to the AlGaInP laser development that occurred in the early 1980s, AlGaInP LED development started at the end of the 1980s (Kuo *et al.*, 1990; Fletcher *et al.*, 1991; Sugawara *et al.*, 1991). In contrast to the AlGaInP laser structures, the LED structures typically employ current-spreading layers so that the entire p-n junction plane of the LED chip lights up and not just the region below the top ohmic contact. Further improvements were attained by using multiple quantum well (MQW) active regions (Huang and Chen, 1997), coherently strained MQW active regions (Chang and Chang, 1998a, 1998b), distributed Bragg reflectors (Huang and Chen, 1997; Chang *et al.*, 1997), transparent GaP substrate technology (Kish and Fletcher, 1997), and chip-shaping (Krames *et al.*, 1999). Comprehensive reviews of the AlGaInP material system and AlGaInP LEDs were published by Stringfellow and Craford (1997), Mueller (2000), and Krames *et al.*, 2002).

1.8 LEDs entering new fields of applications

As devices with higher power capabilities have become available, new application areas emerge constantly. Figure 1.14 shows the use of LEDs integrated into medical goggles worn by a surgeon during an operation (Shimada et al., 2003). The LED-based light source promises substantial weight savings and fulfills the stringent requirements of high-quality color rendition required during medical operations.

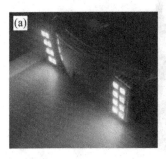

Fig. 1.14. (a) First goggle with integrated white LEDs used for (b) illumination during medical surgery (after Shimada et al., 2001; Shimada et al., 2003)

LED-based automotive headlights were first introduced by Audi Corporation in 2004 using Lumileds Lighting's Luxeon devices. The car is shown in Fig. 1.15.

Fig. 1.15. First automotive daytime running lights based on LEDs.

The use of LEDs in large-scale display and signage applications continues unabated. A seven-story high display and an animated pedestrian traffic signal are shown in Figs. 1.16 and 1.17.

LEDs have also entered the domain of the arts. Figure 1.18 shows the accent-lighted Stone Bridge across the Danube River located in Regensburg, Germany. A line of 21 900 red light-emitting diodes stretches brightly across the bridge, thereby emphasising the link between the two banks of the river. Each LED stands for one month since the city was founded in the year AD 179 (Osram Opto Semiconductors, 2004). Figure 1.19 shows the accent-lighted entrance of the Science Based Industrial Park located in Hsin Chu, Taiwan, an industrial park housing numerous companies active in semiconductor technology.

Fig. 1.16. LED display consisting of 18 million LEDs covering front of building, located in New York City.

Fig. 1.17. Pedestrian sign indicating number of seconds left to cross street, located in Taipei, Taiwan.

Fig. 1.18. Stone Bridge located in Regensburg, Germany, illuminated by LEDs (after *Focus Magazine*, 2004).

Fig. 1.19. Artistic accent lighting at the main gate of the Science Based Industrial Park located in Hsin Chu, Taiwan (courtesy of K. R. Wang and L.-W. Tu, 2005).

References

Akasaki I., Amano H., Itoh K., Koide N., and Manabe K. "GaN based UV/blue light-emitting devices" GaAs and Related Compounds conference, *Inst. Phys. Conf. Ser.* **129**, 851 (1992)

Allen J. W., Moncaster M. E., and Starkiewicz J. "Electroluminescent devices using carrier injection in gallium phosphide" *Solid State Electronics* **6**, 95 (1963)

Amano H., Kito M., Hiramatsu K., Akasaki I. "P-type conduction in Mg-doped GaN treated with low-energy electron beam irradiation (LEEBI)" *Jpn. J. Appl. Phys.* **28**, L2112 (1989)

Chang S. J., Chang C. S., Su Y. K., Chang P. T., Wu Y. R., Huang K. H. and Chen T. P. "AlGaInP multiquantum well light-emitting diodes" *IEE Proc. Optoelectronics* **144**, 1 (1997)

Chang S. J. and Chang C. S. "AlGaInP–GaInP compressively strained multiquantum well light-emitting diodes for polymer fiber applications" *IEEE Photonics Technol. Lett.* **10**, 772 (1998a)

Chang S. J. and Chang C. S. "650 nm AlGaInP/GaInP compressively strained multi-quantum well light-emitting diodes" *Jpn. J. Appl. Phys* **37**, L653 (1998b)

Craford M. G., Shaw R. W., Herzog A. H., and Groves W. O. "Radiative recombination mechanisms in GaAsP diodes with and without nitrogen doping" *J. Appl. Phys.* **43**, 4075 (1972)

Destriau G. "Scintillations of zinc sulfides with alpha-rays" *J. Chimie Physique* **33**, 587 (1936)

Duke C. B. and Holonyak Jr. N. "Advances in light-emitting diodes" *Physics Today*, December issue, p. 23 (1973)

Eastman P. C., Haering R. R., and Barnes P. A. "Injection electroluminescence in metal–semiconductor tunnel diodes" *Solid-State Electronics* **7**, 879 (1964)

Edmond J. A., Kong H. S., and Carter Jr. C. H. "Blue LEDs, UV photodiodes and high-temperature rectifiers in 6 H-SiC" *Physica B* **185**, 453 (1993)

Filsinger D. H. and Bourrie D. B. "Silica to silicon: Key carbothermic reactions and kinetics" *J. Amer. Ceram. Soc.* **73**, 1726 (1990)

Fletcher R. M., Kuo C., Osentowski T. D., and Robbins V. M. "Light-emitting diode with an electrically conductive window" US Patent 5,008,718 (1991)

Foster L. M. and Pilkuhn M. "Electroluminescence near bandgap in GaP containing shallow donor and acceptor levels" *Appl. Phys. Lett.* **7**, 65 (1965)

Grimmeiss H. G. and Scholz H. J. "Efficiency of recombination radiation in GaP" *Phys. Lett.* **8**, 233 (1964)

Groves W. O., Herzog A. H., and Craford M. G. "The effect of nitrogen doping on $GaAs_{1-x}P_x$ electroluminescent diodes" *Appl. Phys. Lett.* **19**, 184 (1971)

Groves W. O. and Epstein A. S. "Epitaxial deposition of III–V compounds containing isoelectronic impurities" US Patent 4,001,056 issued Jan. 4 (1977)

Groves W. O., Herzog A. H., and Craford M. G. "Process for the preparation of electroluminescent III–V materials containing isoelectronic impurities" US Patent Re. 29,648 issued May 30 (1978a)

Groves W. O., Herzog A. H., and Craford M. G. "GaAsP electroluminescent device doped with isoelectronic impurities" US Patent Re. 29,845 issued Nov. 21 (1978b)

Guo X., Graff J. W., and Schubert E. F. "Photon recycling semiconductor light-emitting diode" *IEDM Technical Digest*, **IEDM-99**, 600 (Dec. 1999)

Hall R. N., Fenner G. E., Kingsley J. D., Soltys T. J., and Carlson R. O. "Coherent light emission from GaAs junctions" *Phys. Rev. Lett.* **9**, 366 (1962)

Herzog A. H., Groves W. O., and Craford M. G. "Electroluminescence of diffused $GaAs_{1-x}P_x$ diodes with low donor concentrations" *J. Appl. Phys.* **40**, 1830 (1969)

Holonyak Jr. N. "Active region in visible-light diode laser" *Electronics* **36**, 35 (1963)

Holonyak Jr. N. "Laser action in Ga(AsP) and GaAs" *Proc. IEEE* **52**, 104 (1964)

Holonyak Jr. N., personal communication (2000)

Holonyak Jr. N. and Bevacqua S. F. "Coherent (visible) light emission from $Ga(As_{1-x}P_x)$ junctions" *Appl. Phys. Lett.* **1**, 82 (1962)

Holonyak Jr. N., Bevacqua S. F., Bielan C. V., Carranti F. A., Hess B. G., and Lubowski S. J. "Electrical properties of Ga(AsP) p-n junctions" *Proc. IEEE* **51**, 364 (1963a)

Holonyak Jr. N., Bevacqua S. F., Bielan C. V., and Lubowski S. J. "The "direct–indirect" transition in $Ga(As_{1-x}P_x)$ p-n junctions" *Appl. Phys. Lett.* **3**, 47 (1963b)

Holonyak Jr. N., Nuese C. J., Sirkis M. D., and Stillman G. E. "Effect of donor impurities on the direct–

indirect transition in Ga(AsP)" *Appl. Phys. Lett.* **8**, 83 (1966)
Huang K.-H. and Chen T.-P. "Light-emitting diode structure" US Patent 5,661,742 (1997)
Ikeda M., Nakano K., Mori Y., Kaneko K. and Watanabe N. "MOCVD growth of AlGaInP at atmospheric pressure using triethylmetals and phosphine" *J. Cryst. Growth* **77**, 380 (1986)
Isihamatsu S. and Okuno Y. "High efficiency GaAlAs LED" *Optoelectronics – Dev. Technol.* **4**, 21 (1989)
Itaya K., Ishikawa M., and Uematsu Y. "636 nm room temperature cw operation by heterobarrier blocking structure InGaAlP laser diodes" *Electronics Lett.* **26**, 839 (1990)
Jacobson N. S., Lee K. N., and Fox D. S. "Reactions of SiC and SiO_2 at elevated temperatures" *J. Amer. Ceram. Soc.* **75**, 1603 (1992)
Juza R. and Hahn H. "On the crystal structure of Cu_3N, GaN and InN (translated from German)" *Zeitschrift fuer anorganische und allgemeine Chemie* **239**, 282 (1938)
Kish F. A. and Fletcher R. M. "AlGaInP light-emitting diodes" in *High Brightness Light-Emitting Diodes* edited by G. B. Stringfellow and M. G. Craford, Semiconductors and Semimetals **48**, p. 149 (Academic Press, San Diego, 1997)
Kobayashi K., Kawata S., Gomyo A., Hino I. and Suzuki T. "Room-temperature cw operation of AlGaInP double-heterostructure visible lasers" *Electron. Lett.* **21**, 931 (1985)
Krames M. R., Ochiai-Holcomb M., Höfler G. E., Carter-Coman C., Chen E. I., Tan I.-H., Grillot P., Gardner N. F., Chui H. C., Huang J.-W., Stockman S. A., Kish F. A., Craford M. G.. Tan T. S., Kocot C. P., Hueschen M., Posselt J., Loh B., Sasser G., and Collins D. "High-power truncated-inverted-pyramid $(Al_xGa_{1-x})_{0.5}In_{0.5}P$/GaP light-emitting diodes exhibiting > 50% external quantum efficiency" *Appl. Phys. Lett.* **75**, 2365 (1999)
Krames M. R., Amano H., Brown J. J., and Heremans P. L. "High-efficiency light-emitting diodes" Special Issue of *IEEE J. Sel. Top. Quantum Electron.* **8**, 185 (2002)
Kuo C. P., Fletcher R. M., Osentowski T. D., Lardizabel M. C., Craford M. G., and Robbins V. M. "High performance AlGaInP visible light-emitting diodes" *Appl. Phys. Lett.* **57**, 2937 (1990)
Loebner E. E. "Subhistories of the light-emitting diode" *IEEE Trans. Electron Devices* **ED-23**, 675 (1976)
Logan R. A., personal communication (2000)
Logan R. A., White H. G., and Trumbore F. A. "P-n junctions in compensated solution grown GaP" *J. Appl. Phys.* **38**, 2500 (1967a)
Logan R. A., White H. G., and Trumbore F. A. "P-n junctions in GaP with external electroluminescence efficiencies ~ 2% at 25 °C" *Appl. Phys. Lett.* **10**, 206 (1967b)
Logan R. A., White H. G., Wiegmann W. "Efficient green electroluminescence in nitrogen-doped GaP p-n junctions" *Appl. Phys. Lett.* **13**, 139 (1968)
Logan R. A., White H. G., and Wiegmann W. "Efficient green electroluminescent junctions in GaP" *Solid State Electronics* **14**, 55 (1971)
Lorenz M. R. and Binkowski B. B. "Preparation, stability, and luminescnce of gallium nitride" *J. Electrochem. Soc.* **109**, 24 (1962)
Lossev O. V. "Luminous carborundum detector and detection effect and oscillations with crystals" *Philosophical Magazine* **6**, 1024 (1928)
Maruska H. P., personal communication. The photograph of a GaN MIS LED is gratefully acknowledged (2000)
Maruska H. P. and Pankove J. I. "Efficiency of $GaAs_{1-x}P_x$ electroluminescent diodes" *Solid State Electronics* **10**, 917 (1967)
Maruska H. P. and Tietjen J. J. "The preparation and properties of vapour-deposited single-crystalline GaN" *Appl. Phys. Lett.* **15**, 327 (1969)
Maruska H. P., Rhines W. C., Stevenson D. A. "Preparation of Mg-doped GaN diodes exhibiting violet electroluminescence" *Mat. Res. Bull.* **7**, 777 (1972)
Maruska H. P., Stevenson D. A., Pankove J. I. "Violet luminescence of Mg-doped GaN (light-emitting diode properties)" *Appl. Phys. Lett.* **22**, 303 (1973)
Maruska H. P., Anderson L. J., Stevenson D. A. "Microstructural observations on gallium nitride light-emitting diodes" *J. Electrochem. Soc.* **121**, 1202 (1974a)
Maruska H. P. and Stevenson D. A. "Mechanism of light production in metal–insulator–semiconductor

diodes; GaN:Mg violet light-emitting diodes" *Solid State Electronics* **17**, 1171 (1974b)
Mueller G. (Editor) *Electroluminescence I* Semiconductors and Semimetals **64** (Academic Press, San Diego, 2000)
Mukai, Takashi, personal communication (2005)
Nakamura S., Senoh M., and Mukai T. "Highly p-type Mg doped GaN films grown with GaN buffer layers" *Jpn. J. Appl. Phys.* **30**, L 1708 (1991)
Nakamura S., Senoh M., and Mukai T. "P-GaN/n-InGaN/n-GaN double-heterostructure blue-light-emitting diodes" *Jpn. J. Appl. Phys.* **32**, L8 (1993a)
Nakamura S., Senoh M., and Mukai T. "High-power InGaN/GaN double-heterostructure violet light-emitting diodes" *Appl. Phys. Lett.* **62**, 2390 (1993b)
Nakamura S., Iwasa N., and Senoh M. "Method of manufacturing p-type compound semiconductor" US Patent 5,306,662 (1994a)
Nakamura S., Mukai T., and Senoh M. "Candela-class high-brightness InGaN/AlGaN double-heterostructure blue-light-emitting diodes" *Appl. Phys. Lett.* **64**, 1687 (1994b)
Nakamura S., Senoh M., Iwasa N., Nagahama S. "High-brightness InGaN blue, green, and yellow light-emitting diodes with quantum well structures" *Jpn. J. Appl. Phys.* **34**, L797 (1995)
Nakamura S., Senoh M., Nagahama S., Iwasa N., Yamada T., Matsushita T., Sugimoto Y., and Kiyoku H. "Room-temperature continuous-wave operation of InGaN multi-quantum-well structure laser diodes" *Appl. Phys. Lett.* **69**, 4056 (1996)
Nakamura S. and Fasol G. *The Blue Laser Diode* (Springer, Berlin, 1997)
Nakamura S. *Breakthrough with Anger* (Shueisha, Tokyo, 2001). See also *Compound Semiconductors* **7**, No. 7, 25 (Aug. 2001) and **7**, No. 9, 15 (Oct. 2001)
Nathan M. I., Dumke W. P., Burns G., Dill Jr. F. H., and Lasher G. J. "Stimulated emission of radiation from GaAs p-n junctions" *Appl. Phys. Lett.* **1**, 62 (1962)
Nichia Corporation *Remarkable Technology* edited by I. Matsushita and E. Shibata (Nichia Company, Tokushima, Japan, 2004)
Nuese C. J., Stillman G. E., Sirkis M. D., and Holonyak Jr. N. "Gallium arsenide-phosphide: crystal, diffusion, and laser properties" *Solid State Electronics* **9**, 735 (1966)
Nuese C. J., Tietjen J. J., Gannon J. J., and Gossenberger H. F. "Optimization of electroluminescent efficiencies for vapor-grown GaAsP diodes" *J. Electrochem. Soc.: Solid State Sci.* **116**, 248 (1969)
Ohba Y., Ishikawa M., Sugawara H., Yamamoto T., and Nakanisi T. "Growth of high-quality InGaAlP epilayers by MOCVD using methyl metalorganics and their application to visible semiconductor lasers" *J. Cryst. Growth* **77**, 374 (1986)
Osram Opto Semiconductors "LEDs bridge time and space" press release, June 17 (2004)
Pankove J. I. and Berkeyheiser J. E. "A light source modulated at microwave frequencies" *Proc. IRE*, **50**, 1976 (1962)
Pankove J. I. and Massoulie M. J. "Injection luminescence from GaAs" *Bull. Am. Phys. Soc.* **7**, 88 (1962)
Pankove J. I., Berkeyheiser J. E., Maruska H. P., and Wittke J. "Luminescent properties of GaN" *Solid State Commun.* **8** 1051 (1970a)
Pankove J. I., Maruska H. P., and Berkeyheiser J. E. "Optical absorption of GaN" *Appl. Phys. Lett.* **17**, 197 (1970b)
Pankove J. I., Miller E. A., Richman D., and Berkeyheiser J. E. "Electroluminescence in GaN" *J. Luminescence* **4**, 63 (1971a)
Pankove J. I., Miller E. A., and Berkeyheiser J. E. "GaN electroluminescent diodes" *RCA Review* **32**, 383 (1971b)
Pankove J. I., Miller E. A., and Berkeyheiser J. E. "GaN blue light-emitting diodes" *J. Luminescence* **5**, 84 (1972)
Pankove J. I. and Lampert M. A. "Model for electroluminescence in GaN" *Phys. Rev. Lett.* **33**, 361 (1974)
Perry T. S. "M. George Craford" *IEEE Spectrum*, February issue, p. 52 (1995)
Pilkuhn M. H. and Rupprecht H. "Light emission from $GaAs_xP_{1-x}$ diodes" *Trans. Metallurgical Soc. AIME* **230**, 282 (1964)
Pilkuhn M. H. and Rupprecht H. "Electroluminescence and lasing action in $GaAs_xP_{1-x}$" *J. Appl. Phys.* **36**, 684 (1965)
Pilkuhn M. H. "Light-emitting diodes" in *Handbook of Semiconductors* edited by T. S. Moss, **4**, edited by

C. Hilsum, p. 539 (1981)
Pilkuhn M. H., personal communication. The photograph of GaP:Zn-O LED is gratefully acknowledged (2000)
Potter R. M., Blank J. M., and Addamiano A. "Silicon carbide light-emitting diodes" *J. Appl. Phys.* **40**, 2253 (1969)
Quist T. M., Rediker R. H., Keyes R. J., Krag W. E., Lax B., McWhorter A. L. and Zeigler H. J. "Semiconductor maser of GaAs" *Appl. Phys. Lett.* **1**, 91 (1962)
Rostky G. "LEDs cast Monsanto in unfamiliar role" *Electronic Engineering Times* (*EETimes*) on the internet, see http://eetimes.com/anniversary/designclassics/monsanto.html, Issue 944, March 10 (1997)
Round H. J. "A note on carborundum" *Electrical World*, **19**, 309 (1907)
Rupprecht H., Pilkuhn M., and Woodall J. M. "Continuous stimulated emission from GaAs diodes at 77 K" (First report of 77 K cw laser) *Proc. IEEE* **51**, 1243 (1963)
Rupprecht H., Woodall J. M., Konnerth K., and Pettit D. G. "Efficient electro-luminescence from GaAs diodes at 300 K" *Appl. Phys. Lett.* **9**, 221 (1966)
Rupprecht H., Woodall J. M., and Pettit G. D. "Efficient visible electroluminescence at 300 K from AlGaAs pn junctions grown by liquid phase epitaxy" *Appl. Phys. Lett.* **11**, 81 (1967).
Rupprecht H., Woodall J. M., Pettit G. D., Crowe J. W., and Quinn H. F. "Stimulated emission from AlGaAs diodes at 77 K" *Quantum Electron.* **4**, 35 (1968)
Rupprecht H., personal communication (2000)
Schubert E. F., Wang Y.-H., Cho A. Y., Tu L.-W., and Zydzik G. J. "Resonant cavity light-emitting diode" *Appl. Phys. Lett.* **60**, 921 (1992)
Schubert E. F., Hunt N. E. J., Micovic M., Malik R. J., Sivco D. L., Cho A. Y., and Zydzik G. J. "Highly efficient light-emitting diodes with microcavities" *Science* **265**, 943 (1994)
Schubert E. F., Grieshaber W., and Goepfert I. D. "Enhancement of deep acceptor activation in semiconductors by superlattice doping" *Appl. Phys. Lett.* **69**, 3737 (1996)
Seiko Corporation "Pulsar - it's all in the details" www.pulsarwatches-europe.com (2004)
Shimada J., Kawakami Y., and Fujita S. "Medical lighting composed of LEDs arrays for surgical operation" *SPIE Photonics West: Light Emitting Diodes: Research, Manufacturing, and Applications*, **4278**, 165, San Jose, January 24–25 (2001)
Shimada J., Kawakami Y., and Fujita S. "Development of lighting goggle with power white LED modules" *SPIE Photonics West: Light Emitting Diodes: Research, Manufacturing, and Applications*, **4996**, 174, San Jose, January 28–29 (2003)
Stringfellow G. B. and Craford M. G. (Editors) *High Brightness Light-Emitting Diodes* Semiconductors and Semimetals **48** (Academic Press, San Diego, 1997)
Sugawara H., Ishikawa M., Kokubun Y., Nishikawa Y., and Naritsuka S. "Semiconductor light-emitting device" US Patent 5,048,035 (1991)
Thomas D. G., Hopfield J. J., and Frosch C. J. "Isoelectronic traps due to nitrogen in gallium phosphide" *Phys. Rev. Lett.* **15**, 857 (1965)
Violin E. E., Kalnin A. A., Pasynkov V. V., Tairov Y. M., and Yaskov D. A. "Silicon Carbide – 1968" *2nd International Conference on Silicon Carbide*, published as a special issue of the *Materials Research Bulletin*, p. 231 (1969)
Welker H. "On new semiconducting compounds (translated from German)" *Zeitschrift für Naturforschung* **7a**, 744 (1952)
Welker H. "On new semiconducting compounds II (translated from German)" *Zeitschrift für Naturforschung* **8a**, 248 (1953)
Wolfe C. M., Nuese C. J., and Holonyak Jr. N. "Growth and dislocation structure of single-crystal Ga(AsP)" *J. Appl. Phys.* **36**, 3790 (1965)
Woodall J. M., personal communication (2000)
Woodall J. M., Potemski R. M., Blum S. E., and Lynch R. "AlGaAs LED structures grown on GaP substrates" *Appl. Phys. Lett.* **20**, 375 (1972)

2

Radiative and non-radiative recombination

Electrons and holes in semiconductors recombine either radiatively, i.e. accompanied by the emission of a photon, or non-radiatively. In light-emitting devices, the former is clearly the preferred process. However, non-radiative recombination can, under practical conditions, never be reduced to zero. Thus, there is competition between radiative and non-radiative recombination. Maximization of the radiative process and minimization of the non-radiative process can be attained in a number of ways which will be discussed below.

2.1 Radiative electron–hole recombination

Any undoped or doped semiconductor has two types of free carriers, electrons and holes. Under equilibrium conditions, i.e. without external stimuli such as light or current, the law of mass action teaches that the product of the electron and hole concentrations is, at a given temperature, a constant, i.e.

$$\boxed{n_0 \, p_0 = n_i^2} \qquad (2.1)$$

where n_0 and p_0 are the equilibrium electron and hole concentrations and n_i is the intrinsic carrier concentration. The validity of the law of mass action is limited to non-degenerately doped semiconductors (see, for example, Schubert, 1993).

Excess carriers in semiconductors can be generated either by absorption of light or by an injection current. The total carrier concentration is then given by the sum of equilibrium and excess carrier concentrations, i.e.

$$n = n_0 + \Delta n \quad \text{and} \quad p = p_0 + \Delta p \qquad (2.2)$$

where Δn and Δp are the excess electron and hole concentrations, respectively.

Next, we consider recombination of carriers. The band diagram of a semiconductor with electrons and holes is shown in Fig. 2.1. We are interested in the rate at which the carrier concentration decreases and denote the recombination rate as R. Consider a free electron in the conduction band. The probability that the electron recombines with a hole is proportional to the hole concentration, that is, $R \propto p$. The number of recombination events will also be proportional to the concentration of electrons, as indicated in Fig. 2.1. Thus the recombination rate is proportional to the product of electron and hole concentrations, that is, $R \propto n\,p$. Using a proportionality constant, the recombination rate per unit time per unit volume can be written as

$$R = -\frac{dn}{dt} = -\frac{dp}{dt} = B\,n\,p\;.\qquad(2.3)$$

This equation is the **bimolecular rate equation** and the proportionality constant B is called the **bimolecular recombination coefficient**. It has typical values of 10^{-11}–10^{-9} cm^3/s for direct-gap III–V semiconductors. The bimolecular recombination coefficient will be calculated in a subsequent section using the van Roosbroeck–Shockley model.

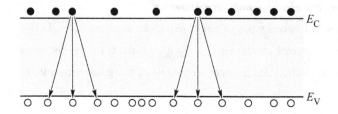

Fig. 2.1. Illustration of electron–hole recombination. The number of recombination events per unit time per unit volume is proportional to the product of electron and hole concentrations, i.e. $R \propto np$.

2.2 Radiative recombination for low-level excitation

Next, we discuss the recombination dynamics as a function of time. Consider a semiconductor subjected to photoexcitation. The equilibrium and excess electron and hole concentrations are n_0, p_0, Δn, and Δp, respectively. Since electrons and holes are generated and annihilated (by recombination) in pairs, the steady-state electron and hole excess concentrations are equal,

$$\Delta n(t) = \Delta p(t)\;.\qquad(2.4)$$

Using the bimolecular rate equation, the recombination rate is given by

$$R = B\,[n_0 + \Delta n(t)]\,[p_0 + \Delta p(t)]\;.\qquad(2.5)$$

2.2 Radiative recombination for low-level excitation

For the case of **low-level excitation**, the photogenerated carrier concentration is much smaller than the majority carrier concentration, i.e. $\Delta n \ll (n_0 + p_0)$. Using $\Delta n(t) = \Delta p(t)$ one obtains from Eq. (2.5)

$$R = B n_i^2 + B(n_0 + p_0)\Delta n(t) \qquad (2.6)$$
$$= R_0 + R_{excess}.$$

The first summand on the right-hand side of the equation can be identified as the *equilibrium recombination rate* (R_0) and the second term as the *excess recombination rate* (R_{excess}).

The time-dependent carrier concentration can be calculated from the rate equation

$$\frac{dn(t)}{dt} = G - R = (G_0 + G_{excess}) - (R_0 + R_{excess}) \qquad (2.7)$$

where G_0 and R_0 are the equilibrium generation and recombination rates, respectively.

Next, we assume that the semiconductor has been illuminated with light and excess carriers are generated. At the time $t = 0$, the illumination is switched off (i.e. $G_{excess} = 0$) as indicated in Fig. 2.2. The recombination rate can then be calculated by insertion of Eq. (2.6) into Eq. (2.7) and using $G_0 = R_0$. This yields the differential equation

$$\frac{d}{dt}\Delta n(t) = -B(n_0 + p_0)\Delta n(t). \qquad (2.8)$$

The solution of the differential equation can be obtained by separation of variables. One obtains

$$\boxed{\Delta n(t) = \Delta n_0 \, e^{-B(n_0+p_0)t}} \qquad (2.9)$$

where $\Delta n_0 = \Delta n(t=0)$. Rewriting the result as

$$\Delta n(t) = \Delta n_0 \, e^{-t/\tau} \qquad (2.10)$$

allows one to identify the **carrier lifetime** τ as

$$\boxed{\tau = \frac{1}{B(n_0 + p_0)}}. \qquad (2.11)$$

For semiconductors with a specific doping type, Eq. (2.11) reduces to

$$\tau_n = \frac{1}{B\,p_0} = \frac{1}{B\,N_A} \qquad \text{for p-type semiconductors} \qquad (2.12)$$

and

$$\tau_p = \frac{1}{B\,n_0} = \frac{1}{B\,N_D} \qquad \text{for n-type semiconductors} \qquad (2.13)$$

where τ_n and τ_p are the electron and hole lifetimes, respectively. Using this result, the rate equation, Eq. (2.8), can be simplified for semiconductors of a specific conductivity type. One obtains the **monomolecular rate equations**:

$$\frac{d}{dt}\Delta n(t) = -\frac{\Delta n(t)}{\tau_n} \qquad \text{for p-type semiconductors} \qquad (2.14)$$

and

$$\frac{d}{dt}\Delta p(t) = -\frac{\Delta p(t)}{\tau_p} \qquad \text{for n-type semiconductors.} \qquad (2.15)$$

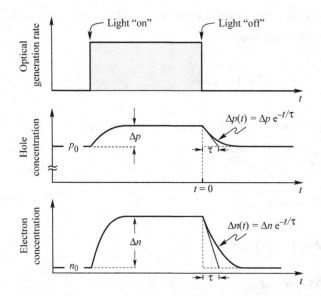

Fig. 2.2. Carrier concentration as a function of time before, during, and after an optical excitation pulse. The semiconductor is assumed to be p-type and thus $p_0 \gg n_0$. Electrons and holes are generated in pairs, thus $\Delta p = \Delta n$. Under low-level excitation as shown here, it is $\Delta n \ll p_0$. In most practical cases the equilibrium minority carrier concentration is extremely small so that $n_0 \ll \Delta n$.

Figure 2.2 shows the majority and minority carrier concentrations in a p-type semiconductor as a function of time (similar considerations apply if an n-type semiconductor is chosen). Note that Fig. 2.2 shows the case of *low-level excitation* in which the photogenerated carrier concentration is much smaller than the majority carrier concentration. However, the photogenerated carrier concentration is much larger than the minority carrier concentration.

Once photoexcitation is terminated, the minority carrier concentration decays exponentially with a characteristic time constant denoted as the **minority carrier lifetime** τ. It is the mean time between generation and recombination of a minority carrier.

Note that the majority carrier concentration also decays with the same time constant τ. However, only a very *small fraction* of the majority carriers disappear by recombination, as illustrated in Fig. 2.2. Thus, for low-level excitation, the average time it takes for a majority carrier to recombine is much longer than the minority carrier lifetime. For many practical purposes, the majority carrier lifetime can be assumed to be infinitely long.

Theoretical and experimental values for the minority carrier lifetime in GaAs as a function of the doping concentration are shown in Fig. 2.3 (Hwang, 1971; Nelson and Sobers, 1978a, 1978b; Ehrhardt et al., 1991; Ahrenkiel, 1993). The theoretical line in the figure is calculated from Eq. (2.10) using $B = 10^{-10}$ cm^3/s. In nominally undoped material, minority carrier lifetimes as long as 15 µs have been measured in GaAs at room temperature (Nelson and Sobers, 1978a, 1978b).

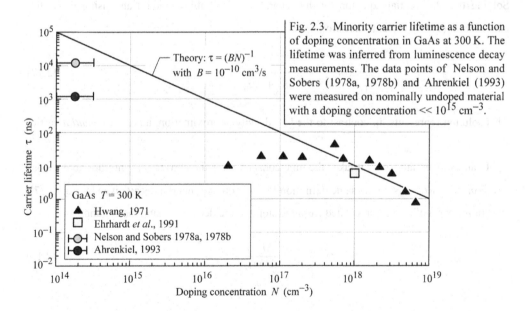

Fig. 2.3. Minority carrier lifetime as a function of doping concentration in GaAs at 300 K. The lifetime was inferred from luminescence decay measurements. The data points of Nelson and Sobers (1978a, 1978b) and Ahrenkiel (1993) were measured on nominally undoped material with a doping concentration $\ll 10^{15}$ cm^{-3}.

2 Radiative and non-radiative recombination

Exercise: *Minority carrier lifetimes*. Calculate the minority carrier lifetime in p-type GaAs at doping concentrations of 10^{15} and 10^{18} cm^{-3} using a bimolecular recombination coefficient of $B = 10^{-10}$ cm^3/s. Assume that one could fabricate GaAs without any impurities. What would be the carrier lifetime in intrinsic GaAs with a carrier concentration of 2×10^6 cm^{-3}?

Solution: $\tau_n = 10$ μs   for $N_A = 10^{15}$ cm^{-3}
              $\tau_n = 10$ ns   for $N_A = 10^{18}$ cm^{-3}
              $\tau = 2500$ s   for undoped GaAs.

Discuss how the modulation speed of communication LEDs is affected by the radiative lifetime and the doping concentration.

2.3 Radiative recombination for high-level excitation

For the case of **high-level excitation**, the photogenerated carrier concentration is larger than the equilibrium carrier concentration, i.e. $\Delta n \gg (n_0 + p_0)$. The bimolecular rate equation (Eq. 2.3) is then given by

$$\frac{d\Delta n(t)}{dt} = -B \Delta n^2 . \tag{2.16}$$

Solving this differential equation by the separation-of-variables method and using the initial condition $\Delta n(0) = \Delta n_0$ yields the solution

$$\boxed{\Delta n(t) = \frac{1}{Bt + \Delta n_0^{-1}} .} \tag{2.17}$$

This solution represents, in contrast to the low-density approximation, a *non-exponential* carrier decay.

In an exponential decay, it takes the time constant τ for the carrier concentration to decrease from Δn_0 to $\Delta n_0\, e^{-1}$. Using the same definition for the non-exponential decay given by Eq. (2.17), the "time constant" can be calculated from the slope of the decay by using the equation

$$\tau(t) = -\frac{\Delta n(t)}{\dfrac{d\Delta n(t)}{dt}} . \tag{2.18}$$

Using this definition for the non-exponential decay of Eq. (2.17), one obtains the "time constant"

$$\tau(t) = t + \frac{1}{B \Delta n_0}. \tag{2.19}$$

Thus, for non-exponential decays, the "time constant" depends on time. Equation (2.19) shows that the minority carrier lifetime increases with time. For sufficiently long times, low-level excitation conditions will be reached and τ will approach the low-level value.

2.4 Bimolecular rate equations for quantum well structures

Quantum wells provide a means of confining the free carriers to a narrow quantum well region by using the two barrier regions cladding the quantum well. Assume that the well region has a thickness of L_{QW}. Assume further that the conduction band and valence band wells have carrier densities of n^{2D} and p^{2D}, respectively. The effective three-dimensional (3D) carrier concentration for electrons and holes can be approximated by n^{2D}/L_{QW} and p^{2D}/L_{QW}, respectively. Using these values as the 3D carrier concentration, the recombination rate can be inferred from Eq. (2.5), and it is given by

$$R = B \frac{n^{2D}}{L_{QW}} \frac{p^{2D}}{L_{QW}}. \tag{2.20}$$

This equation illustrates one of the essential advantages of quantum well and double heterostructures. A decrease of the quantum well thickness allows one to attain high 3D carrier concentrations (carriers per cm^3). As a result, the carrier lifetime for radiative recombination is reduced, as inferred from Eq. (2.11), and the radiative efficiency is increased.

For sufficiently small quantum well thicknesses, the wave function no longer scales with the physical well width. L_{QW} must be replaced by the carrier distribution width, which for sufficiently small well thicknesses is larger than L_{QW}, since the wave function will extend into the barriers. This effect should be considered for well thicknesses < 100 Å in the AlGaAs/GaAs material system.

2.5 Luminescence decay

The carrier decay in semiconductors can be measured by the decay of the luminescence after a short optical excitation pulse. The luminescence intensity is proportional to the recombination rate. Calculating the recombination rate for the low and high excitation cases (Eqs. 2.9 and 2.17), one obtains

2 Radiative and non-radiative recombination

$$R = -\frac{dn(t)}{dt} = \frac{\Delta n_0}{\tau} e^{-t/\tau} \qquad \text{for low excitation} \qquad (2.21)$$

and

$$R = -\frac{dn(t)}{dt} = \frac{-B}{\left(Bt + \Delta n_0^{-1}\right)^2} \qquad \text{for high excitation.} \qquad (2.22)$$

Figure 2.4 illustrates schematically the decay of the luminescence after optical excitation by a short pulse. For the case of low excitation density, the decay is exponential with a time constant τ. For the case of high excitation, the decay is non-exponential. All non-exponential decay functions can be expressed by an exponential function *with a time-dependent time constant*, i.e. $\exp[-t/\tau(t)]$. In most cases, the time constant τ increases with time. This type of decay function is frequently called a **stretched exponential decay function** which describes a *slower-than-exponential* decay.

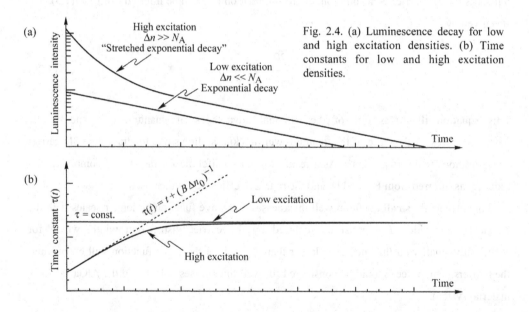

Fig. 2.4. (a) Luminescence decay for low and high excitation densities. (b) Time constants for low and high excitation densities.

A particularly well-known stretched exponential function is given by $\exp\{-[t/\tau(t)]^\beta\}$, where β, the ***disorder parameter***, represents the disorder of the radiative material. For $\beta = 1$, no disorder exists in the material. For $\beta \approx 1/2$, a strong disorder exists in the material and such values of β have been found in glasses (Phillips, 1996) as well as semiconductors. Non-exponential decays were first discovered and discussed by Friedrich Kohlrausch in the late 1800s and such decays are therefore also referred to as **Kohlrausch decays**.

The recombination dynamics of carriers in LEDs is one of the factors that limits the time it takes to switch an LED on and off. The modulation speed of LEDs used for communication applications can be limited by the minority carrier lifetime. The carrier lifetime can be reduced by either a high doping of the active region or a high concentration of injected carriers in the active region. Heterostructures that confine free carriers to the small well region are frequently employed to obtain high carrier concentrations and thus short carrier lifetimes.

2.6 Non-radiative recombination in the bulk

There are two basic recombination mechanisms in semiconductors, namely *radiative* recombination and *non-radiative* recombination. In a radiative recombination event, one photon with energy equal to the bandgap energy of the semiconductor is emitted, as illustrated in Fig. 2.5. During non-radiative recombination, the electron energy is converted to vibrational energy of lattice atoms, i.e. phonons. Thus, the electron energy is converted to heat. For obvious reasons, non-radiative recombination events are unwanted in light-emitting devices.

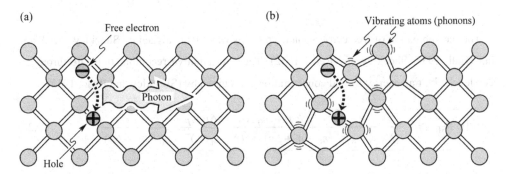

Fig. 2.5. (a) Radiative recombination of an electron–hole pair accompanied by the emission of a photon with energy $h\nu \approx E_g$. (b) In non-radiative recombination events, the energy released during the electron–hole recombination is converted to phonons (adapted from Shockley, 1950).

There are several physical mechanisms by which non-radiative recombination can occur. Defects in the crystal structure are the most common cause for non-radiative recombination. These defects include unwanted foreign atoms, native defects, dislocations, and any complexes of defects, foreign atoms, or dislocations. In compound semiconductors, native defects include interstitials, vacancies, and antisite defects (Longini and Greene, 1956; Baraff and Schluter,

2 Radiative and non-radiative recombination

1985). All such defects have energy level structures that are different from substitutional semiconductor atoms. It is quite common for such defects to form one or several energy levels within the forbidden gap of the semiconductor.

Energy levels within the gap of the semiconductor are efficient recombination centers; in particular, if the energy level is close to the middle of the gap. The recombination of carriers via a trap level is shown schematically in Fig. 2.6. Owing to the promotion of non-radiative processes, such deep levels or traps are called *luminescence killers*.

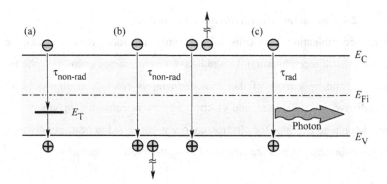

Fig. 2.6. Band diagram illustrating recombination: (a) non-radiative via deep level, (b) non-radiative via Auger process and (c) radiative.

The recombination of free carriers via deep levels was first analyzed by Shockley, Read, and Hall (Hall, 1952; Shockley and Read, 1952). The non-radiative recombination rate through a deep level with trap energy E_T and concentration N_T is given by (Shockley and Read, 1952)

$$R_{SR} = \frac{p_0 \Delta n + n_0 \Delta p + \Delta n \Delta p}{(N_T v_p \sigma_p)^{-1}(n_0 + n_1 + \Delta n) + (N_T v_n \sigma_n)^{-1}(p_0 + p_1 + \Delta p)} \qquad (2.23)$$

where $\Delta n = \Delta p$; v_n and v_p are the electron and hole thermal velocities, and σ_n and σ_p are the capture cross sections of the traps. The quantities n_1 and p_1 are the electron and hole concentrations if the Fermi energy is located at the trap level. These quantities are given by

$$n_1 = n_i \exp\left(\frac{E_T - E_{Fi}}{kT}\right) \quad \text{and} \quad p_1 = n_i \exp\left(\frac{E_{Fi} - E_T}{kT}\right) \qquad (2.24)$$

where E_{Fi} is the Fermi level in the intrinsic semiconductor.

The non-radiative lifetime of excess electrons can be deduced from the equation $R_{SR} = \Delta n / \tau$. Consequently, the lifetime is given by

2.6 Non-radiative recombination in the bulk

$$\frac{1}{\tau} = \frac{p_0 + n_0 + \Delta n}{(N_T v_p \sigma_p)^{-1}(n_0 + n_1 + \Delta n) + (N_T v_n \sigma_n)^{-1}(p_0 + p_1 + \Delta p)}. \quad (2.25)$$

We now differentiate between majority carriers and minority carriers and assume that the semiconductor is p-type. Then holes are in the majority, i.e. $p_0 \gg n_0$ and $p_0 \gg p_1$. If we further assume a small deviation from equilibrium, i.e. $\Delta n \ll p_0$, then the minority carrier lifetime is given by

$$\frac{1}{\tau} = \frac{1}{\tau_{n_0}} = N_T v_n \sigma_n. \quad (2.26)$$

If electrons were the majority carriers, the lifetime would be obtained in an analogous way, i.e.

$$\frac{1}{\tau} = \frac{1}{\tau_{p_0}} = N_T v_p \sigma_p. \quad (2.27)$$

The results show that the Shockley–Read recombination rate is limited by the rate of capture of minority carriers. This result suggests itself since the capture of *majority* carriers is a much more *likely* event than the capture of *minority* carriers. Equation (2.25) can then be written as

$$\frac{1}{\tau} = \frac{p_0 + n_0 + \Delta n}{\tau_{p_0}(n_0 + n_1 + \Delta n) + \tau_{n_0}(p_0 + p_1 + \Delta p)}. \quad (2.28)$$

For small deviations from equilibrium, i.e. $\Delta n \ll p_0$, the equation simplifies to

$$\tau = \tau_{n_0}\frac{p_0 + p_1}{p_0 + n_0} + \tau_{p_0}\frac{n_0 + n_1 + \Delta n}{p_0 + n_0} \approx \tau_{n_0}\frac{p_0 + p_1}{p_0 + n_0}. \quad (2.29)$$

Inspection of the equation reveals that the lifetime does not change for small deviations from equilibrium in an extrinsic semiconductor.

For further insight, we assume that the trap captures electrons and holes at the same rate, i.e. $v_n \sigma_n = v_p \sigma_p$ and $\tau_{n_0} = \tau_{p_0}$. One obtains from Eq. (2.29)

$$\tau = \tau_{n_0}\left(1 + \frac{p_0 + p_1}{p_0 + n_0}\right). \quad (2.30)$$

For the special case of intrinsic material, i.e. $n_0 = p_0 = n_i$, the equation simplifies to

$$\tau_i = \tau_{n_0}\left(1 + \frac{p_1 + n_1}{2n_i}\right) = \tau_{n_0}\left[1 + \cosh\left(\frac{E_T - E_{Fi}}{kT}\right)\right] \quad (2.31)$$

where E_{Fi} is the intrinsic Fermi level, which is typically close to the middle of the gap. The cosh function has a minimum when the argument of the function is zero. Thus the non-radiative lifetime is minimized if $E_T - E_{Fi}$ is zero; i.e. when the trap level is at or close to the midgap energy. For such midgap levels, the lifetime is given by $\tau = 2\tau_{n0}$. This result demonstrates that deep levels are effective recombination centers if they are near the middle of the gap.

Inspection of Eq. (2.31) also reveals the temperature dependence of Shockley–Read recombination. As T increases the non-radiative recombination lifetime *decreases*. As a result, the radiative band-to-band recombination efficiency *decreases* at high temperatures. The highest band-to-band radiative efficiencies of direct-gap semiconductors can be obtained at cryogenic temperatures.

However, some devices are based on radiative recombination through a deep state. A well-known example of radiative recombination mediated by a deep level is N-doped GaP. It follows from the Shockley–Read model that the deep-level recombination rate increases with increasing temperature.

In *indirect-gap* semiconductors such as GaP, radiative transitions are mediated by phonons. That is, radiative recombination must be accompanied by absorption or emission of a phonon. Since phonons are more abundant at high temperatures, radiative recombination (mediated by the absorption of a phonon) can increase with temperature.

In the vicinity of a deep level, the luminescence intensity decreases. A single **point defect** will be difficult to observe due to its relatively small effect. Frequently, however, defects group into clusters of defects or **extended defects**. Such extended defects are, for example, threading dislocations and misfit dislocations occurring when epitaxial semiconductors are grown on mismatched substrates, i.e. substrates with a different lattice constant from that of the epitaxial layer. There are also many other types of extended defects. The luminescence-killing nature of extended defects is illustrated in Fig. 2.7, which shows a cathodoluminescence micrograph of a GaAs layer measured at room temperature. The figure reveals several dark spots. Luminescence in the vicinity of the defects is reduced due to the non-radiative recombination channels so that the defects manifest themselves as dark spots. The size of the dark spots depends on the size of the defect and the minority carrier diffusion length.

While most deep-level transitions are non-radiative, some deep-level transitions are radiative.

An example of a radiative deep-level transition in GaN is shown in Fig. 2.8 (Grieshaber *et al.*, 1996). The luminescence spectrum shows the band-to-band transition at 365 nm and a broad deep-level transition around 550 nm. The deep-level transition occurs near the yellow range of the visible spectrum. The yellow luminescence line has been shown to be due to Ga vacancies (Neugebauer and Van de Walle, 1996; Schubert *et al.* 1997; Saarinen *et al.* 1997), a common point defect in n-type GaN.

Fig. 2.7. Cathodoluminescence micrograph of a GaAs epitaxial layer. The dark spots are due to large clusters of non-radiative recombination centers (after Schubert, 1995).

GaAs
$T = 295$ K   10 µm

Deep levels can be caused by **native defects** (group-III vacancies, group-V vacancies, group-III interstitials, and group-V interstitials), unwanted foreign impurities, dislocations, impurity–defect complexes, and combinations of different types of defects. Frequently it takes many years to unambiguously identify the atomic nature of a defect. For a review of defects in semiconductors see, for example, Pantelides (1992).

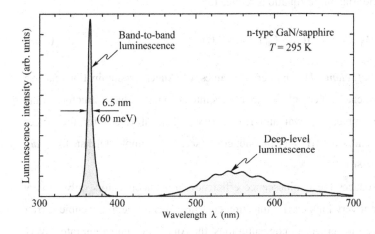

Fig. 2.8. Room-temperature photoluminescence spectrum of GaN with a band-to-band optical transition at 365 nm and a second transition at 550 nm identified as an optically active deep-level transition (after Grieshaber *et al.*, 1996).

Another important non-radiative recombination mechanism is **Auger recombination**. In this process, the energy becoming available through electron–hole recombination (approximately E_g), is dissipated by the excitation of a free electron high into the conduction band, or by a hole deeply excited into the valence band. The processes are shown schematically in Fig. 2.6 (b). The highly excited carriers will subsequently lose energy by multiple phonon emission until they are close to the band edge.

The recombination rates due to the two Auger processes shown in Fig. 2.6 (b) are given by

$$R_{\text{Auger}} = C_p n p^2 \quad (2.32)$$

and

$$R_{\text{Auger}} = C_n n^2 p \ . \quad (2.33)$$

Auger recombination is proportional to the square of the carrier concentration (either p^2 or n^2) since two carriers of the same type (either two holes or two electrons) are required for the recombination process. The first process (see Eq. 2.32) is more likely to happen in p-type semiconductors due to the abundance of holes. The second process (see Eq. 2.33) is more likely in n-type semiconductors due to the abundance of electrons.

During Auger recombination, energy and momentum must be conserved. Owing to the differences in conduction and valence band structure in semiconductors, the two Auger coefficients C_p and C_n are generally different.

In the high-excitation limit in which the non-equilibrium carriers have a higher concentration than equilibrium carriers, the Auger rate equations reduce to

$$R_{\text{Auger}} = (C_p + C_n) n^3 = C n^3 \quad (2.34)$$

where C is the **Auger coefficient**. The numerical values of Auger recombination can be determined by a quantum mechanical calculation that takes into account the band structure of the semiconductor (see, for example, Agrawal and Dutta, 1986). Typical values for the Auger coefficient are 10^{-28}–10^{-29} cm^6/s for III–V semiconductors (see, for example, Olshansky et al., 1984; Agrawal and Dutta, 1986).

Auger recombination reduces the luminescence efficiency in semiconductors only at very high excitation intensity or at very high carrier injection currents. This is due to the cubic carrier concentration dependence. At lower carrier concentrations, the Auger recombination rate is very small and can be neglected for practical purposes.

2.7 Non-radiative recombination at surfaces

Substantial non-radiative recombination can occur at semiconductor surfaces. Surfaces are a strong perturbation of the periodicity of a crystal lattice. Recall that the band diagram model is based on the strict periodicity of a lattice. Since this periodicity ends at a surface, the band diagram will need to be modified at a semiconductor surface. This modification includes the addition of electronic states within the forbidden gap of the semiconductor.

Next, we consider a semiconductor surface from a chemical point of view. Atoms at the surface cannot have the same bonding structure as bulk atoms due to the lack of neighboring atoms. Thus some of the valence orbitals do not form a chemical bond. These partially filled electron orbitals, or **dangling bonds**, are electronic states that can be located in the forbidden gap of the semiconductor where they act as recombination centers. Depending on the charge state of these valence orbitals, the states can be acceptor-like or donor-like states.

The dangling bonds may also rearrange themselves and form bonds between neighboring atoms in the same surface plane. This **surface reconstruction** can lead to a *locally* new atomic structure with state energies different from bulk atomic states. The surface bonding structure depends on the specific nature of the semiconductor surface. The energetic location of surface states is very difficult to predict, even with powerful theoretical models. Thus phenomenological models of surface recombination are commonly used.

It has been shown that electronic states within the forbidden gap appear at semiconductor surfaces. Bardeen and Shockley (Shockley, 1950) pioneered the understanding of surface states and their role as recombination centers.

Next, we calculate the effect of surface recombination on the carrier distribution in a p-type semiconductor subjected to illumination. Assume that the illumination causes a uniform steady-state generation rate G. The one-dimensional continuity equation must be fulfilled at any point in the semiconductor. The continuity equation for electrons is given by

$$\frac{\partial \Delta n(x,t)}{\partial t} = G - R + \frac{1}{e}\frac{\partial}{\partial x} J_n \tag{2.35}$$

where J_n is the current density caused by electrons flowing to the surface. In the bulk of the uniform semiconductor, there is no dependence on space and thus the continuity equation reduces to $G = R$ under steady-state conditions. Using the recombination rate in the bulk as given by Eq. (2.14), the excess carrier concentration in the bulk is given by $\Delta n_\infty = G\,\tau_n$ as indicated in Fig. 2.9. Assuming that the electron current is a diffusion current of the form

2 Radiative and non-radiative recombination

$$J_n = eD_n \frac{\partial \Delta n(x,t)}{\partial x} \tag{2.36}$$

and inserting the diffusion current into Eq. (2.35) yields the continuity equation for diffusive currents, i.e.

$$\frac{\partial \Delta n(x,t)}{\partial t} = G - \frac{\Delta n(x,t)}{\tau_n} + D_n \frac{\partial^2 \Delta n(x,t)}{\partial x^2}. \tag{2.37}$$

(a)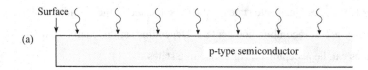

(b)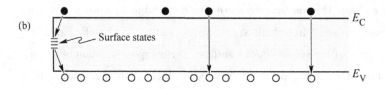

(c)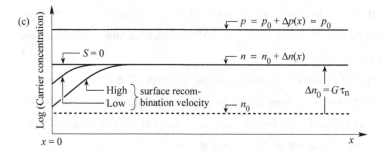

Fig. 2.9. (a) Illuminated p-type semiconductor, (b) band diagram, and (c) minority and majority carrier concentrations near the surface assuming uniform carrier generation due to illumination. The excess carrier concentrations are Δn and Δp.

At the semiconductor surface, carriers will recombine rapidly due to surface states. The **boundary condition** at the surface is given by

$$eD_n \frac{\partial \Delta n(x,t)}{\partial x}\bigg|_{x=0} = eS\, \Delta n(x,t)\big|_{x=0} \tag{2.38}$$

where S is the surface recombination velocity. The boundary condition states that minority carriers diffusing to the surface will recombine at the surface. We assume that the generation rate is constant with time, and thus the minority carrier concentration has no time dependence. The steady-state solution to the differential equation with the above boundary condition is given by

$$n(x) = n_0 + \Delta n(x) = n_0 + \Delta n_\infty \left[1 - \frac{\tau_n S \exp(-x/L_n)}{L_n + \tau_n S} \right]. \quad (2.39)$$

The carrier concentration near a semiconductor surface is shown in Fig. 2.9 for different surface recombination velocities. For $S \to 0$, the minority carrier concentration at the surface is identical to the bulk value, i.e. $n(0) \to n_0 + \Delta n_\infty$. For $S \to \infty$, the minority carrier concentration at the surface approaches the equilibrium value, i.e. $n(0) \to n_0$.

Surface recombination leads to a reduced luminescence efficiency and also to heating of the surface due to non-radiative recombination at the surface. Both effects are unwanted in electroluminescent devices. The surface recombination velocities for several semiconductors are summarized in Table 2.1. The data shown in the table show that GaAs has a particularly high surface recombination velocity.

Table 2.1. Surface recombination velocities of several semiconductors (GaN data after Tu et al., 2000; Aleksiejunas et al., 2003).

Semiconductor	Surface recombination velocity
GaAs	$S = 10^6$ cm/s
GaN	$S = 5 \times 10^4$ cm/s
InP	$S = 10^3$ cm/s
Si	$S = 10^1$ cm/s

Experimental evidence of surface recombination is illustrated in Fig. 2.10, which shows the luminescence emanating from a stripe-like current-injection contact on a GaAs laser chip. The luminescence is viewed from the substrate side so that the stripe-like metal contact is "behind" the light-emitting region. Figure 2.10 clearly reveals that the luminescence decreases in the near-surface region.

Surface recombination can occur only when both types of carriers are present. It is important in the design of LEDs that the carrier-injected active region, in which naturally both types of carriers are present, *be far removed from any surface*. This can be achieved, for example, by carrier injection under a contact that is much smaller than the semiconductor die. Furthermore, the contact must be sufficiently far away from the side surfaces of the die. If the current flow is confined to the region below the contact, carriers will not "see" any semiconductor surfaces.

2 Radiative and non-radiative recombination

Note that unipolar regions of a semiconductor device, e.g. the confinement regions, are not affected by surface recombination due to the lack of minority carriers.

Several passivation techniques have been developed to reduce the surface recombination in semiconductors, including treatments with sulfur, and other chemicals (Lipsanen *et al.*, 1999).

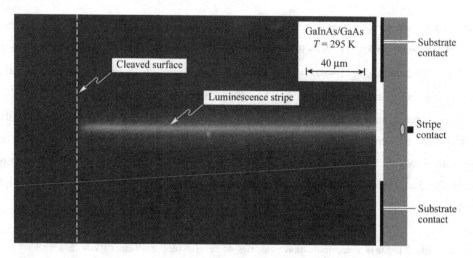

Fig. 2.10. (a) Micrograph and (b) schematic frontal view of a GaInAs/GaAs structure with a stripe-shaped top contact and contact window on substrate side under current injection conditions. The luminescence emanating from active region located below the stripe clearly decreases in the vicinity of the surface due to surface recombination.

2.8 Competition between radiative and non-radiative recombination

So far we have seen that several mechanisms for non-radiative recombination exist, including Shockley–Read, Auger, and surface recombination. Even though non-radiative recombination can be reduced, it can never be totally eliminated. For example, surface recombination can be drastically reduced by device designs that spatially separate the active region from any surfaces. However, even if the separation is large, a few carriers will still diffuse to the surface and recombine there.

Just as for surface recombination, non-radiative bulk recombination and Auger recombination can never be totally avoided. Any semiconductor crystal will have some native defects. Even though the concentration of these native defects can be low, it is never zero. Thermodynamic considerations predict that if an energy E_a is needed to create a specific point defect in a crystal lattice, the probability that such a defect does indeed form at a specific lattice site, is given by the Boltzmann factor, i.e. $\exp(-E_a/kT)$. The product of the concentration of lattice sites and the Boltzmann factor gives the concentration of defects. A native point defect or extended defect may form a deep state in the gap and thus be a non-radiative recombination center.

Exercise: *Concentration of point defects*. Assume that the energy required to move a substitutional lattice atom into an interstitial position is $E_a = 1.1$ eV. What is the equilibrium concentration of interstitial defects of a simple cubic lattice with lattice constant $a_0 = 2.5$ Å?
Solution: The concentration of lattice atoms of a simple cubic lattice is given by $N = a_0^{-3} = 6.4 \times 10^{22}$ cm^{-3}. The concentration of interstitial defects under equilibrium conditions at room temperature is then given by

$$N_{\text{defect}} = N \exp(-E_a/kT) = 2.7 \times 10^4 \text{ cm}^{-3}.$$

Note that the calculated concentration of defects is small when compared to the typical concentrations of electrons and holes. If the defect discussed here forms a level in the gap, non-radiative recombination through the defect level can occur.

Another issue is the chemical purity of semiconductors. It is difficult to fabricate materials with impurity levels lower than the *parts per billion* (ppb) range. Thus, even the purest semiconductors contain impurities in the 10^{12} cm^{-3} range. Some elements may form deep levels and thus reduce the luminescence efficiency.

In the 1960s, when the first III–V semiconductors had been demonstrated, the internal luminescence efficiencies at room temperature were very low, typically a fraction of 1%. At the present time, high-quality bulk semiconductors and quantum well structures can have internal efficiencies exceeding 90%, and in some cases even 99%. This remarkable progress is due to improved crystal quality, and reduced defect and impurity concentrations.

Next, we calculate the internal quantum efficiency in a semiconductor with non-radiative recombination centers. If the radiative lifetime is denoted as τ_r and the non-radiative lifetime is denoted as τ_{nr}, then the total probability of recombination is given by the sum of the radiative and non-radiative probabilities:

$$\tau^{-1} = \tau_r^{-1} + \tau_{nr}^{-1}. \tag{2.40}$$

The relative probability of radiative recombination is given by the radiative probability over the total probability of recombination. Thus the probability of radiative recombination or **internal quantum efficiency** is given by

$$\eta_{\text{int}} = \frac{\tau_r^{-1}}{\tau_r^{-1} + \tau_{nr}^{-1}}. \qquad (2.41)$$

The internal quantum efficiency gives the ratio of the number of light quanta emitted inside the semiconductor to the number of charge quanta undergoing recombination. Note that not all photons emitted internally may escape from the semiconductor due to the light-escape problem, reabsorption in the substrate, or other reabsorption mechanisms.

References

Agrawal G. P. and Dutta N. K. *Long Wavelength Semiconductor Lasers* (John Wiley and Sons, New York, 1986)

Ahrenkiel R. K. "Minority-carrier lifetime in III–V semiconductors" in *Minority Carriers in III–V Semiconductors: Physics and Applications* edited by R. K. Ahrenkiel and M. S. Lundstrom, Semiconductors and Semimetals **39**, 40 (Academic Press, San Diego, 1993)

Aleksiejunas R., Sudzius M., Malinauskas T., Vaitkus J., Jarasiunas K. and Sakai S. "Determination of free carrier bipolar diffusion coefficient and surface recombination velocity of undoped GaN epilayers" *Appl. Phys. Lett.* **83**, 1157 (2003)

Baraff G. A. and Schluter M. "Electronic structure, total energies, and abundances of the elementary point defects in GaAs" *Phys. Rev. Lett.* **55**, 1327 (1985)

Ehrhardt A., Wettling W., and Bett A. "Transient photoluminescence decay study of minority carrier lifetime in GaAs heteroface solar cell structures" *Appl. Phys. A* (*Solids and Surfaces*) **A53**, 123 (1991)

Grieshaber W., Schubert E. F., Goepfert I. D., Karlicek R. F. Jr., Schurman M. J. and Tran C. "Competition between band gap and yellow luminescence in GaN and its relevance for optoelectronic devices" *J. Appl. Phys.* **80**, 4615 (1996)

Hall R. N. "Electron–hole recombination in germanium" *Phys. Rev.* **87**, 387 (1952)

Hwang C. J. "Doping dependence of hole lifetime in n-type GaAs" *J. Appl. Phys.* **42**, 4408 (1971)

Lipsanen H., Sopanen M., Ahopelto J., Sandman J., and Feldmann J. "Effect of InP passivation on carrier recombination in $In_xGa_{1-x}As$/GaAs surface quantum wells" *Jpn. J. Appl. Phys.* **38**, 1133 (1999)

Longini R. L. and Greene R. F. "Ionization interaction between impurities in semiconductors and insulators" *Phys. Rev.* **102**, 992 (1956)

Nelson R. J. and Sobers R. G. "Interfacial recombination velocity in GaAlAs/GaAs heterostructures" *Appl. Phys. Lett.* **32**, 761 (1978a)

Nelson R. J. and Sobers R. G. "Minority-carrier lifetime and internal quantum efficiency of surface-free GaAs" *Appl. Phys. Lett.* **49**, 6103 (1978b)

Neugebauer J. and Van de Walle C. "Gallium vacancies and the yellow luminescence in GaN" *Appl. Phys. Lett.* **69**, 503 (1996)

Olshansky R., Su C. B., Manning J., and Powazinik W. "Measurement of radiative and non-radiative recombination rates in InGaAsP and AlGaAs light sources" *IEEE J. Quantum Electronics* **QE-20**, 838 (1984)

Pantelides S. T. (Editor) *Deep Centers in Semiconductors* (Gordon and Breach, Yverdon, Switzerland, 1992)

Phillips J. C. "Stretched exponential relaxation in molecular and electronic glasses" *Rep. Prog. Phys.* **59**, 1133 (1996)

Saarinen K. *et al.* "Observation of native Ga vacancies in GaN by positron annihilation" *Phys. Rev. Lett.* **79**, 3030 (1997)

Schubert E. F. *Doping in III–V Semiconductors* (Cambridge University Press, Cambridge UK, 1993)

Schubert E. F., unpublished (1995)

Schubert E. F., Goepfert I., and Redwing J. M. "Evidence of compensating centers as origin of yellow

luminescence in GaN" *Appl. Phys. Lett.* **71**, 3224 (1997)
Shockley W. *Electrons and Holes in Semiconductors* (D. Van Nostrand Company, New York, 1950)
Shockley W. and Read W. T. "Statistics of the recombinations of holes and electrons" *Phys. Rev.* **87**, 835 (1952)
Tu L. W., Kuo W. C., Lee K. H., Tsao P. H., and Lai C. M., Chu A. K., and Sheu J. K. "High-dielectric-constant Ta_2O_5 / n-GaN metal-oxide-semiconductor structure" *Appl. Phys. Lett.* **77**, 3788 (2000)

3

Theory of radiative recombination

In this chapter, the theory of radiative recombination is first discussed in terms of a rigorous quantum mechanical model. Subsequently recombination is discussed in terms of a semi-classical model based on equilibrium generation and recombination. This model was developed by van Roosbroeck and Shockley (1954). Finally, the Einstein model of spontaneous and stimulated transitions in a two-level atom is discussed.

3.1 Quantum mechanical model of recombination

Spontaneous recombination based on quantum mechanics was discussed by Bebb and Williams (1972), Agrawal and Dutta (1986), Dutta (1993), Thompson (1980), and others. The quantum mechanical calculation for the spontaneous emission rate is based on the induced emission rate given by **Fermi's Golden Rule** which gives the transition probability per unit time (called transition rate) from a quantum mechanical state j to a state m

$$W_{j \to m} = \frac{d}{dt}|a'_m(t)|^2 = \frac{2\pi}{\hbar}|H'_{mj}|^2 \rho(E = E_j + \hbar\omega_0) \quad (3.1)$$

where H_{mj}' is the **transition matrix element**. For the one-dimensional case with a dependence on only the spatial variable x, the matrix element that connects the (initial) jth state with the (final) mth state via the perturbation hamiltonian H', is given by

$$H'_{mj} = \langle \psi_m^0 | H' | \psi_j^0 \rangle = \int_{-\infty}^{\infty} \psi_m^{0*}(x)\, A(x)\, \psi_j^0(x)\, dx \; . \quad (3.2)$$

For the derivation of Fermi's Golden Rule, the perturbation hamiltonian H' is assumed to have a harmonic time dependence, i.e. $H' = A(x)[\exp(i\omega_0 t) + \exp(-i\omega_0 t)]$, as expected for the excitation by a harmonic wave of a photon. Equation (3.2) indicates that a necessary condition for recombination is **spatial overlap** between electron and hole wave functions. This is intuitively clear, because spatially separated electrons and holes will not be able to recombine.

For optical transitions between the conduction and valence band, the electron momentum must be conserved because the photon momentum ($p = \hbar k$) is negligibly small. The conservation of momentum condition is known as the ***k*-selection rule**. For the remainder of this section, we closely follow the quantum mechanical analysis given by Agrawal and Dutta (1986) and the reader is referred to that reference for a comprehensive discussion.

The average matrix element for the Bloch states, $|M_b|$ can be derived using the four-band Kane model (Kane, 1957), which takes into account the conduction, heavy-hole, light-hole, and split-off band. In bulk semiconductors $|M_b|^2$ is given by (Kane, 1957; Casey and Panish, 1978)

$$|M_b|^2 = \frac{m_e^2 E_g (E_g + \Delta)}{12 m_e^* (E_g + 2\Delta/3)} \qquad (3.3)$$

where m_e is the free-electron mass, E_g is the band gap, and Δ is the spin-orbit splitting. For GaAs, using $E_g = 1.424$ eV, $\Delta = 0.33$ eV, $m_e^* = 0.067 \, m_e$, we get $|M_b|^2 = 1.3 \, m_e E_g$.

Taking into account the *k*-selection rule, the total spontaneous emission rate per unit volume is given by

$$r_{sp}(E) = \frac{4\pi \bar{n} e^2 E}{m_e^2 \varepsilon_0 h^2 c^3} |M_b|^2 \frac{(2\pi)^3}{V} 2 \left(\frac{V}{(2\pi)^3}\right)^2 \frac{1}{V}$$

$$\times \sum \int \cdots \int f_c(E_c) f_v(E_v) \, d^3 \vec{k}_c \, d^3 \vec{k}_v \, \delta(\vec{k}_c - \vec{k}_v) \, \delta(E_i - E_f - E) \qquad (3.4)$$

where f_c and f_v are the Fermi factors for electrons and holes, and the term $\delta(\vec{k}_c - \vec{k}_v)$ ensures satisfaction of the *k*-selection rule. The factor 2 arises from the two spin states. In Eq. (3.4), Σ stands for the sum over the three valence bands (heavy-hole, light-hole, and split-off bands). For definiteness, we first consider transitions involving electrons and heavy holes. The integrals in Eq. (3.4) can be evaluated with the following result:

$$r_{sp}(E) = \frac{2\bar{n} e^2 E |M_b|^2}{\pi m_e^2 \varepsilon_0 h^2 c^3} \left(\frac{2 m_r}{\hbar^2}\right)^{3/2} \sqrt{E - E_g} \, f_c(E_c) f_v(E_v) \qquad (3.5)$$

where

3 Theory of radiative recombination

$$E_c = (m_r/m_e^*)(E - E_g), \qquad (3.6)$$

$$E_v = (m_r/m_{hh}^*)(E - E_g), \qquad (3.7)$$

$$m_r = \frac{m_e^* \, m_{hh}^*}{m_e^* + m_{hh}^*}, \qquad (3.8)$$

and m_{hh}^* is the effective mass of the heavy hole. Equation (3.5) gives the spontaneous emission rate at the photon energy E. To obtain the total spontaneous emission rate, a final integral should be carried out over all possible energies. Thus the total spontaneous emission rate per unit volume due to electron–heavy-hole transitions is given by

$$R = \int_{E_g}^{\infty} r_{sp}(E)\,dE = A\,|M_b|^2\,I \qquad (3.9)$$

where

$$I = \int_{E_g}^{\infty} \sqrt{E - E_g}\, f_c(E_c)\, f_v(E_v)\, dE \qquad (3.10)$$

and A represents the remaining constants in Eq. (3.5). A similar equation holds for the electron–light-hole transitions if we replace m_{hh}^* by the effective light-hole mass m_{lh}^*.

The quantum mechanical absorption coefficient $\alpha(E)$ derived by Agrawal and Dutta (1986) using a similar analysis is given by

$$\alpha(E) = \frac{e^2 h |M_b|^2}{4\pi^2\,\varepsilon_0\,m_e^2\,c\,\bar{n}\,E} \left(\frac{2m_r}{\hbar^2}\right)^{3/2} \sqrt{E - E_g}\,[1 - f_c(E_c) - f_v(E_v)]. \qquad (3.11)$$

Although the quantum mechanical model of recombination is most appropriate and accurate, it can be time-consuming and awkward to deal with. The following sections analyze recombination in semi-classical terms that usually are more convenient to work with.

3.2 The van Roosbroeck–Shockley model

The van Roosbroeck–Shockley model allows one to calculate the spontaneous radiative recombination rate under equilibrium and non-equilibrium conditions. To calculate the recombination rate, the model requires the knowledge of only a few basic parameters, namely the

3.2 The van Roosbroeck–Shockley model

bandgap energy, the absorption coefficient, and the refractive index. All of these parameters can be determined by simple, well-known experimental methods.

Consider a semiconductor with an absorption coefficient $\alpha(\nu)$ given in units of cm^{-1}. A photon is generated in the semiconductor by electron–hole recombination and is subsequently absorbed, as illustrated in Fig. 3.1. The mean distance a photon of frequency ν travels before being absorbed is simply $\alpha(\nu)^{-1}$. The time it takes for a photon to be absorbed is given by

$$\tau(\nu) = \frac{1}{\alpha(\nu)\, v_{gr}} \tag{3.12}$$

where v_{gr} is the group velocity of photons propagating in the semiconductor. The group velocity of photons is given by

$$v_{gr} = \frac{d\omega}{dk} = \frac{d\nu}{d(1/\lambda)} = c\, \frac{d\nu}{d(\bar{n}\nu)} \tag{3.13}$$

where $\bar{n}$ is the refractive index. Inserting the group velocity into Eq. (3.12) yields

$$\boxed{\frac{1}{\tau(\nu)} = \alpha(\nu)\, v_{gr} = \alpha(\nu)\, c\, \frac{d\nu}{d(\bar{n}\nu)}} \tag{3.14}$$

This equation gives the *inverse photon lifetime* or *photon absorption probability* per unit time. The product of the absorption probability and the photon density yields the photon absorption rate per unit time per unit volume.

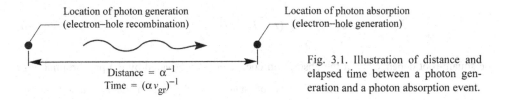

Fig. 3.1. Illustration of distance and elapsed time between a photon generation and a photon absorption event.

Under equilibrium conditions, the density of photons per unit volume in a medium with refractive index $\bar{n}$ is given by Planck's **black-body radiation formula**

$$N(\lambda)\, d\lambda = \frac{8\pi}{\lambda^4}\, \frac{1}{e^{h\nu/kT} - 1}\, d\lambda \tag{3.15}$$

from which we can readily obtain $N(\nu)\,d\nu$, the number of photons having frequencies in the interval ν and $\nu + d\nu$. We have $\lambda = c/(\bar{n}\nu)$, so that

$$d\lambda = -\frac{c}{(\bar{n}\nu)^2}\frac{d(\bar{n}\nu)}{d\nu}\,d\nu. \tag{3.16}$$

Inserting this value in Eq. (3.15) yields Planck's black-body photon distribution as a function of frequency

$$\boxed{N(\nu)\,d\nu = \frac{8\pi\nu^2\bar{n}^2}{c^3}\frac{d(\bar{n}\nu)}{d\nu}\frac{1}{e^{h\nu/kT}-1}\,d\nu}. \tag{3.17}$$

The absorption rate per unit volume in the frequency interval ν and $\nu + d\nu$ is given by the photon density divided by the mean lifetime of photons

$$R_0(\nu) = \frac{N(\nu)}{\tau(\nu)} = \frac{8\pi\nu^2\bar{n}^2}{c^3}\frac{d(\bar{n}\nu)}{d\nu}\frac{1}{e^{h\nu/kT}-1}\,\alpha(\nu)\,c\,\frac{d\nu}{d(\bar{n}\nu)}. \tag{3.18}$$

Integration over all frequencies yields the absorption rate per unit volume

$$\boxed{R_0 = \int_0^\infty R_0(\nu)\,d\nu = \int_0^\infty \frac{8\pi\nu^2\bar{n}^2}{c^2}\frac{\alpha(\nu)}{e^{h\nu/kT}-1}\,d\nu} \tag{3.19}$$

which is the celebrated **van Roosbroeck–Shockley equation**. The van Roosbroeck–Shockley equation can be simplified by writing the absorption coefficient as

$$\alpha = \alpha_0\sqrt{(E-E_\text{g})/E_\text{g}}. \tag{3.20}$$

The square-root dependence of the absorption coefficient is motivated by the proportionality of the absorption coefficient and the density of states, which in turn follows a square-root dependence on energy. Note that α_0 is the absorption coefficient at $h\nu = 2E_\text{g}$. Approximate values of α_0 for several semiconductors are given in Table 3.1.

The van Roosbroeck–Shockley equation can be simplified further by neglecting the frequency dependence of the refractive index and using the refractive index value at the band edge. One obtains

3.2 The van Roosbroeck–Shockley model

$$R_0 = 8\pi c \bar{n}^2 \alpha_0 \sqrt{\frac{kT}{E_g}} \left(\frac{kT}{ch}\right)^3 \int_{x_g}^{\infty} \frac{x^2 \sqrt{x-x_g}}{e^x - 1} dx \qquad (3.21)$$

where $x = h\nu/(kT) = E/(kT)$ and $x_g = E_g/(kT)$. Owing to the strong increase of the exponential function with x, only a small range of energies close to the bandgap contributes to the integral. The integral has no simple analytical solution and it needs to be evaluated by a numerical method.

Table 3.1. Bimolecular recombination coefficient at 300 K for different semiconductors as calculated from the energy gap, absorption coefficient, and refractive index at the bandgap energy. The spontaneous lifetime is given by $B^{-1} N_{D,A}^{-1}$ and it is calculated for a majority carrier concentration of 10^{18} cm^{-3}.

Material	E_g (eV)	α_0 (cm^{-1})	$\bar{n}$ (–)	R_0 (cm^{-3} s^{-1})	n_i (cm^{-3})	B (cm^3 s^{-1})	τ_{spont} (s)
GaAs	1.42	2×10^4	3.3	7.9×10^2	2×10^6	2.0×10^{-10}	5.1×10^{-9}
InP	1.35	2×10^4	3.4	1.2×10^4	1×10^7	1.2×10^{-10}	8.5×10^{-9}
GaN	3.4	2×10^5	2.5	8.9×10^{-30}	2×10^{-10}	2.2×10^{-10}	4.5×10^{-9}
GaP	2.26	2×10^3	3.0	1.0×10^{-12}	1.6×10^0	3.9×10^{-13}	2.6×10^{-6}
Si	1.12	1×10^3	3.4	3.3×10^6	1×10^{10}	3.2×10^{-14}	3.0×10^{-5}
Ge	0.66	1×10^3	4.0	1.1×10^{14}	2×10^{13}	2.8×10^{-13}	3.5×10^{-6}

Under equilibrium conditions, the carrier generation rate (photon absorption rate) is equal to the carrier recombination rate (photon emission rate). Thus, the van Roosbroeck–Shockley model provides the *equilibrium* recombination rate. As discussed earlier, the bimolecular rate equation, which applies to both equilibrium and non-equilibrium conditions, gives the number of recombination events occurring per unit volume per unit time:

$$R = B n p . \qquad (3.22)$$

Next, we use the van Roosbroeck–Shockley model to calculate the **bimolecular recombination coefficient** B. Under equilibrium conditions, it is $R = R_0 = B n_i^2$. Thus the bimolecular recombination coefficient is related to the equilibrium recombination rate by

$$B = \frac{R_0}{n_i^2} . \qquad (3.23)$$

3 Theory of radiative recombination

Table 3.1 shows the bimolecular recombination coefficient for different semiconductors as calculated from Eqs. (3.21) and (3.23). All material parameters used in the calculation are given in the table. The calculated results reveal that $B = 10^{-9}$–10^{-11} cm³/s for direct-gap III–V semiconductors. This calculated result agrees well with experimental results. GaP, Si, and Ge, all indirect-gap semiconductors, have a much smaller bimolecular recombination coefficient compared with direct-gap III–V semiconductors.

There are several other ways to calculate the bimolecular recombination coefficient. An early calculation by Hall (1960) using a two-band model produced a bimolecular recombination coefficient of

$$B = 5.8 \times 10^{-13} \frac{\text{cm}^3}{\text{s}} \left(\frac{m_h^*}{m_e} + \frac{m_e^*}{m_e} \right)^{-3/2} \left(1 + \frac{m_e}{m_h^*} + \frac{m_e}{m_e^*} \right) \left(\frac{300 \text{ K}}{T} \right)^{3/2} \left(\frac{E_g}{1 \text{ eV}} \right)^2 \bar{n} \qquad (3.24)$$

where m_e^*, m_h^*, and m_e are the effective electron mass, effective hole mass, and free electron mass, respectively. Garbuzov (1982) described a simple quantum mechanical calculation for direct-gap semiconductors and obtained the following expression for the bimolecular recombination coefficient:

$$B = 3.0 \times 10^{-10} \frac{\text{cm}^3}{\text{s}} \left(\frac{300 \text{ K}}{T} \right)^{3/2} \left(\frac{E_g}{1.5 \text{ eV}} \right)^2. \qquad (3.25)$$

All of the methods described here to calculate B give reasonably similar results. The B coefficient is in the 10^{-10} cm³/s range when calculated for GaAs at 300 K using Eqs. (3.21) and (3.23) – (3.25).

3.3 Temperature and doping dependence of recombination

The temperature dependence of the recombination probability is elucidated in Fig. 3.2, which shows a parabolic $E(k)$ relationship at low and high temperatures. Inspection of the figure reveals that the number of carriers per dk interval decreases with increasing temperature. As radiative recombination requires momentum conservation and the recombination probability of an electron is proportional to the number of holes available at equal momentum, the recombination probability decreases with increasing temperature. This trend is confirmed by Eqs. (3.24) and (3.25), which display a $T^{-3/2}$ dependence of the bimolecular recombination coefficient.

The doping-concentration dependence of the recombination probability is elucidated in

3.3 Temperature and doping dependence of recombination

Fig. 3.3, which shows $E(k)$ for non-degenerate and degenerate doping concentrations. Inspection of the figure reveals that the number of holes per dk interval remains constant in the degenerate doping regime. Thus the recombination probability does not increase in the degenerate doping regime.

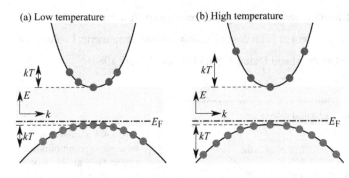

Fig. 3.2. Carrier distribution at (a) low and (b) high temperatures. Recombination probability decreases at high temperatures due to reduced number of carriers per dk interval.

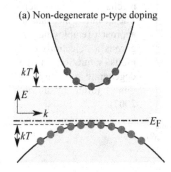 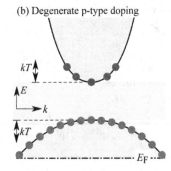

Fig. 3.3. Carrier distribution in (a) non-degenerately and (b) degenerately doped p-type semiconductor. Degenerate doping does not increase the overlap between electrons and holes with equal momentum.

This contention is confirmed by the quantum mechanical calculation of the bimolecular recombination coefficient that is shown versus doping concentration in Fig. 3.4 (Waldron, 2002). Inspection of the figure reveals that the bimolecular recombination coefficient saturates in the degenerate doping regime. The van Roosbroeck–Shockley model does not exhibit this characteristic as its validity is limited to the non-degenerate case.

The bimolecular rate equation, $R = B n p$, applies to *dilute* carrier concentrations, i.e. to non-degenerately doped semiconductors. Thus the bimolecular recombination coefficient applies to semiconductors with non-degenerate carrier concentrations. In this case, the bimolecular recombination coefficient is independent of the carrier concentration. For very high carrier concentrations, however, the bimolecular recombination coefficient decreases due to an

3 Theory of radiative recombination

increasing momentum mismatch between electrons and holes. The bimolecular recombination coefficient in the degenerate regime can be expressed as

$$B|_{\text{high concentrations}} = B - \frac{n}{N_c} B^*. \tag{3.26}$$

That is, the recombination coefficient is reduced at high concentrations. A detailed discussion of the bimolecular recombination coefficient at high concentrations including numerical values for B^* can be found in the literature (Agrawal and Dutta, 1986; Olshansky *et al.*, 1984).

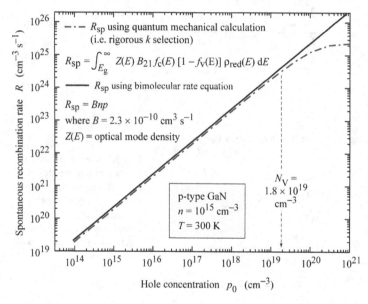

Fig. 3.4. Calculated spontaneous recombination rate in GaN at 300 K as a function of p-type doping concentration using a classical and quantum mechanical approach. The quantum mechanical approach (employing rigorous k selection) exhibits saturation in the degenerate doping regime (after Waldron, 2002).

Exercise: *Radiative efficiency*. Analyze the temperature dependence of the radiative lifetime based on the van Roosbroeck–Shockley model and the non-radiative lifetime based on the Shockley–Read model and predict the temperature dependence of the radiative efficiency in semiconductors.

3.4 The Einstein model

The first theory of optical transitions was developed by Albert Einstein. The Einstein model includes **spontaneous** and **stimulated** or **induced transitions**. *Spontaneous transitions* occur spontaneously, that is, without an external stimulus. In contrast, *stimulated transitions* are induced by an external stimulus, namely a photon. Thus the induced transition rates are proportional to the photon density or radiation density.

3.4 The Einstein model

The coefficients A and B describe spontaneous and stimulated transitions in an atom with two quantized levels. These transitions are illustrated schematically in Fig. 3.5. Denoting the two levels as "1" and "2", Einstein postulated the probability per unit time for the downward transition (2 → 1) and upward transition (1 → 2) as

$$W_{2 \to 1} = B_{2 \to 1}\, \rho(\nu) + A \tag{3.27}$$

and

$$W_{1 \to 2} = B_{1 \to 2}\, \rho(\nu). \tag{3.28}$$

The downward transition probability (per atom) has *two* terms, namely the induced term and the spontaneous term. The induced term, $B_{2 \to 1}\, \rho(\nu)$, is proportional to the radiation density $\rho(\nu)$. The spontaneous downward transition probability is a constant A. Note that the spontaneous lifetime is given by $\tau_{\text{spont}} = A^{-1}$. The upward probability is just $B_{1 \to 2}\, \rho(\nu)$.

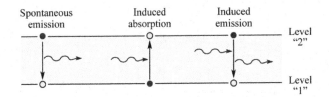

Fig. 3.5. Spontaneous emission, induced absorption, and induced emission events in the two-level atom model. Initially occupied and unoccupied levels are indicated by solid and open circles, respectively.

The Einstein A coefficient in an atom corresponds to the bimolecular recombination coefficient in a semiconductor. In an atom, the concentration terms (i.e. n and p in the bimolecular rate equation $R = Bnp$) do not come into play, since for a downward transition to occur, the upper level must be occupied (one electron, "$n = 1$") and the lower level must be unoccupied (one hole, "$p = 1$").

Einstein showed that $B = B_{2 \to 1} = B_{1 \to 2}$. Thus stimulated absorption and stimulated emission are complementary processes. He also showed that the ratio of the coefficients at a frequency ν in an isotropic and homogeneous medium with refractive index $\bar{n}$ is a constant given by $A/B = 8\pi\, \bar{n}^3 h \nu^3 / c^3$.

The equivalence of $B_{2 \to 1}$ and $B_{1 \to 2}$ can also be shown by quantum mechanical considerations, namely by using Fermi's Golden Rule. However, a detailed discussion of the Einstein model goes beyond the scope of this discussion.

References

Agrawal G. P. and Dutta N. K. *Long-Wavelength Semiconductor Lasers* Chapter 3 (Van Nostrand Reinhold, New York, 1986)

Casey Jr. H. C. and Panish M. B. *Heterostructure Lasers*, Part A, Cap. 3 (Academic Press, New York, 1978)

Bebb H. B. and Williams E. W. "Photoluminescence I: Theory" in *Transport and Optical Phenomena* edited by R. K. Willardson and A. C. Beer, Semiconductors and Semimetals **8**, p. 181 (Academic Press, New York, 1972)

Dutta N. K. "Radiative transitions in GaAs and other III–V compounds" in *Minority Carriers in III–V Semiconductors: Physics and Applications* edited by R. K. Ahrenkiel and M. S. Lundstrom, Semiconductors and Semimetals **39**, p. 1 (Academic Press, San Diego, 1993)

Garbuzov D. Z. "Radiation effects, lifetimes and probabilities of band-to-band transitions in direct A_3B_5 compounds of GaAs type" *J. Luminescence* **27**, 109 (1982)

Hall R. N. "Recombination processes in semiconductors" *Proc. Inst. Electr. Eng.* **106 B**, Suppl. 17, 983 (1960)

Kane E. O. "Band structure of indium antimonide" *J. Phys. Chem. Solids* **1**, 249 (1957)

Olshansky R., Su C. B., Manning J., and Powazinik W. "Measurement of radiative and non-radiative recombination rates in InGaAsP and AlGaAs light sources" *IEEE J. Quantum Electron.* **QE-20**, 838 (1984)

Thompson G. H. B. *Physics of Semiconductor Laser Devices* (John Wiley and Sons, New York, 1980)

Van Roosbroeck W. and Shockley W. "Photon-radiative recombination of electrons and holes in germanium" *Phys. Rev.* **94**, 1558 (1954)

Waldron E. L. "Optoelectronic properties of AlGaN/GaN superlattices" Ph.D. Dissertation, Boston University (2002)

4

LED basics: Electrical properties

4.1 Diode current–voltage characteristic

The electrical characteristics of p-n junctions will be summarized, however, a detailed derivation of the results will not be provided in this chapter. We consider an *abrupt p-n junction* with a donor concentration of N_D and an acceptor concentration of N_A. All dopants are assumed to be fully ionized so that the free electron concentration is given by $n = N_D$ and the free hole concentration is given by $p = N_A$. It is further assumed that no *compensation* of the dopants occurs by unintentional impurities and defects.

In the vicinity of an unbiased p-n junction, electrons originating from donors on the n-type side diffuse over to the p-type side where they encounter many holes with which they recombine. A corresponding process occurs with holes that diffuse to the n-type side. As a result, a region near the p-n junction is depleted of free carriers. This region is known as the ***depletion region***.

(a) p-n junction under zero bias

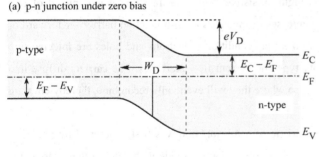

(b) p-n junction under forward bias

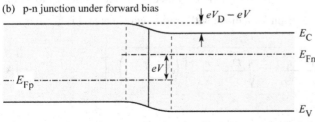

Fig. 4.1. P-n junction under (a) zero bias and (b) forward bias. Under forward-bias conditions, minority carriers diffuse into the neutral regions where they recombine.

59

4 LED basics: electrical properties

In the absence of free carriers in the depletion region, the only charge in the depletion region is from ionized donors and acceptors. These dopants form a space charge region, i.e. donors on the n-type side and acceptors on the p-type side. The space charge region produces a potential that is called the **diffusion voltage**, V_D. The diffusion voltage is given by

$$V_D = \frac{kT}{e} \ln \frac{N_A N_D}{n_i^2} \tag{4.1}$$

where N_A and N_D are the acceptor and donor concentrations, respectively, and n_i is the intrinsic carrier concentration of the semiconductor. The diffusion voltage is shown in the band diagram of Fig. 4.1. The diffusion voltage represents the barrier that free carriers must overcome in order to reach the neutral region of opposite conductivity type.

The width of the depletion region, the charge in the depletion region, and the diffusion voltage are related by the Poisson equation. It is therefore possible to determine the depletion layer width from the diffusion voltage. The depletion layer width is given by

$$W_D = \sqrt{\frac{2\varepsilon}{e}(V_D - V)\left(\frac{1}{N_A} + \frac{1}{N_D}\right)} \tag{4.2}$$

where $\varepsilon = \varepsilon_r \varepsilon_0$ is the dielectric permittivity of the semiconductor and V is the diode bias voltage.

Upon application of the bias voltage to the p–n junction, the voltage is going to drop across the depletion region. This region is highly resistive due to the fact that it is depleted of free carriers. An external bias therefore decreases or increases the p–n junction barrier for forward or reverse bias, respectively. Under *forward*-bias conditions, electrons and holes are injected into the region with opposite conductivity type and current flow *increases*. The carriers diffuse into the regions of opposite conductivity type where they will eventually recombine, thereby emitting a photon.

The current–voltage (*I–V*) characteristic of a p–n junction was first developed by Shockley and the equation describing the *I–V* curve of a p–n junction diode is therefore referred to as the **Shockley equation**. The Shockley equation for a diode with cross-sectional area A is given by

$$\boxed{I = eA\left(\sqrt{\frac{D_p}{\tau_p}}\frac{n_i^2}{N_D} + \sqrt{\frac{D_n}{\tau_n}}\frac{n_i^2}{N_A}\right)\left(e^{eV/kT} - 1\right)} \tag{4.3}$$

where $D_{n,p}$ and $\tau_{n,p}$ are the electron and hole diffusion constants and the electron and hole minority-carrier lifetimes, respectively.

Under reverse-bias conditions, the diode current saturates and the saturation current is given by the factor preceding the exponential function in the Shockley equation. The diode I–V characteristic can be written as

$$I = I_s \left(e^{eV/kT} - 1 \right) \quad \text{with} \quad I_s = eA \left(\sqrt{\frac{D_p}{\tau_p}} \frac{n_i^2}{N_D} + \sqrt{\frac{D_n}{\tau_n}} \frac{n_i^2}{N_A} \right). \qquad (4.4)$$

Under typical forward-bias conditions, the diode voltage is $V \gg kT/e$, and thus $[\exp(eV/kT) - 1] \approx \exp(eV/kT)$. Using Eq. (4.1), the Shockley equation can be rewritten, for forward-bias conditions, as

$$I = eA \left(\sqrt{\frac{D_p}{\tau_p}} N_A + \sqrt{\frac{D_n}{\tau_n}} N_D \right) e^{e(V-V_D)/kT}. \qquad (4.5)$$

The exponent of the exponential function in Eq. (4.5) illustrates that the current strongly increases as the diode voltage approaches the diffusion voltage, i.e. $V \approx V_D$. The voltage at which the current strongly increases is called the **threshold voltage** and this voltage is given by $V_{th} \approx V_D$.

The band diagram of a p-n junction, shown in Fig. 4.1, also illustrates the separation of the Fermi level from the conduction and valence band edge. The difference in energy between the Fermi level and the band edges can be inferred from Boltzmann statistics and is given by

$$E_C - E_F = -kT \ln \frac{n}{N_c} \qquad \text{for the n-type side} \qquad (4.6)$$

and

$$E_F - E_V = -kT \ln \frac{p}{N_v} \qquad \text{for the p-type side.} \qquad (4.7)$$

The band diagram shown in Fig. 4.1 illustrates that the following sum of energies is zero:

$$eV_D - E_g + (E_F - E_V) + (E_C - E_F) = 0. \qquad (4.8)$$

In highly doped semiconductors, the separation between the band edges and the Fermi level is small compared with the bandgap energy, i.e. $(E_C - E_F) \ll E_g$ on the n-type side and $(E_F - $

4 LED basics: electrical properties

E_V) $\ll E_g$ on the p-type side. Furthermore, these quantities depend only weakly (logarithmic dependence) on the doping concentration as inferred from Eqs. (4.6) and (4.7). Thus, the third and fourth summand of Eq. (4.8) can be neglected and the diffusion voltage can be approximated by the bandgap energy divided by the elementary charge

$$V_{th} \approx V_D \approx E_g/e \,. \tag{4.9}$$

Several diode I–V characteristics of semiconductors made from different materials are shown in Fig. 4.2 along with the bandgap energy of these materials. The experimental threshold voltages shown in the figure, and the comparison with the bandgap energy of these materials, indicates that the energy gap and the threshold voltage indeed agree reasonably well.

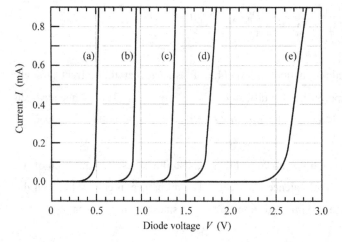

Fig. 4.2. Room-temperature current–voltage characteristics of p-n junctions made from different semiconductors.

The forward diode voltage at a diode current of 20 mA versus bandgap energy for LEDs emitting in the ultraviolet, visible, and infrared wavelength range is shown in Fig. 4.3 (Krames et al., 2000; Emerson et al., 2002). The solid line illustrates the expected forward diode voltage. The line equals the bandgap energy divided by the elementary charge. Inspection of the figure reveals that most semiconductor LEDs follow the solid line, except for LEDs based on III–V nitrides. This peculiarity is due to several reasons. Firstly, large bandgap discontinuities occur in the nitride material system, which cause an additional voltage drop. Secondly, the contact technology is less mature in the nitride material system, which causes an additional voltage drop at the ohmic contacts. Thirdly, the p-type conductivity in bulk GaN is generally low. Lastly, a parasitic voltage drop can occur in the n-type buffer layer.

4.2 Deviations from the ideal I–V characteristic

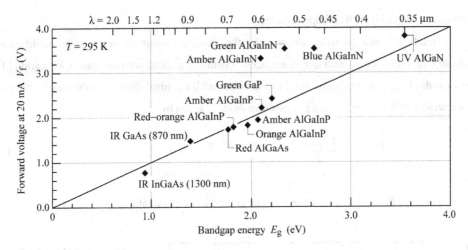

Fig. 4.3. Diode forward voltage versus bandgap energy for LEDs made from different materials (after Krames *et al.*, 2000; updated with UV LED data of Emerson *et al.*, 2002).

Assuming a chip area of 250 µm × 250 µm and a current of 20 mA, the current density used in Fig. 4.3 to characterize the forward voltage is 32 A/cm². Typical current densities in LEDs range from 30 A/cm² in low-power devices to 100 A/cm² in high-power devices.

4.2 Deviations from the ideal I–V characteristic

The Shockley equation gives the expected theoretical *I–V* characteristic of a p-n junction. To describe experimentally measured characteristics, the following equation is used:

$$I = I_s e^{eV/(n_{ideal} kT)} \tag{4.10}$$

where n_{ideal} is the ***ideality factor*** of the diode. For a perfect diode, the ideality factor has a value of unity (n_{ideal} = 1.0). For real diodes, the ideality factor assumes values of typically n_{ideal} = 1.1–1.5. However, values as high as n_{ideal} = 2.0 have been found for III–V arsenide and phosphide diodes. Values as high as n_{ideal} = 7.0 have been found for GaN/GaInN diodes. For a detailed analysis of the diode ideality factor see, for example, Rhoderick and Williams (1988), and Shah *et al.* (2003).

Frequently a diode has unwanted or ***parasitic resistances***. The effect of a series resistance and a parallel resistance is shown in Fig. 4.4 (a). A series resistance can be caused by excessive contact resistance or by the resistance of the neutral regions. A parallel resistance can be caused by any channel that bypasses the p-n junction. This bypass can be caused by damaged regions of

4 LED basics: electrical properties

the p–n junction or by surface imperfections.

The diode I–V characteristic, as given by the Shockley equation, needs to be modified in order to take into account parasitic resistances. Assuming a shunt with resistance R_p (parallel to the ideal diode) and a series resistance R_s (in series with the ideal diode *and* the shunt), the I–V characteristic of a forward-biased p–n junction diode is given by

$$I - \frac{(V - IR_s)}{R_p} = I_s e^{e(V-IR_s)/(n_{\text{ideal}} kT)} . \qquad (4.11)$$

For $R_p \to \infty$ and $R_s \to 0$, this equation reduces to the Shockley equation.

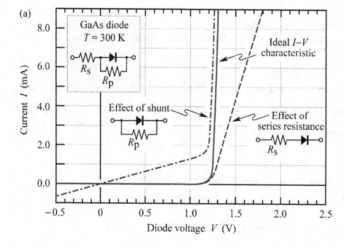
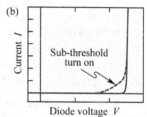

Fig. 4.4. (a) Effect of series and parallel resistance (shunt) on I–V characteristic. (b) I–V with clearly discernable sub-threshold turn-on, caused by defects or surface states.

Occasionally, the diode turn-on is distributed over a range of voltages rather than occurring abruptly at the threshold voltage. Both types of turn-on are shown in Fig. 4.4 (b). The non-abrupt turn-on is referred to as **sub-threshold turn-on** or **premature turn-on**. The sub-threshold current can be caused by carrier transport through surface states or deep levels in the bulk of the semiconductor.

Detailed inspection of the diode I–V characteristic on a linear as well as logarithmic scale allows for the diagnosis of potential problems such as shunts, series resistances, premature turn-on, and parasitic diodes. Figure 4.5 shows a number of parasitic effects that can occur in diodes. The diagrams may allow the reader to diagnose and identify specific problems in diodes.

4.2 Deviations from the ideal I–V characteristic

Fig. 4.5(a)

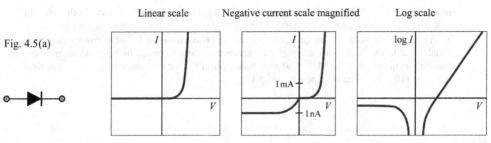

Ideal diode: The ideal diode *I-V* characteristic is given by the Shockley equation.

Fig. 4.5(b)

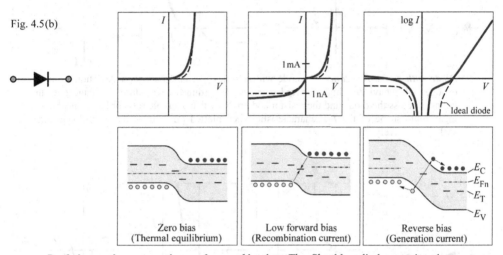

Depletion region generation and recombination: The Shockley diode equation does not account for carrier generation and recombination events in the depletion region. However, in practical diodes, there are trap levels in the depletion region, which make such events likely. Carrier generation and recombination causes an excess current for both forward and reverse bias. In the forward-bias regime, the excess current is due to the recombination of minority carriers in the depletion region. This recombination current dominates only at low voltages and gives an ideality factor of 2.0. At higher voltages, the diffusion current dominates resulting in an ideality factor of 1.0. In the reverse-bias regime, the excess current is due to the generation of carriers in the depletion region. Under the influence of the electric field in the depletion region, generated carriers drift to the neutral regions. This generation current keeps increasing with reverse voltage due to the increasing depletion-layer width.

Fig. 4.5(c)

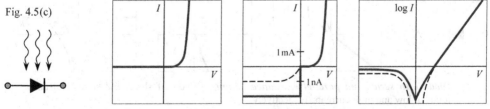

Photocurrent: In a practical measurement within a lighted room, a diode located in a transparent package generates a photocurrent. Therefore measurements need to be carried out in the dark. Switching off room lights or covering the setup with a dark cloth helps in reducing the photocurrent. In the dark, at zero voltage, there should be zero current. However, a very small non-zero current (e.g. 10^{-12} A) is frequently measured. The non-zero current is usually due to the limited accuracy of the measurement instrument. The best instruments will measure a current of about 10^{-15} A at zero bias, even if the measurement is carried out in total darkness (10^{-15} A = 1 atto ampere = 10^{-3} pA).

Fig. 4.5(d)

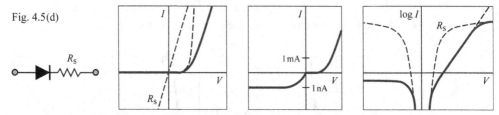

Diode with series resistance: A diode with a series resistance shows a deviation from the exponential behavior at high forward currents. According to Kirchhoff's voltage law, the voltages across the diode and the resistor add up. Note that a simple resistor has a linear and a logarithmic shape of the *I–V* characteristic when plotted on a linear and semi-logarithmic scale, respectively.

Fig. 4.5(e)

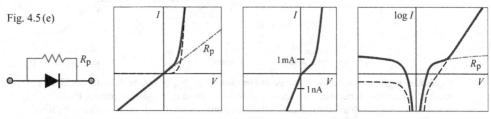

Diode with parallel resistance (shunt): According to Kirchhoff's current law, the currents through diode and resistor add up. Note that the forward "hump" seen on the semi-logarithmic plot has about the same level as the reverse saturation current. This is a characteristic by which a shunt can be identified.

Fig. 4.5(f)

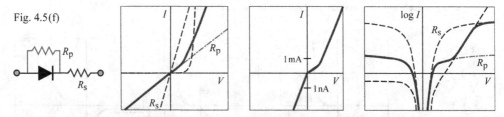

Diode with series and parallel resistance (shunt): Effects of shunt and series resistance found at low and high currents, respectively.

4.3 Evaluation of diode parasitic resistances

Fig. 4.5(g)

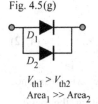

$V_{th1} > V_{th2}$
Area$_1$ >> Area$_2$

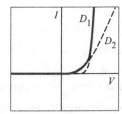
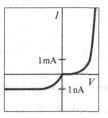
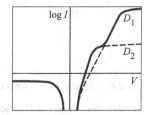

Parasitic diode with lower barrier height and smaller area than main diode: Such diodes display premature turn-on caused by leakage through either surface states at the perimeter of the diode chip or defective regions within the p-n junction plane that have a lower barrier height than the main p-n junction. Note that the forward "hump" on the semi-logarithmic plot has much higher level than the reverse saturation current, which is not the case for diodes with a shunt.

Exercise: *Critical points of diode current–voltage characteristics.* The *I–V* characteristics of diodes are frequently characterized in terms of **four critical points**, namely *forward voltage one*, V_{f1}, *forward voltage two*, V_{f2}, *forward voltage three*, V_{f3}, and *reverse saturation current*, I_s, specified at the operating current (e.g. 100 mA), a small forward current (e.g. 10 µA), a very small forward current (e.g. 1 µA), and at negative bias (e.g. – 5 V), respectively. The critical points are shown in Fig. 4.6.
(a) Explain the relevance of the critical points.
(b) Two GaInN diodes have the following data: (1) V_{f1} = 3.2 V, V_{f2} = 2.5 V, V_{f3} = 2.3 V, I_s = 0.8 µA; (2) V_{f1} = 3.4 V, V_{f2} = 2.0 V, V_{f3} = 1.8 V, I_s = 0.8 µA. Which device has the more favorable characteristics?

Solution: (a) For devices emitting at the same peak wavelength, V_{f1} should be as *low* as possible, as high values indicate a high series resistance. The forward voltage two, V_{f2}, should be as *high* as possible (as close to V_{f1} as possible), as low values of V_{f2} indicate excessive sub-threshold leakage. The same argument applies to V_{f3}. The reverse saturation current should be as low as possible as high values of I_s indicate excessive leakage paths (e.g. surface leakage or bulk leakage mediated by surface states, bulk point defects and dislocations). Low values of V_{f1}, high values of V_{f2} and V_{f3}, and low values of I_s are consistently correlated with high device reliability. (b) Device (1) has more favorable characteristics due to lower series resistance and lower sub-threshold leakage.

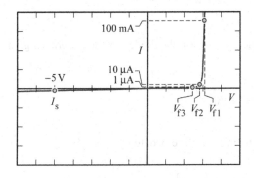

Fig. 4.6. Critical points of diode *I–V* characteristic, namely "forward voltage one", V_{f1} (measured at operating current, e.g. 100 mA), "forward voltage two", V_{f2} (measured at low current, e.g. 10 µA), "forward voltage three", V_{f3} (measured at very low current, e.g. 1 µA), and reverse saturation current (measured at e.g. – 5.0 V).

4.3 Evaluation of diode parasitic resistances

The diode parallel resistance can be evaluated near the origin of the *I–V* diagram where $V \ll E_g / e$. For this voltage range, the p-n junction current can be neglected and the parallel

4 LED basics: electrical properties

resistance is given by

$$R_p \approx dV/dI \big|_{\text{near origin}}. \tag{4.12}$$

Note that in any reasonable diode, the parallel resistance is much larger than the series resistance so that the series resistance need not be taken into account when evaluating the parallel resistance.

The series resistance can be evaluated at a high voltage where $V > E_g/e$. For sufficiently large voltages, the diode I–V characteristic becomes linear and the series resistance is given by the tangent to the I–V curve, as shown in Fig. 4.7 (a).

$$R_s = dV/dI \big|_{\text{at voltages exceeding turn-on}}. \tag{4.13}$$

However, it may not be practical to evaluate the diode resistance at high voltages due to device heating effects. For this case, the following procedure will be suitable.

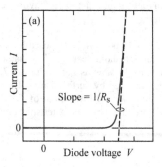
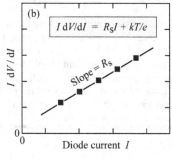

Fig. 4.7. Methods for evaluating diode series resistance. (a) Tangent for $V > V_{th}$ provides R_s. (b) Equation shown as inset is valid for forward bias ($V \gg kT/e$).

For devices with a high parallel resistance ($R_p \to \infty$), the diode I–V characteristic, given in Eq. (4.11), can be written as

$$I = I_s \, e^{e(V - IR_s)/(n_{\text{ideal}} kT)}. \tag{4.14}$$

Solving the equation for V and then differentiating V with respect to I yields

$$\frac{dV}{dI} = R_s + \frac{n_{\text{ideal}} kT}{e} \frac{1}{I} \tag{4.15}$$

where the second summand on the right-hand side of the equation represents the differential p-n junction resistance. Multiplication of the equation by I allows one to identify the series resistance of the diode as the slope of a $(I\,dV/dI)$-versus-I plot, as shown in Fig. 4.7 (b).

4.4 Emission energy

The energy of photons emitted from a semiconductor with energy gap E_g is given by the bandgap energy, i.e.

$$h\nu \approx E_g. \tag{4.16}$$

In an ideal diode, every electron injected into the active region will generate a photon. Conservation of energy thus requires that the energy with which an electron is injected is equal to the photon energy. Thus energy conservation requires

$$eV = h\nu. \tag{4.17}$$

That is, the voltage applied to the LED multiplied by the elementary charge is equal to the photon energy. There are several effects that can change the diode voltage from the ideal value given by Eq. (4.17). These effects will be discussed below.

4.5 Carrier distribution in p-n homojunctions

The carrier distribution in p-n homojunctions, i.e. p-n junctions consisting of a single material, depends on the *diffusion constant* of the carriers. The diffusion constant of carriers is not easily measured. Much more common is the measurement of the *carrier mobility*; for example, by the Hall effect. The diffusion constant can be inferred from the carrier mobility by the **Einstein relation**, which, for non-degenerate semiconductors, is given by

$$D_n = \frac{kT}{e}\mu_n \quad \text{and} \quad D_p = \frac{kT}{e}\mu_p. \tag{4.18}$$

Carriers injected into a neutral semiconductor, with no external electric field applied, propagate by diffusion. If carriers are injected into a region with opposite conductivity type, the minority carriers will eventually recombine. The mean distance a minority carrier diffuses before recombination is the **diffusion length**. Electrons injected into the p-type region will, on average, diffuse over the diffusion length L_n, before recombining with holes. The diffusion length is given by

$$L_n = \sqrt{D_n \tau_n} \quad \text{and} \quad L_p = \sqrt{D_p \tau_p} \tag{4.19}$$

4 LED basics: electrical properties

where τ_n and τ_p are the electron and hole minority carrier lifetimes, respectively. In typical semiconductors, the diffusion length is of the order of a several micrometers. For example, the diffusion length of electrons in p-type GaAs is given by $L_n = (220 \text{ cm}^2/\text{s} \times 10^{-8} \text{ s})^{1/2} \approx 15 \text{ μm}$. Thus, minority carriers are distributed over a region several micrometers thick.

The distribution of carriers in a p-n junction under zero bias and under forward bias is shown in Figs. 4.8 (a) and (b), respectively. Note that minority carriers are distributed over quite a large distance. Furthermore, the minority carrier concentration decreases as these carriers diffuse further into the adjacent region. Thus recombination occurs over a large region, with a strongly changing minority carrier concentration. As will be shown below, the *large recombination region in homojunctions* is not beneficial for efficient recombination.

(a) Homojunction under zero bias

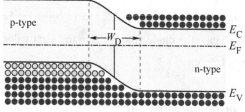

(b) Homojunction under forward bias

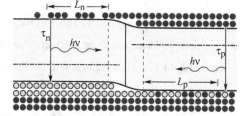

(c) Heterojunction under forward bias

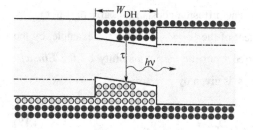

Fig. 4.8. P-n homojunction under (a) zero and (b) forward bias. (c) P-n heterojunction under forward bias. In homojunctions, carriers diffuse, on average, over the diffusion lengths L_n and L_p before recombining. In heterojunctions, carriers are confined by the heterojunction barriers.

4.6 Carrier distribution in p-n heterojunctions

All high-intensity light-emitting diodes do *not* use the homojunction design but rather employ heterojunctions, which have clear advantages over homojunction devices. Heterojunction devices employ two types of semiconductors, namely a small-bandgap active region and a large-bandgap barrier region. If a structure consists of *two* barriers, i.e. two large-bandgap semiconductors, then the structure is called a **double heterostructure** (frequently abbreviated as **DH**).

The effect of heterojunctions on the carrier distribution is shown in Fig. 4.8 (c). Carriers injected into the active region of the double heterostructure are confined to the active region by means of the barriers. As a result, the thickness of the region in which carriers recombine *is given by the thickness of the active region rather than the diffusion length.*

The consequences of this change are significant. We assume that the thickness of the active region is much smaller than the typical diffusion length. Diffusion lengths may range from 1 to 20 µm. The active region of double heterojunctions may range from 0.01 to 1.0 µm. Thus, carriers in the active region of a double heterostructure have a much higher concentration than carriers in homojunctions, which are distributed over several diffusion lengths. Recalling that the radiative recombination rate is given by the bimolecular recombination equation, i.e.

$$R = Bnp \qquad (4.20)$$

it is clear that a high concentration of carriers in the active region *increases* the radiative recombination rate and decreases the recombination lifetime. For this reason, all high-efficiency LED designs employ double heterostructure or quantum well designs.

4.7 Effect of heterojunctions on device resistance

The employment of heterostructures allows one to improve the efficiency of LEDs by confining carriers to the active region, thereby avoiding diffusion of minority carriers over long distances. Heterostructures can also be used to confine light to waveguide regions; in particular, in edge-emitting LEDs. Generally, modern semiconductor LEDs and lasers have many heterojunctions, e.g. for contact layers, active regions, and waveguiding regions. Although heterostructures allow for improved LED designs, there are also problems associated with heterojunctions.

One of the problems introduced by heterostructures is the resistance caused by the heterointerface. The origin of the resistance is illustrated in Fig. 4.9 (a), which shows the band diagram of a heterostructure. The heterostructure consists of two semiconductors with different bandgap energy and it is assumed that both sides of the heterostructure are of n-type conductivity. Carriers in the large-bandgap material will diffuse over to the small-bandgap material where they occupy conduction band states of lower energy. As a result of the electron transfer, an electrostatic dipole forms, consisting of a positively charged depletion layer with ionized donors in the large-bandgap material, and a negatively charged electron accumulation layer in the small-bandgap material. The charge transfer leads to the band bending illustrated in Fig. 4.9 (a). Carriers transferring from one semiconductor to the other must overcome this barrier by either tunneling or by thermal emission over the barrier. The resistance caused by heterojunctions can have a strong deleterious effect on device performance, especially in high-power devices. The thermal power produced by heterostructure resistances leads to heating of the active region, thereby decreasing the radiative efficiency.

It has been shown that heterostructure band discontinuities can be completely eliminated by *grading* of the chemical composition of the semiconductor in the vicinity of the heterostructure (Schubert *et al.*, 1992). The band diagram of a graded heterostructure is shown in Fig. 4.9 (b). Inspection of the figure reveals that there is no longer a spike in the conduction band which hinders the electron flow. It has been shown that the resistance of parabolically graded heterostructures is comparable to bulk material resistance. Thus, the additional resistance introduced by abrupt heterostructures can be completely eliminated by parabolic grading.

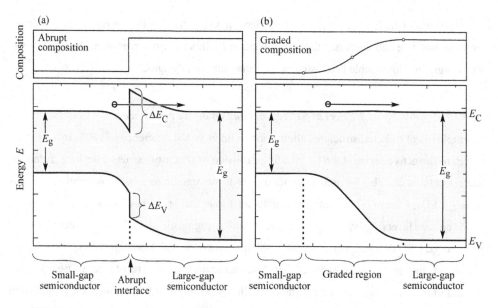

Fig. 4.9. Band diagram of (a) an abrupt n-type–n-type heterojunction and (b) a graded heterojunction of two semiconductors with different bandgap energy. The abrupt junction is more resistive than the graded junction due to the electron barrier forming at the abrupt junctions (after Schubert *et al.*, 1992).

The shape of the graded region should be parabolic for the following reason. The large-bandgap material will be depleted of free carriers due to electron transfer to the small-bandgap material. Thus the charge concentration in the large-bandgap material will be the donor concentration. Assuming that the donor concentration N_D is a constant throughout the heterostructure, the solution of the Poisson equation yields the electrostatic potential

$$\Phi = \frac{eN_D}{2\varepsilon}x^2. \qquad (4.21)$$

The equation reveals that the potential depends *quadratically* on the spatial coordinate x, i.e. the potential has a *parabolic* shape. In order to compensate for the parabolic shape of the depletion potential, the composition of the semiconductor *is varied parabolically as well*, so that an overall *flat potential* results. It is assumed here that the parabolic variation of the chemical composition results in a parabolic change of the bandgap energy, i.e. that the bandgap energy depends linearly on the chemical composition and that bandgap *bowing* can be neglected.

Next, an approximate design rule for the grading of a heterostructure is given. Assume that the conduction band discontinuity of an abrupt heterojunction is given by ΔE_C and that the structure is uniformly doped with doping concentration N_D. Let us assume that carriers have transferred to the small-bandgap semiconductor, thus causing a depletion region of thickness W_D in the large-bandgap semiconductor. If the potential created in the depletion region is equal to $\Delta E_C / e$, then electrons will no longer transfer to the small-bandgap material. The thickness of the depletion region can be inferred from Eq. (4.21) to be

$$W_D = \sqrt{\frac{2\varepsilon \Delta E_C}{e^2 N_D}}. \qquad (4.22)$$

A heterostructure interface should be graded over the distance W_D in order to minimize the resistance introduced by an abrupt heterostructure. Although the result of Eq. (4.22) is an approximation it provides excellent guidance for device design. Steps can be taken to refine the calculation. For example, the potential change due to the electron accumulation layer in the small-bandgap material can be taken into account. Several software packages are available that allow for the numerical calculation of semiconductor heterostructures, for example the software package *Atlas* by the Silvaco Corporation.

Exercise: *Grading of heterostructures*. Assume that the conduction band discontinuity of an AlGaAs/GaAs heterostructure is given by ΔE_C = 300 meV and that the structure is uniformly doped with donors of concentration $N_D = 5 \times 10^{17}$ cm^{-3}. Over what distance should the interface be graded in order to minimize the resistance occurring in abrupt heterostructures?

Solution: Calculating the depletion layer thickness from Eq. (4.22) yields W_D = 30 nm. Thus the heterostructure should be graded over 30 nm to minimize the heterostructure resistance. The graded region should have *two* parabolic regions as shown in Fig. 4.9 (b).

Grading is useful for all heterostructures, including the heterostructures adjoining the active region. The effect of grading in a double heterostructure is shown in Fig. 4.10. The composition and the band diagram of an ungraded structure are shown in Fig. 4.10 (a). At both heterointerfaces, barriers develop as a result of free charge transferring to the active region. These barriers increase the device resistance under forward-bias conditions.

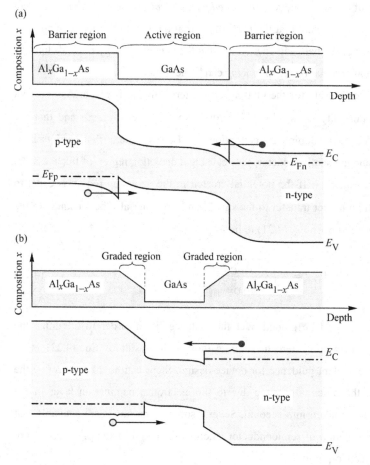

Fig. 4.10. Band diagram of (a) an abrupt double heterostructure and (b) a graded double heterostructure. The barrier–well interface of the abrupt junction is more resistive than the graded junction due to barriers forming at the interfaces.

The case of graded heterointerfaces is shown in Fig. 4.10 (b). The figure shows two linearly graded regions cladding the active region. The band diagram illustrates that barriers at the heterointerfaces can be effectively reduced or completely eliminated by grading. Note that the linear grading shown in Fig. 4.10 (b) results in small "spikes" at the interfaces between the linearly graded and the non-graded regions. These "spikes" are a result of the linear grading assumed, and would not result for parabolic grading.

Generally, the transport of carriers in heterostructures should be as *adiabatic* as possible, i.e. the carrier transport within the semiconductor device should not generate unnecessary heat. This is particularly true for high-power devices where additional heat generated inside the device leads to a performance loss due to increased operating temperature.

Finally, note that *lattice matching* is desirable in all heterostructure devices. It is also desirable in graded structures in order to minimize the number of misfit dislocations that act as non-radiative recombination centers.

4.8 Carrier loss in double heterostructures

In an ideal LED, the injected carriers are confined to the active region by the barrier layers adjoining the active regions. By means of confinement of carriers to the active region, a high carrier concentration is attained resulting in a high radiative efficiency of the recombination process.

The energy barriers confining the carriers to the active region are typically of the order of several hundred meV, i.e. much larger than kT. Nevertheless, some carriers will succeed in escaping from the active region into the barrier layers. The concentration of the escaping carriers in the barrier layers will be rather low, resulting in a low radiative efficiency of carriers in the barrier layers.

Free carriers in the active region are distributed according to the Fermi–Dirac distribution and, as a result, some carriers have a higher energy than the height of the confining barrier. Thus some of the carriers escape from the active region into the barrier regions as illustrated in Fig. 4.11.

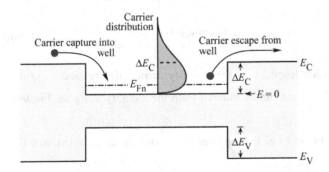

Fig. 4.11. Carrier capture and escape in a double heterostructure. Also shown is the carrier distribution in the active layer.

Consider electrons in the active region of a double heterostructure and assume that the active region is clad by a barrier with height ΔE_C, as shown in Fig. 4.11. The energy distribution of carriers is given by the Fermi–Dirac distribution. Thus, a certain fraction of the carriers residing in the active region have an energy higher than the energy of the barrier. The concentration of electrons with energy higher than the barrier is given by

$$n_B = \int_{E_B}^{\infty} \rho_{DOS}\, f_{FD}(E)\, dE \tag{4.23}$$

where ρ_{DOS} is the density of states, f_{FD} is the Fermi–Dirac distribution function, and E_B is the height of the barrier. For a bulk-type density of states, the concentration of carriers with energy higher than E_B is given by

$$n_B = \frac{1}{2\pi^2}\left(\frac{2m^*}{\hbar^2}\right)^{3/2} \int_{E_B}^{\infty} \frac{\sqrt{E-E_C}}{1+e^{(E-E_{Fn})/kT}}\, dE . \tag{4.24}$$

Taking into account that we are interested in carriers with energies much higher than the Fermi energy, the Fermi–Dirac distribution can be approximated by the Boltzmann distribution. One obtains

$$n_B = N_c\, e^{(E_{Fn}-E_B)/kT} \tag{4.25}$$

where N_c is the effective density of states in the active region. Equation (4.25) gives the concentration of free carriers at the active-region–cladding-region interface. Minority carriers at the edge of the cladding layer will diffuse into the cladding layer. The diffusion process is governed by the initial concentration n_B and the electron diffusion length L_n. Taking the location of the origin ($x = 0$) at the edge of the barrier, the carrier distribution can be written as

$$n_B(x) = n_B(0)\, e^{-x/L_n} = N_c\, e^{-(E_B-E_{Fn})/kT}\, e^{-x/L_n} \tag{4.26}$$

where $L_n = (D_n \tau_n)^{1/2}$ is the diffusion length, τ_n is the minority carrier lifetime, and D_n is the diffusion constant. The diffusion constant can be inferred from the mobility using the Einstein relation $D = \mu kT/e$.

The diffusion current density of electrons leaking over the barrier can be obtained from the carrier concentration gradient at $x = 0$, i.e.

$$J_n\big|_{x=0} = -eD_n \frac{dn_B(x)}{dx}\bigg|_{x=0} = -eD_n \frac{n_B(0)}{L_n} . \tag{4.27}$$

4.8 Carrier loss in double heterostructures

The leakage current depends on the carrier concentration at the edge of the barrier. Thus a high barrier height is required to minimize the leakage current. Clearly, barriers must be much larger than kT for efficient confinement of carriers. Some material systems such as AlGaN/GaN or AlGaAs/GaAs have relatively high barriers and therefore lower leakage currents over the barrier. Other material systems such as AlGaInP/AlGaInP emitting at 600–650 nm have lower barriers and therefore have a stronger carrier leakage over the barriers.

Note that the leakage increases exponentially with temperature. Thus, a decrease of the radiative efficiency of LEDs results as the temperature increases. To reduce the temperature dependence of the emission, high barriers are required. In addition to carrier leakage, other effects, such as Shockley–Read recombination also contribute to the lower radiative efficiency at high temperatures.

Exercise: *Carrier leakage over a barrier*. Electrons in the active region of a GaAs structure have a concentration of 2×10^{18} cm^{-3}. Calculate the current density of the carrier loss over the barrier for barrier heights of 200 and 300 meV, assuming an electron mobility of 2000 cm^2/(V s) and a minority carrier lifetime of 5 ns. Compare the calculated leakage current to LED injection currents of 0.1–1.0 kA/cm^2.

Solution: The Fermi level in GaAs with electron density of 2×10^{18} cm^{-3} is 77 meV above the conduction band edge. Assuming that the effective density of states in the barrier is the same as in the GaAs active region, the carrier concentrations at the edge of the barrier are 3.9×10^{15} cm^{-3} for a 200 meV barrier and 8.3×10^{13} cm^{-3} for a 300 meV barrier. The diffusion constant, as inferred from the Einstein relation, is $D_n = 51.7$ cm^2/s. The diffusion length is then given by $L_n = (D_n \tau_n)^{1/2} = 5.1$ µm. The leakage current is calculated using Eq. (4.27), and one obtains 63 A/cm^2 for the 200 meV barrier and 1.3 A/cm^2 for the 300 meV barrier. Comparison with diode current densities of 0.1–1.0 kA/cm^2 suggests that leakage currents can be a significant loss mechanism, particularly for small barrier heights.

In the consideration above, we have assumed that electrons diffuse in the p-type region and have neglected any drift. However, if the p-type region has a substantial resistance, electron drift cannot be neglected. This drift will enhance the electron current. Furthermore, electrical contacts have been neglected. The minority carrier concentration at the contact–semiconductor interface can be assumed to be zero due to the high surface recombination velocity of such interfaces. Taking these effects into account, the leakage current was calculated by Ebeling (1993). If the distance of the contact from the active–barrier interface is denoted by x_p, the leakage current is given by

$$J_n = -e D_n n_B(0) \left(\sqrt{\frac{1}{L_n^2} + \frac{1}{L_{nf}^2}} \coth \sqrt{\frac{1}{L_n^2} + \frac{1}{L_{nf}^2}} \, x_p + \frac{1}{L_{nf}} \right) \quad (4.28)$$

where

4 LED basics: electrical properties

$$L_{nf} = \frac{kT}{e} \frac{\sigma_p}{J_{tot}}, \quad (4.29)$$

σ_p is the conductivity of the p-type cladding region, and J_{tot} is the total diode current density.

4.9 Carrier overflow in double heterostructures

The **overflow of carriers** from the active region into the confinement regions is another loss mechanism. Carrier overflow occurs at high injection current densities. As the injection current increases, the carrier concentration in the active region increases and the Fermi energy rises. For sufficiently high current densities, the Fermi energy will rise to the top of the barrier. The active region is flooded with carriers and a further increase in injection current density will *not* increase the carrier concentration in the active region. As a result, the optical intensity saturates. At high injection current densities, carrier overflow occurs, even if the barriers are sufficiently high, so that carrier leakage over the barriers at low injection current densities can be neglected.

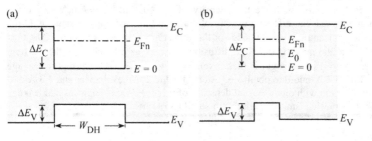

Fig. 4.12. Fermi level (E_{Fn}) and subband level (E_0) in (a) a double heterostructure and (b) a quantum well structure.

Consider a double heterostructure LED with an active region of thickness W_{DH}, as shown in Fig. 4.12. The rate equation of carrier supply to (by injection) and removal from (by recombination) the active region is given by

$$\frac{dn}{dt} = \frac{J}{eW_{DH}} - B n p, \quad (4.30)$$

where B is the bimolecular recombination coefficient. For high injection densities, it is $n = p$. Solving Eq. (4.30) for n under steady-state conditions ($dn/dt = 0$), yields

$$n = \sqrt{\frac{J}{eBW_{DH}}}. \quad (4.31)$$

The carrier density increases with the current injected into the device. As a result, the Fermi energy rises. In the high-density approximation, the Fermi energy is given by

$$\frac{E_F - E_C}{kT} = \left(\frac{3\sqrt{\pi}}{4} \frac{n}{N_c} \right)^{2/3}. \tag{4.32}$$

At high injection levels, the Fermi energy rises and will eventually reach the top of the barrier. At that point, it is $E_F - E_C = \Delta E_C$. Using this value, the current density at which the active region overflows can be calculated from Eqs. (4.31) and (4.32) and one obtains

$$\boxed{J|_{\text{overflow}} = \left(\frac{4 N_c}{3\sqrt{\pi}} \right)^2 \left(\frac{\Delta E_C}{kT} \right)^3 e B W_{\text{DH}}}. \tag{4.33}$$

Either the conduction band or the valence band well may overflow first depending on the effective density of states (N_c, N_v) and the band discontinuities (ΔE_C, ΔE_V).

Exercise: *Carrier overflow in a double heterostructure.* Consider electrons in a GaAs double heterostructure with a barrier height of $\Delta E_C = 200$ meV and an active region thickness of $W_{\text{DH}} = 500$ Å. Calculate the current level at which the electron well overflows.
  Solution: Using $N_c = 4.4 \times 10^{17}$ cm^{-3} and $B = 10^{-10}$ cm^3/s, one obtains from Eq. (4.33) a current level of $J_{\text{max}} = 3990$ A/cm^2.

Generally, the problem of carrier overflow is more severe in structures with a small active-region volume. In particular, single-quantum-well structures and quantum-dot active regions have an inherently small volume. At a certain current density, the active region is filled with carriers, and the injection of additional carriers will not lead to an increase in the emitted light intensity.

Experimental results of an LED structure with one, four, six, and eight quantum wells (QWs) are shown in Fig. 4.13 (Hunt *et al.*, 1992). The light intensity for the single QW structure saturates at a low current level. As the number of quantum wells is increased, the current level at which saturation occurs increases, and the optical saturation intensity increases as well. The saturation of the light intensity displayed in Fig. 4.13 is caused by the overflow of carriers.

The calculation of the overflow current level is different for quantum well structures and bulk active regions. For quantum well structures, we must employ the two-dimensional (2D) density of states, rather than the 3D density of states that was used in the above calculation. The Fermi

4 LED basics: electrical properties

level in a QW with one quantized state with energy E_0 is given by

$$\frac{E_F - E_0}{kT} = \ln\left[\exp\left(\frac{n^{2D}}{N_c^{2D}}\right) - 1\right] \tag{4.34}$$

where n^{2D} is the 2D carrier density per cm² and N_c^{2D} is the effective 2D density of states given by

$$N_c^{2D} = \frac{m^*}{\pi \hbar^2} kT. \tag{4.35}$$

Because we are dealing with high carrier densities, the high-degeneracy approximation can be employed and one obtains

$$E_F - E_0 = \frac{\pi \hbar^2}{m^*} n^{2D}. \tag{4.36}$$

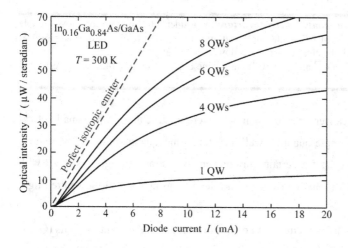

Fig. 4.13. Optical intensity emitted by $In_{0.16}Ga_{0.84}As$/ GaAs LEDs with active regions consisting of one, four, six, and eight quantum wells and theoretical intensity of a perfect isotropic emitter (dashed line) (after Hunt et al., 1992).

Next, we write the rate equation for the quantum well. The rate equation of carrier supply to (by injection) and removal from (by recombination) the active region is given by

$$\frac{dn^{2D}}{dt} = \frac{J}{e} - B^{2D} n^{2D} p^{2D} \tag{4.37}$$

where $B^{2D} \approx B/W_{QW}$ is the bimolecular recombination coefficient for a 2D structure. For high injection densities, it is $n^{2D} = p^{2D}$. Solving Eq. (4.37) for n^{2D} under steady-state conditions ($dn^{2D}/dt = 0$), yields

$$n^{2D} = \sqrt{\frac{J}{eB^{2D}}} = \sqrt{\frac{JW_{QW}}{eB}}. \tag{4.38}$$

At high injection levels, the Fermi energy will reach the top of the barrier. At that point, $E_F - E_0 = \Delta E_C - E_0$. The use of this value in Eq. (4.36) and subsequent elimination of n^{2D} from Eqs. (4.36) and (4.38) yields the current density at which the active region overflows

$$J|_{\text{overflow}} = \left[\frac{m^*}{\pi\hbar^2}(\Delta E_C - E_0)\right]^2 \frac{eB}{W_{QW}}. \tag{4.39}$$

Thus overflow of the active region is a potential problem in double heterostructures as well as quantum well structures. In order to avoid this problem, high-current LEDs must employ *thick* double heterostructure active regions, or *many* QWs of multiple QW (MQW) active regions, or a large injection (contact) area. By choosing these parameters, the volume of the active region can be designed in such a way that at the intended current density of operation, carrier overflow does not occur.

4.10 Electron-blocking layers

Carriers tend to escape from the active layer of an LED into the confinement layers. The carrier escape can be substantial in double heterostructures with a low barrier height at the active–confinement interface. In addition, high temperatures promote carrier loss out of the active region due to the increase in carrier thermal energy.

The electron leakage current is larger than the hole leakage current due to the usually larger diffusion constant of electrons compared with holes in III–V semiconductors. To reduce carrier leakage out of the active region, carrier-blocking layers are used. In particular, *electron-blocking layers* or *electron blockers* are used in many LED structures to reduce electron escape out of the active region. Such electron-blocking layers are regions with a high bandgap energy located at the confinement–active interface.

The band diagram of a GaInN LED with an electron-blocking layer is shown in Fig. 4.14. The LED has AlGaN confinement layers and a GaInN/GaN multiple quantum well active region. An AlGaN electron-blocking layer is included in the p-type confinement layer at the

4 LED basics: electrical properties

confinement–active interface. Figure 4.14 (a), showing the undoped structure, illustrates that the AlGaN electron-blocking layer creates a barrier to current flow in both the conduction band as well as the valence band.

However, Fig. 4.14 (b), showing the doped structure, illustrates that the barrier in the valence band is screened by free carriers so that there is *no barrier* to the flow of *holes* in the p-type confinement layer. That is, the entire band discontinuity is located in the conduction band, i.e.

$$\text{barrier height for electrons} = E_{C,\text{confinement}} - E_{C,\text{active}} + \Delta E_g \qquad (4.40)$$

where ΔE_g is the difference in bandgap energy between the confinement and the electron-blocking layer.

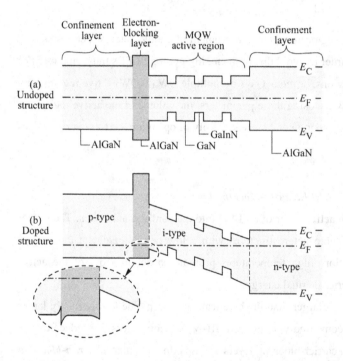

Fig. 4.14. AlGaN electron-blocking layer in an AlGaN/GaN/GaInN multi-quantum well structure. (a) Band diagram without doping. (b) Band diagram with doping. The Al content in the electron-blocking layer is higher than in the p-type confinement layer.

The inset of Fig. 4.14 shows the valence band edge of the electron-blocking layer in greater detail. A potential spike (hole depletion layer in the electron blocker) and notch (hole accumulation layer in the p-type confinement layer) occur at the confinement–blocking layer interface. Holes must tunnel through the potential spike when propagating towards the active region. Note that the valence band edge can be completely smoothed out by compositional grading at the confinement–blocking layer interface so that the electron blocking layer does not impede the hole flow at all.

4.11 Diode voltage

The energy of an injected electron is converted into optical energy upon electron–hole recombination. Thus, conservation of energy requires that the **drive voltage** or **forward voltage** of a light-emitting device is equal to (or larger than) the bandgap energy divided by the elementary charge. The diode voltage is thus given by

$$V = h\nu/e \approx E_g/e. \tag{4.41}$$

There are several mechanisms causing the drive voltage to be slightly different from this value and these mechanisms will be discussed below.

Firstly, if the diode has a significant series resistance, an additional voltage drop occurs. The additional resistance can be caused by (*i*) *contact resistance*, (*ii*) *resistances caused by abrupt heterostructures*, and (*iii*) *bulk resistance* occurring particularly in materials with low carrier concentrations or low carrier mobilities. A voltage drop of magnitude $I R_s$ occurs at the series resistance thereby increasing the drive voltage.

Secondly, carrier energy may be lost upon injection into a quantum well structure or double heterostructure. An example of non-adiabatic injection is shown in Fig. 4.15, which shows a thin quantum well under forward-bias conditions. The figure illustrates that upon injection into the quantum well, the electron loses energy $\Delta E_C - E_0$, where ΔE_C is the band discontinuity and E_0 is the energy of the lowest quantized state in the conduction-band quantum well. Similarly, the energy lost by holes is given by $\Delta E_V - E_0$, where ΔE_V is the band discontinuity and E_0 is the energy of the lowest state in the valence-band quantum well. Upon injection of carriers into the well, the carrier energy is dissipated by **phonon emission**, i.e. by conversion of the carrier energy to heat. The energy loss due to **non-adiabatic injection** of carriers is relevant in semiconductors with large band discontinuities, for example GaN and other group-III nitride materials.

Thus, the total voltage drop across a forward-biased LED is given by

$$V = \frac{E_g}{e} + I R_s + \frac{\Delta E_C - E_0}{e} + \frac{\Delta E_V - E_0}{e} \tag{4.42}$$

where the first summand on the right-hand side of the equation is the theoretical voltage minimum, the second summand is due to the series resistance in the device, and the third and fourth summands are due to non-adiabatic injection of carriers into the active region.

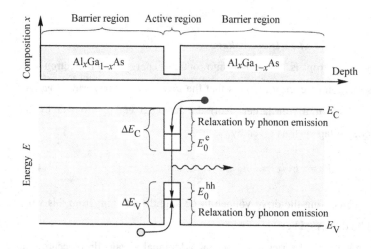

Fig. 4.15. Chemical composition and energy band diagram of a quantum well structure, illustrating the energy loss of carriers as they are captured into the quantum well.

One finds experimentally that the diode voltage can be *slightly lower than the minimum* value predicted by Eq. (4.42), i.e. can be slightly lower than $E_g/e \approx h\nu/e$. Both electrons and holes carry, on average, the thermal energy kT. In a forward-biased p-n junction, high-energy carriers are more likely than low-energy carriers to diffuse over to the side of opposite conductivity type where they recombine. At room temperature, $4kT/e$ amounts to a voltage of about 100 mV. In low-resistance devices, the diode voltage can be 100–200 mV *lower* than $h\nu/e$. For example, in forward-biased GaAs LEDs ($E_g = 1.42$ eV), some photon emission with $h\nu = 1.42$ eV is observed at diode voltages of about 1.32 V, i.e. lower than the photon energy.

Exercise: *Drive voltages of LEDs*. Calculate the approximate forward diode voltage of LEDs emitting in the blue, green, and red parts of the visible spectrum. Also calculate the forward diode voltage of LEDs emitting at 870 nm and 1.55 μm.
Solution:

Emission color	Wavelength	Photon energy	Drive voltage
Blue	470 nm	2.6 eV	2.6 V
Green	550 nm	2.2 eV	2.2 V
Red	650 nm	1.9 eV	1.9 V
IR	870 nm	1.4 eV	1.4 V
IR	1550 nm	0.8 eV	0.8 V

References

Ebeling K. J. *Integrated Opto-Electronics* Chapter 9 (Springer, Berlin, 1993)
Emerson D., Abare A., Bergmann M., Slater D., and Edmond J. "Development of deep UV III–N optical sources" *7th International Workshop on Wide-Bandgap III–Nitrides*, Richmond VA, March (2002)

Hunt N. E. J., Schubert E. F., Sivco D. L., Cho A. Y., and Zydzik G. J. "Power and efficiency limits in single-mirror light-emitting diodes with enhanced intensity" *Electron. Lett.* **28**, 2169 (1992)

Krames M. R. *et al.* "High-brightness AlGaInP light-emitting diodes" *Proceedings of SPIE* **3938**, 2 (2000)

Rhoderick E. H. and Williams R. H. *Metal–Semiconductor Contacts* (Clarendon Press, Oxford, UK, 1988)

Schubert E. F., Tu L.-W., Zydzik G. J., Kopf R. F., Benvenuti A., and Pinto M. R. "Elimination of heterojunction band discontinuities by modulation doping" *Appl. Phys. Lett.* **60**, 466 (1992)

Shah J. M., Li Y.-L., Gessmann Th., and Schubert E. F. "Experimental analysis and theoretical model for anomalously high ideality factors ($n \gg 2.0$) in AlGaN/GaN p-n junction diodes" *J. Appl. Phys.* **94**, 2627 (2003)

5

LED basics: Optical properties

5.1 Internal, extraction, external, and power efficiencies

The active region of an ideal LED emits *one photon* for *every electron* injected. Each charge quantum-particle (electron) produces one light quantum-particle (photon). Thus the ideal active region of an LED has a *quantum efficiency* of unity. The **internal quantum efficiency** is defined as

$$\eta_{int} = \frac{\text{number of photons emitted from active region per second}}{\text{number of electrons injected into LED per second}} = \frac{P_{int}/(h\nu)}{I/e} \qquad (5.1)$$

where P_{int} is the optical power emitted from the active region and I is the injection current.

Photons emitted by the active region should escape from the LED die. In an ideal LED, all photons emitted by the active region are also emitted into free space. Such an LED has unity *extraction efficiency*. However, in a real LED, not all the power emitted from the active region is emitted into free space. Some photons may never leave the semiconductor die. This is due to several possible loss mechanisms. For example, light emitted by the active region can be reabsorbed in the substrate of the LED, assuming that the substrate is absorbing at the emission wavelength. Light may be incident on a metallic contact surface and be absorbed by the metal. In addition, the phenomenon of *total internal reflection*, also referred to as the *trapped light phenomenon*, reduces the ability of the light to escape from the semiconductor. The light **extraction efficiency** is defined as

$$\eta_{extraction} = \frac{\text{number of photons emitted into free space per second}}{\text{number of photons emitted from active region per second}} = \frac{P/(h\nu)}{P_{int}/(h\nu)} \qquad (5.2)$$

where P is the optical power emitted into free space.

The extraction efficiency can be a severe limitation for high-performance LEDs. It is quite difficult to increase the extraction efficiency beyond 50% without resorting to highly sophisticated and costly device processes.

The ***external quantum efficiency*** is defined as

$$\eta_{\text{ext}} = \frac{\text{number of photons emitted into free space per second}}{\text{number of electrons injected into LED per second}} = \frac{P/(h\nu)}{I/e} = \eta_{\text{int}} \eta_{\text{extraction}} \cdot \quad (5.3)$$

The external quantum efficiency gives the ratio of the number of useful light particles to the number of injected charge particles.

The ***power efficiency*** is defined as

$$\eta_{\text{power}} = \frac{P}{IV} \quad (5.4)$$

where IV is the electrical power provided to the LED. Informally, the power efficiency is also called the *wallplug efficiency*.

Exercise: *LED efficiency*. Consider an LED with a threshold voltage of $V_{\text{th}} = E_g/e = 2.0$ V with a differential resistance of $R_s = 20\,\Omega$, so that the I–V characteristic in the forward direction is given by $V = V_{\text{th}} + IR_s$. When the device is operated at 20 mA it emits a light power of 4 mW of energy $h\nu = E_g$. Determine the internal quantum efficiency, the external quantum efficiency, and the power efficiency, assuming that the extraction efficiency is 50%.

5.2 Emission spectrum

The physical mechanism by which semiconductor LEDs emit light is spontaneous recombination of electron–hole pairs and simultaneous emission of photons. The spontaneous emission process is fundamentally different from the stimulated emission process occurring in semiconductor lasers and superluminescent LEDs. Spontaneous recombination has certain characteristics that determine the optical properties of LEDs. The properties of spontaneous emission in LEDs will be discussed in this section.

An electron–hole recombination process is illustrated schematically in Fig. 5.1. Electrons in the conduction band and holes in the valence band are assumed to have the parabolic dispersion relations

$$E = E_C + \frac{\hbar^2 k^2}{2m_e^*} \quad \text{(for electrons)} \quad (5.5)$$

and

$$E = E_V - \frac{\hbar^2 k^2}{2m_h^*} \qquad \text{(for holes)} \qquad (5.6)$$

where m_e^* and m_h^* are the electron and hole effective masses, $\hbar$ is Planck's constant divided by 2π, k is the carrier wave number, and E_V and E_C are the valence and conduction band edges, respectively.

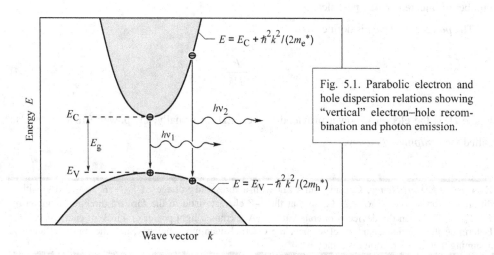

Fig. 5.1. Parabolic electron and hole dispersion relations showing "vertical" electron–hole recombination and photon emission.

The requirement of energy and momentum conservation leads to further insight into the radiative recombination mechanism. It follows from the Boltzmann distribution that electrons and holes have an average kinetic energy of kT. Energy conservation requires that the photon energy is given by the difference between the electron energy, E_e, and the hole energy, E_h, i.e.

$$h\nu = E_e - E_h \approx E_g. \qquad (5.7)$$

The photon energy is approximately equal to the bandgap energy, E_g, if the thermal energy is small compared with the bandgap energy $kT \ll E_g$. Thus the desired emission wavelength of an LED can be attained by choosing a semiconductor material with an appropriate bandgap energy. For example, GaAs has a bandgap energy of 1.42 eV at room temperature and thus GaAs LEDs emit at the infrared wavelength of 870 nm.

It is helpful to compare the average carrier momentum with the photon momentum. A carrier with kinetic energy kT and effective mass m^* has the momentum

$$p = m^*v = \sqrt{2m^*\tfrac{1}{2}m^*v^2} = \sqrt{2m^*kT} \ . \tag{5.8}$$

The momentum of a photon with energy E_g can be derived from the de Broglie relation

$$p = \hbar k = \frac{h\nu}{c} = \frac{E_g}{c} \ . \tag{5.9}$$

Calculation of the carrier momentum (using Eq. 5.8) and the photon momentum (using Eq. 5.9) yields that the carrier momentum is *orders of magnitude larger* than the photon momentum. Therefore the electron momentum cannot change significantly during the transition from the conduction to the valence band. The transitions are therefore "vertical" as shown in Fig. 5.1, i.e. electrons only recombine with holes that have the same momentum or k value.

Using the requirement that electron and hole momenta are the same, the photon energy can be written as the *joint dispersion relation*

$$h\nu = E_C + \frac{\hbar^2 k^2}{2m_e^*} - E_V + \frac{\hbar^2 k^2}{2m_h^*} = E_g + \frac{\hbar^2 k^2}{2m_r^*} \tag{5.10}$$

where m_r^* is the reduced mass given by

$$\frac{1}{m_r^*} = \frac{1}{m_e^*} + \frac{1}{m_h^*} \ . \tag{5.11}$$

Using the joint dispersion relation, the joint density of states can be calculated and one obtains

$$\rho(E) = \frac{1}{2\pi^2} \left(\frac{2m_r^*}{\hbar^2} \right)^{3/2} \sqrt{E - E_g} \ . \tag{5.12}$$

The distribution of carriers in the allowed bands is given by the Boltzmann distribution, i.e.

$$f_B(E) = e^{-E/(kT)} \ . \tag{5.13}$$

The *emission intensity* as a function of energy is proportional to the product of Eqs. (5.12) and (5.13),

$$\boxed{I(E) \propto \sqrt{E - E_g}\ e^{-E/(kT)}} \ . \tag{5.14}$$

5 LED basics: optical properties

The lineshape of an LED, as given by Eq. (5.14), is shown in Fig. 5.2. The maximum emission intensity occurs at

$$E = E_g + \tfrac{1}{2} kT \quad . \tag{5.15}$$

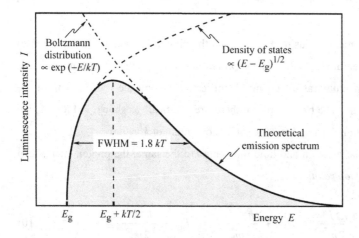

Fig. 5.2. Theoretical emission spectrum of an LED. The full-width at half-maximum (FWHM) of the emission line is $1.8kT$.

The full-width at half-maximum of the emission is

$$\Delta E = 1.8 kT \quad \text{or} \quad \Delta\lambda = \frac{1.8 kT \lambda^2}{hc} \quad . \tag{5.16}$$

For example, the theoretical room-temperature linewidth of a GaAs LED emitting at 870 nm is ΔE = 46 meV or $\Delta\lambda$ = 28 nm.

The spectral linewidth of LED emission is important in several respects. Firstly, the linewidth of an LED emitting in the visible range is relatively narrow compared with the range of the entire visible spectrum. The LED emission is even narrower than the spectral width of a single color as perceived by the human eye. For example, *red* colors range in wavelength from 625 to 730 nm, which is much wider than the typical emission spectrum of an LED. Therefore, LED emission is perceived by the human eye as *monochromatic*.

Secondly, optical fibers are dispersive, which leads to a range of propagation velocities for a light pulse consisting of a range of wavelengths. The material dispersion in optical fibers limits the "*bit rate × distance product*" achievable with LEDs.

The spontaneous lifetime of carriers in LEDs in direct-gap semiconductors is of the order of 1–100 ns depending on the active region doping concentration (or carrier concentrations) and the material quality. Thus, modulation speeds up to 1 Gbit/s are attainable with LEDs.

5.3 The light escape cone

Light generated inside a semiconductor cannot escape from the semiconductor if it is totally internally reflected at the semiconductor–air interface. If the angle of incidence of a light ray is close to normal incidence, light can escape from the semiconductor. However, total internal reflection occurs for light rays with oblique and grazing-angle incidence. Total internal reflection reduces the external efficiency significantly, in particular for LEDs consisting of high-refractive-index materials.

Assume that the angle of incidence in the semiconductor at the semiconductor–air interface is given by ϕ. Then the angle of incidence of the refracted ray, Φ, can be inferred from Snell's law

$$\bar{n}_s \sin \phi = \bar{n}_{air} \sin \Phi \tag{5.17}$$

where $\bar{n}_s$ and $\bar{n}_{air}$ are the refractive indices of the semiconductor and air, respectively. The ***critical angle for total internal reflection*** is obtained using $\Phi = 90°$, as illustrated in Fig. 5.3 (a). Using Snell's law, one obtains

$$\sin \phi_c = \frac{\bar{n}_{air}}{\bar{n}_s} \sin 90° = \frac{\bar{n}_{air}}{\bar{n}_s} \tag{5.18a}$$

and

$$\phi_c = \arcsin \frac{\bar{n}_{air}}{\bar{n}_s}. \tag{5.18b}$$

The refractive indices of semiconductors are usually quite high. For example, GaAs has a refractive index of 3.4. Thus, according to Eq. (5.18), the critical angle for total internal reflection is quite small. In this case, we can use the approximation $\sin \phi_c \approx \phi_c$. The critical angle for total internal reflection is then given by

$$\phi_c \approx \frac{\bar{n}_{air}}{\bar{n}_s}. \tag{5.19}$$

The angle of total internal reflection defines the ***light escape cone***. Light emitted into the cone can escape from the semiconductor, whereas light emitted outside the cone is subject to total internal reflection.

5 LED basics: optical properties

Next, we calculate the surface area of the spherical cone with radius r in order to determine the total fraction of light that is emitted into the light escape cone. The surface area of the calotte-shaped surface shown in Figs. 5.3 (b) and (c) is given by the integral

$$A = \int dA = \int_{\phi=0}^{\phi_c} 2\pi r \sin\phi \, r \, d\phi = 2\pi r^2 (1 - \cos\phi_c) . \tag{5.20}$$

Let us assume that light is emitted from a point-like source in the semiconductor with a total power of P_{source}. Then the power that can escape from the semiconductor is given by

$$P_{\text{escape}} = P_{\text{source}} \frac{2\pi r^2 (1 - \cos\phi_c)}{4\pi r^2} \tag{5.21}$$

where $4\pi r^2$ is the entire surface area of the sphere with radius r.

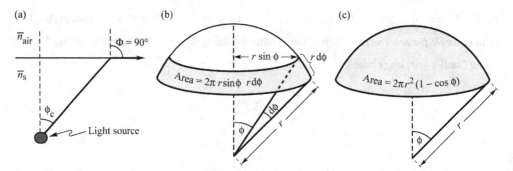

Fig. 5.3. (a) Definition of the escape cone by the critical angle ϕ_c. (b) Area element dA. (c) Area of calotte-shaped section of the sphere defined by radius r and angle ϕ_c.

The calculation indicates that only a fraction of the light emitted inside a semiconductor can escape from the semiconductor. This fraction is given by

$$\boxed{\frac{P_{\text{escape}}}{P_{\text{source}}} = \frac{1}{2}(1 - \cos\phi_c)} . \tag{5.22}$$

Because the critical angle of total internal reflection for high-index materials is relatively small, the cosine term can be expanded into a power series. Neglecting higher-than-second-order terms yields

$$\frac{P_{escape}}{P_{source}} \approx \frac{1}{2}\left[1 - \left(1 - \frac{\phi_c^2}{2}\right)\right] = \frac{1}{4}\phi_c^2. \tag{5.23}$$

Using the approximation of Eq. (5.19), one obtains

$$\boxed{\frac{P_{escape}}{P_{source}} \approx \frac{1}{4}\frac{\bar{n}_{air}^2}{\bar{n}_s^2}}. \tag{5.24}$$

The escape problem is a significant problem for high-efficiency LEDs. In most semiconductors, the refractive index is quite high (> 2.5) and thus only a small percentage of the light generated in the semiconductor can escape from a planar LED. The problem is less significant in semiconductors with a small refractive index and for polymers, which have refractive indices of the order of 1.5.

Exercise: *Light escape from planar GaAs, GaN, and polymer LED structures*. The refractive indices of GaAs, GaN, and light-emitting polymers are 3.4, 2.5, and 1.5, respectively. Calculate the critical angle of total internal reflection for GaAs, GaN, and for polymers. Also calculate the fraction of light power that can escape from a planar GaAs and GaN semiconductor structures and a polymer LED structure.
What improvement can be attained if a planar GaAs LED is encapsulated in a transparent polymer of refractive index 1.5, if the reflection at the polymer–air interface is neglected?
Solution:
Critical angle for total internal reflection:
GaAs $\phi_c = 17.1°$ GaN $\phi_c = 23.6°$ Polymer $\phi_c = 41.8°$.
Fraction of light that can escape:
GaAs 2.21%   GaN 4.18%   Polymer 12.7%.
Improvement of the GaAs planar LED due to polymer encapsulation: 232%.

5.4 Radiation pattern

All LEDs have a certain **radiation pattern** or **far-field pattern**. The intensity, measured in W/cm^2, depends on the longitudinal and azimuth angle and the distance from the LED. The total optical power emitted by the LED is obtained by integration over the area of a sphere.

$$P = \int_A \int_\lambda I(\lambda)\, d\lambda\, dA \tag{5.25}$$

where $I(\lambda)$ is the **spectral light intensity** (measured in W per nm per cm^2) and A is the surface area of the sphere. The integration is carried out over the entire surface area.

5.5 The lambertian emission pattern

The index contrast between the light-emitting material and the surrounding material leads to a non-isotropic emission pattern. For high-index light-emitting materials with a planar surface, a lambertian emission pattern is obtained. Figure 5.4 illustrates a point-like light source located a short distance below a semiconductor–air interface. Consider a light ray emitted from the source at an angle ϕ with respect to the surface normal. The light ray is refracted at the semiconductor–air interface and the refracted light ray has an angle Φ with respect to the surface normal. The two angles are related by Snell's law, which, for small angles of ϕ (for which $\sin \phi \approx \phi$), can be written as

$$\bar{n}_s \phi = \bar{n}_{air} \sin \Phi . \quad (5.26)$$

Light emitted into the angle $d\phi$ in the semiconductor is emitted into the angle $d\Phi$ in air as shown in Fig. 5.4 (a). Differentiating the equation with respect to Φ and solving the resulting equation for $d\Phi$ yields

$$d\Phi = \frac{\bar{n}_s}{\bar{n}_{air}} \frac{1}{\cos \Phi} d\phi . \quad (5.27)$$

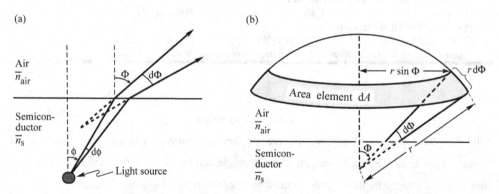

Fig. 5.4. Geometrical model used to derive the lambertian emission pattern. (a) The light emitted into angle $d\phi$ inside the semiconductor is emitted into the angle $d\Phi$ in air. (b) Illustration of the area element dA of the calotte-shaped section of the sphere.

Power conservation requires that the optical power emitted into the angle $d\phi$ in the semiconductor be equal to the optical power emitted into the angle $d\Phi$ in air, i.e.

5.5 The lambertian emission pattern

$$I_s \, dA_s = I_{air} \, dA_{air} \tag{5.28}$$

where I_s and I_{air} are the light intensities (measured in units of W/m²) in the semiconductor and air, respectively. Owing to the cylindrical symmetry of the emission pattern we choose the area element shown in Fig. 5.4 (b). The area element in air is given by

$$dA_{air} = 2\pi r \sin\Phi \, r \, d\Phi \, . \tag{5.29}$$

Using Eqs. (5.27) and (5.28) yields

$$dA_{air} = 2\pi r^2 \frac{\bar{n}_s^2}{\bar{n}_{air}^2} \frac{1}{\cos\Phi} \phi \, d\phi \, . \tag{5.30}$$

Similarly, the surface element in the semiconductor is given by

$$dA_s = 2\pi r \sin\phi \, r \, d\phi \approx 2\pi r^2 \phi \, d\phi \, . \tag{5.31}$$

The light intensity in the semiconductor at a distance r from the light source is given by the total source power divided by the surface area of a sphere with radius r, i.e.

$$I_s = \frac{P_{source}}{4\pi r^2} \, . \tag{5.32}$$

The light intensity in air can then be inferred from Eqs. (5.28), (5.30), (5.31), and (5.32). One obtains the **lambertian emission pattern** given by

$$\boxed{I_{air} = \frac{P_{source}}{4\pi r^2} \frac{\bar{n}_{air}^2}{\bar{n}_s^2} \cos\Phi} \, . \tag{5.33}$$

The lambertian emission pattern follows a cosine dependence on the angle Φ. The intensity is highest for emission normal to the semiconductor surface, i.e. for $\Phi = 0°$. At an angle of $\Phi = 60°$, the intensity decreases to half of its maximum value. The lambertian emission pattern is shown schematically in Fig. 5.5.

Several other surface shapes are also shown in Fig. 5.5. These non-planar surfaces exhibit various emission patterns. An isotropic emission pattern is obtained for hemispherically shaped LEDs, which have the light-emitting region in the center of the sphere. A strongly directed

5 LED basics: optical properties

emission pattern can be obtained in LEDs with parabolically shaped surfaces. However, both hemispherical as well as parabolic surfaces are difficult to fabricate.

The total power emitted into air can be calculated by integrating the intensity over the entire hemisphere. The total power is then given by

$$P_{air} = \int_{\Phi=0°}^{90°} I_{air} \, 2\pi r \sin\Phi \, r \, d\Phi \; . \tag{5.34}$$

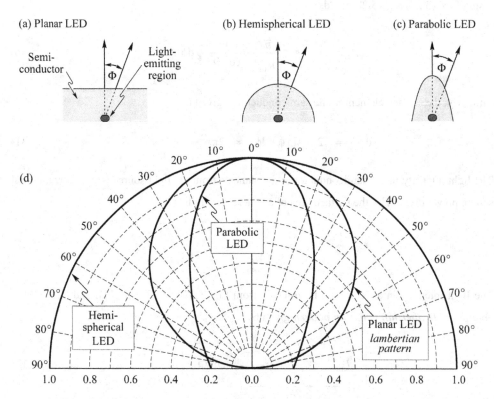

Fig. 5.5. Light-emitting diodes with (a) planar, (b) hemispherical, and (c) parabolic surfaces. (d) Far-field patterns of the different types of LEDs. At an angle of $\Phi = 60°$, the lambertian emission pattern decreases to 50% of its maximum value occurring at $\Phi = 0°$. The three emission patterns are normalized to unity intensity at $\Phi = 0°$.

By using the lambertian emission pattern for I_{air} in Eq. (5.34) and using $\cos\Phi \sin\Phi = (1/2) \sin(2\Phi)$, the integral can be calculated to yield

$$P_{air} = \frac{P_{source}}{4} \frac{\bar{n}_{air}^2}{\bar{n}_s^2}. \qquad (5.35)$$

This result is identical to Eq. (5.24). This is not surprising because the light power that escapes from the semiconductor (P_{escape}) must be identical to the power in air (P_{air}).

In the calculation above, **Fresnel reflection** at the semiconductor–air interface has been neglected. At normal incidence, the Fresnel power transmittance is given by

$$T = 1 - R = 1 - \left(\frac{\bar{n}_s - \bar{n}_{air}}{\bar{n}_s + \bar{n}_{air}}\right)^2 = \frac{4\,\bar{n}_s\,\bar{n}_{air}}{(\bar{n}_s + \bar{n}_{air})^2}. \qquad (5.36)$$

Fresnel reflection losses must be taken into account in a rigorous calculation.

Exercise: *LED-to-fiber coupling efficiency*. Consider a GaAs LED with a point-like light-emitting region located in close proximity to the planar GaAs LED surface. An optical fiber has an acceptance angle of 12° in air. What fraction of the light emitted by the active region can be coupled into the fiber? Assume a GaAs refractive index of 3.4. Neglect Fresnel reflection losses at the semiconductor–air and air–fiber interfaces.

Solution: The acceptance angle in the semiconductor is obtained from Snell's law and is 3.5°. Thus 0.093% of the power emitted by the active region can be coupled into the fiber.

5.6 Epoxy encapsulants

The light extraction efficiency can be enhanced by using dome-shaped encapsulants with a large refractive index. As a result of the encapsulation, the angle of total internal reflection through the top surface of the semiconductor is increased (Nuese *et al.*, 1969). It follows from Eq. (5.22) that the ratio of extraction efficiency with and without epoxy encapsulant is given by

$$\frac{\eta_{epoxy}}{\eta_{air}} = \frac{1 - \cos\phi_{c,epoxy}}{1 - \cos\phi_{c,air}} \qquad (5.37)$$

where $\phi_{c,epoxy}$ and $\phi_{c,air}$ are the critical angles for total internal reflection at the semiconductor–epoxy and semiconductor–air interface, respectively. Figure 5.6 shows the calculated ratio of the extraction efficiency with and without an epoxy dome. Inspection of the figure yields that the efficiency of a typical semiconductor LED increases by a factor of 2–3 upon encapsulation with an epoxy having a refractive index of 1.5.

The inset of Fig. 5.6 shows that light is incident at an angle of approximately 90° at the epoxy–air interface due to the dome-shape of the epoxy. Thus, total internal reflection losses do not occur at the epoxy–air interface. Besides improving the external efficiency of an LED, the encapsulant can also be used as a spherical lens for applications requiring a directed emission pattern. In polymer LEDs, encapsulants increase the extraction efficiency by only a small amount due to the inherently small refractive index of polymers.

Advanced encapsulants including graded-index encapsulants, encapsulants with a high refractive index ($\bar{n} > 2.0$), and encapsulants containing mineral diffusers will be discussed in the chapter on packaging.

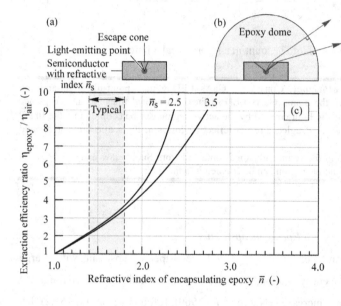

Fig. 5.6. LED (a) without and (b) with dome-shaped epoxy encapsulant. A larger escape angle is obtained for the LED with an epoxy dome. (c) Calculated ratio of light extraction efficiency emitted through the top surface of a planar LED with and without an epoxy dome. The refractive indices of typical epoxies range between 1.4 and 1.8 (adapted from Nuese et al., 1969).

5.7 Temperature dependence of emission intensity

The emission intensity of LEDs decreases with increasing temperature. This decrease of the emission intensity is due to several temperature-dependent factors including (*i*) non-radiative recombination via deep levels, (*ii*) surface recombination, and (*iii*) carrier loss over heterostructure barriers.

Near room temperature, the temperature dependence of the LED emission intensity is frequently described by the phenomenological equation

5.7 Temperature dependence of emission intensity

$$I = I|_{300K} \exp - \frac{T - 300\,K}{T_1} \tag{5.38}$$

where T_1 is the **characteristic temperature**. A *high* characteristic temperature, implying a *weak* temperature dependence, is desirable.

It is interesting to note that both LEDs as well as semiconductor lasers have a distinct temperature dependence of the emission intensity. In LEDs, the decrease is expressed in terms of the "T_1 equation". In semiconductor lasers, the threshold current, i.e. the electrical current required for the onset of lasing, increases. In lasers the increase in threshold current is expressed in terms of the well-known T_0 equation. This equation is given by

$$I_{th} = I_{th}|_{300K} \exp \frac{T - 300\,K}{T_0} \tag{5.39}$$

where I_{th} is the threshold current of the laser. Note the formal similarity of the "T_1 equation" (Eq. 5.38) and the "T_0 equation" (Eq. 5.39). Both equations are purely phenomenological equations intended to describe the experimental results without a strong theoretical framework allowing the derivation of the equations from basic principles.

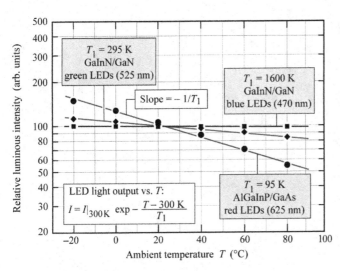

Fig. 5.7. Characteristic temperature T_1 of GaInN/GaN blue, GaInN/GaN green, and AlGaInP/GaAs red LEDs near room temperature (after data from Toyoda Gosei Corp., 2000). More recent data (Toyoda Gosei Corp., 2004) show the following values for T_1: Blue GaInN LED, 460 nm, T_1 = 1600 K; Cyan GaInN LED, 505 nm, T_1 = 832 K; Green GaInN LED, 525 nm, T_1 = 341 K; Red AlGaInP LED, 625 nm, T_1 = 199 K.

Experimental results of the temperature dependence of the emission intensity are shown in Fig. 5.7 (Toyoda Gosei Corporation, 2000). The figure shows the temperature dependence of the emission intensity at a constant current for a blue GaInN/GaN, a green GaInN/GaN, and a red AlGaInP/GaAs LED. Inspection of Fig. 5.7 reveals that the blue LED has the highest T_1 and the red LED has the lowest T_1. III–V nitride LEDs have deeper wells so that carrier confinement is more effective in III–V nitride structures than in the III–V phosphide structures.

References
Nuese C. J., Tietjen J. J., Gannon J. J., and Gossenberger H. F. "Optimization of electroluminescent efficiencies for vapor-grown GaAsP diodes" *J. Electrochem Soc.: Solid State Sci.* **116**, 248 (1969)
Toyoda Gosei Corporation, Japan, General LED catalogue (2000)
Toyoda Gosei Corporation, Japan, General LED catalogue (2004)

6

Junction and carrier temperatures

The temperature of the active region crystal lattice, frequently referred to as the *junction temperature*, is a critical parameter. The junction temperature is relevant for several reasons. Firstly, the internal quantum efficiency depends on the junction temperature. Secondly, high-temperature operation shortens the device lifetime. Thirdly, a high device temperature can lead to degradation of the encapsulant. It is therefore desirable to know the junction temperature as a function of the drive current.

Heat can be generated in the contacts, cladding layers, and the active region. At low current levels, heat generation in the parasitic resistances of contacts and cladding layers is small due to the I^2R dependence of Joule heating. The dominant heat source at low current levels is the active region, where heat is created by non-radiative recombination. At high current levels, the contribution of parasitics becomes increasingly important and can even dominate.

There are several different ways to measure the junction temperature, which include micro-Raman spectroscopy (Todoroki *et al.*, 1985), threshold voltage (Abdelkader *et al.*, 1992), thermal resistance (Murata and Nakada, 1992), photothermal reflectance microscopy (Epperlein, 1990), electroluminescence (Epperlein and Bona, 1993), photoluminescence (Hall *et al.*, 1992) and a non-contact method based on the peak ratio of a dichromatic source (Gu and Narendran, 2003). Most methods are *indirect* methods that infer the junction temperature from an easily measurable parameter. In this chapter, we discuss a method based on the shift of the peak emission wavelength with the temperature and a method based on the shift of the diode forward voltage with temperature. We also discuss the carrier temperature as inferred from the high-energy slope of the emission spectrum.

6.1 Carrier temperature and high-energy slope of spectrum

The Boltzmann distribution of carriers, applicable to the high-energy part of the emission spectrum, results in an exponential dependence of the emission intensity on energy, i.e.

6 Junction and carrier temperature

$$I \propto \exp\left[-h\nu/(kT_c)\right] \qquad (6.1)$$

where T_c is the **carrier temperature**. The high-energy slope of the spectrum is given by

$$\frac{d(\ln I)}{d(h\nu)} \propto \frac{-1}{kT_c}. \qquad (6.2)$$

Thus, the carrier temperature can be directly inferred from the slope. Because the carrier temperature is generally higher than the junction temperature, e.g. due to high-energy injection of carriers into the active region, this method gives an *upper limit* for the actual junction temperature.

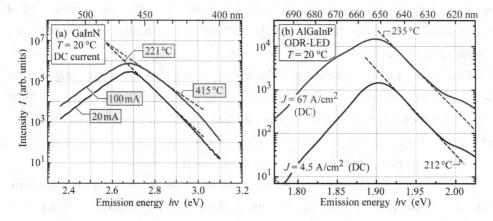

Fig. 6.1. Carrier temperatures in (a) GaInN blue and (b) AlGaInP red LEDs inferred from the high-energy slope of emission spectrum. Due to the alloy-broadening effect, the measured carrier temperatures overestimate the true carrier temperature (after Chhajed et al., 2005; Gessmann et al., 2003).

Figure 6.1 shows the evaluation of the carrier temperature from the emission spectrum of a GaInN and an AlGaInP LED (Chhajed et al., 2005; Gessmann et al., 2003). Inspection of the figure reveals that the carrier temperature increases along with the current level. At low current levels, the GaInN device has a carrier temperature of 221 °C and the AlGaInP device has a carrier temperature of 212 °C. At high current levels, the carrier temperature increases to 415 °C and 235 °C for the GaInN and AlGaInP LED, respectively. Due to the alloy-broadening effect occurring in ternary and quaternary semiconductor alloys, these temperatures overestimate the true carrier temperature.

Semiconductor alloys exhibit substantial broadening of the emission spectrum (and its high-energy slope) due to alloy broadening, i.e. the statistical fluctuation of the chemical composition occurring in ternary and quaternary semiconductors (Schubert *et al.*, 1984). De-convolution of the alloy-broadening effect and the kT-broadening effect allows for a more accurate estimate of the carrier temperature.

The determination of the carrier temperature using the high-energy slope works best for binary compounds such as GaAs or InP. Such semiconductors do not exhibit alloy broadening and thus the high-energy slope is more representative of the true carrier temperature.

6.2 Junction temperature and peak emission wavelength

This method makes use of the dependence of the bandgap energy (and thus the peak emission wavelength) on temperature. The method consists of a calibration measurement and a junction-temperature measurement. In the calibration measurement, the peak energy is measured at different ambient temperatures, typically in the range 20 °C to 120 °C, by placing the device in a temperature-controlled oven. The device is injected with a range of pulsed currents with a duty cycle << 1 to minimize additional heating. As a consequence, the ambient temperature in the oven and the junction temperature can be assumed to be identical. The calibration measurement establishes the junction-temperature versus emission-peak-energy relation for a range of currents. Calibration data for a deep UV LED are shown in Fig. 6.2 (a) (Xi *et al.*, 2004; 2005).

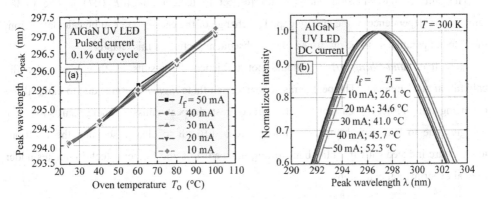

Fig. 6.2. (a) Peak emssion wavelength versus oven temperature of an AlGaN UV LED for pulsed current injection with 0.1% duty cycle. (b) Emssion spectra and junction temperatures for different DC currents (after Xi *et al.*, 2005).

Subsequent to the calibration, the peak emission energy is measured as a function of the DC injection current with the device in a room-temperature ambient. The junction temperature for each current level can then be determined by using the calibration data. Figure 6.2 (b) shows the emission spectra of the UV LED for different injection currents. The junction temperatures inferred from the calibration measurement are shown in Fig. 6.3 (Xi et al., 2005).

The accuracy of the method is limited by the ability to determine the peak wavelength. As a rule of thumb, the error bar of the peak wavelength is about 5–10% of the full-width at half-maximum of the luminescence line. Alloy-broadening effects and kT broadening impose a limitation on the accuracy of the method.

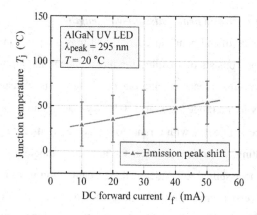

Fig. 6.3. Junction temperature inferred from emission peak energy as a function of DC injection current for a 300 μm × 300 μm deep UV LED emitting at 295 nm. The error bar stems from an uncertainty in the peak energy (after Xi et al., 2005).

The shift of the emission energy with respect to temperature (dE_g/dT) is due to the temperature dependence of the energy gap. The change of the energy gap can be calculated from the Varshni parameters to be discussed in the subsequent section of this chapter.

Note that **band-filling effects** should not influence the results as they also enter the calibration measurement. The peak emission energy shifts to *higher* energies due to band filling occurring at high injection current densities. In contrast, the peak emission energy shifts to *lower* energies due to bandgap shrinkage. Although it is difficult to clearly separate the junction-temperature-induced shift from the band-filling-induced shift, the former effect dominates under typical experimental conditions.

6.3 Theory of temperature dependence of diode forward voltage

The derivation of the temperature dependence of the forward voltage presented here follows the analysis first given by Xi et al. (2004; 2005). The I–V characteristic of an ideal p-n junction diode is given by the Shockley equation

6.3 Theory of temperature dependence of diode forward voltage

$$J = J_s \left(e^{eV_f/(n_{ideal}kT)} - 1 \right) \tag{6.3}$$

where J_s is the saturation current density. For non-degenerate semiconductors and under forward bias conditions, $V_f \gg kT/e$, one obtains

$$\frac{dV_f}{dT} = \frac{d}{dT}\left[\frac{n_{ideal} kT}{e} \ln\left(\frac{J_f}{J_s}\right)\right]. \tag{6.4}$$

The saturation current density depends on the diffusion constants of electrons and holes, the lifetimes of electrons and holes, the effective density of states at the conduction band and valence band edges, and the bandgap energy, all of which depend on the junction temperature. The temperature dependence of the effective density of states is given by $N_{c,v} \propto T^{3/2}$. Assuming phonon scattering, the temperature dependence of the carrier mobility is $\mu \propto T^{-3/2}$. Using the Einstein relation, the diffusion constant depends on temperature according to $D \propto T^{-1/2}$. The minority carrier lifetime can either decrease (non-radiative recombination) or increase (radiative recombination) with temperature. Due to this uncertainty, the minority carrier lifetime is assumed to be independent of temperature. Using these temperature dependences in Eq. (6.4) and executing the derivative yields

$$\boxed{\frac{dV_f}{dT} = \frac{eV_f - E_g}{eT} + \frac{1}{e}\frac{dE_g}{dT} - \frac{3k}{e}}. \tag{6.5}$$

This equation gives the fundamental temperature dependence of the forward voltage. The first, second, and third summands on the right-hand side of the equation are due to the temperature dependence of the intrinsic carrier concentration, bandgap energy, and effective densities of states, respectively. The equation includes the temperature dependence of the bandgap energy, which had not been taken into account in earlier derivations of dV_f/dT (Millman and Halkias, 1972).

LEDs are typically operated at forward voltages close to the built-in voltage, i.e. $V_f \approx V_{bi}$. Thus, for non-degenerate doping concentrations, we can write

$$eV_f - E_g \approx kT \ln\left(\frac{N_D N_A}{n_i^2}\right) - kT \ln\left(\frac{N_c N_v}{n_i^2}\right) = kT \ln\left(\frac{N_D N_A}{N_c N_v}\right). \tag{6.6}$$

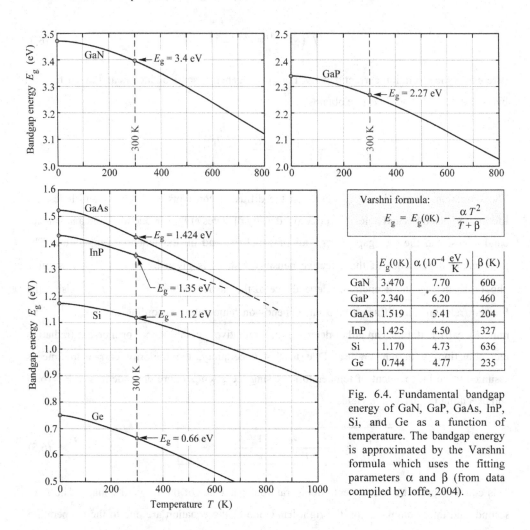

Fig. 6.4. Fundamental bandgap energy of GaN, GaP, GaAs, InP, Si, and Ge as a function of temperature. The bandgap energy is approximated by the Varshni formula which uses the fitting parameters α and β (from data compiled by Ioffe, 2004).

The second summand on the right-hand side of Eq. (6.5) is due to the changes in bandgap energy. As the temperature increases, the energy gap of semiconductors generally decreases. The temperature dependence of the energy gap of a semiconductor can be expressed by the **Varshni formula** (Varshni, 1967)

$$E_g = E_g\big|_{T=0K} - \frac{\alpha T^2}{T + \beta} \qquad (6.7)$$

where α and β are fitting parameters, frequently called the *Varshni parameters*. The bandgap energy versus temperature for several semiconductors is shown in Fig. 6.4 along with the values for α and β. The Varshni parameters for several semiconductors, compiled by the Ioffe (2004), are given in Table 6.1. Substituting Eqs. (6.6) and (6.7) into Eq. (6.5) yields

$$\frac{dV_f}{dT} \approx \underbrace{\frac{k}{e} \ln\left(\frac{N_D N_A}{N_c N_v}\right)}_{\text{due to } T \text{ dependence of } n_i} - \underbrace{\frac{\alpha T(T+2\beta)}{e(T+\beta)^2}}_{\frac{1}{e}\frac{dE_g}{dT}} - \underbrace{\frac{3k}{e}}_{\text{due to } T \text{ dependence of DOS}}. \quad (6.8)$$

This equation is a very useful expression for the temperature coefficient of the forward voltage.

Table 6.1. Varshni parameters of common semiconductors (from data compiled by Ioffe, 2004).

Semiconductor	E_g at 0 K (eV)	α (10^{-4} eV/K)	β (K)	Validity range
AlN (wurtzite)	6.026	18.0	1462	$T \leq 300$ K
GaN (wurtzite)	3.47	7.7	600	$T \leq 600$ K
GaP	2.34	6.2	460	$T \leq 1\,200$ K
GaAs	1.519	5.41	204	$T \leq 1\,000$ K
GaSb	0.813	3.78	94	$T \leq 300$ K
InN (wurtzite)	1.994	2.45	624	$T \leq 300$ K
InP	1.425	4.50	327	$T \leq 800$ K
InAs	0.415	2.76	83	$T \leq 300$ K
InSb	0.24	6.0	500	$T \leq 300$ K
Si	1.170	4.73	636	$T \leq 1\,000$ K
Ge	0.744	4.77	235	$T \leq 700$ K

For GaN diodes, Xi et al. (2004; 2005) reported a calculated dV_f/dT of –1.76 mV/K, which is in good agreement with the experimental value of –2.3 mV/K. Deviations between theory and experiment were attributed to the temperature coefficient of the resistivity in the neutral regions which decreases with increasing temperature due to a higher doping activation (Xi et al., 2005).

The temperature dependence of a GaPAs/GaAs LED is illustrated in Fig. 6.5, which shows the I–V characteristic at 77 K and at room temperature. Inspection of the figure reveals that the threshold voltage as well as the series resistance of the diode increases as the diode is cooled. If the device were driven at a constant voltage, e.g. 1.9 V, a large current change would result from the change in temperature.

6 Junction and carrier temperature

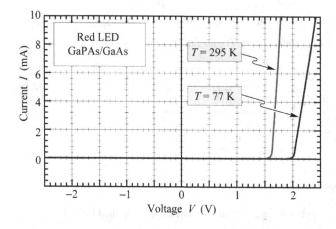

Fig. 6.5. Current–voltage characteristic of GaAsP/GaAs LED emitting in the red part of the visible spectrum, measured at 77 and 295 K. The threshold voltages are 2.0 and 1.6 V, at 77 and 300 K, respectively.

Exercise: *Temperature dependence of diode forward voltage*. Experimentally determined linear temperature coefficients (dV_f/dT) for GaAs diodes range from 1.2 to 1.4 mV/K. Calculate the linear temperature coefficient of the forward voltage of a GaAs diode with $N_A = N_D = 2 \times 10^{17}$ cm^{-3} at room temperature. What is the decrease in forward voltage if the ambient temperature is increased from 20 to 40 °C and the internal heating in the diode can be neglected?
Solution: For GaAs with $\alpha = 5.41 \times 10^{-4}$ eV/K and $\beta = 204$ K, one obtains at room temperature $dV_f/dT = -1.09$ mV/K. The decrease in diode voltage for the 20 °C temperature increase is $\Delta V_f = 21.9$ mV.

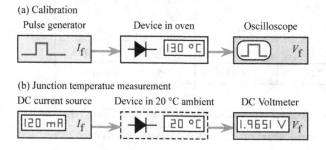

Fig. 6.6. (a) Pulsed calibration procedure establishing the forward voltage versus junction temperature (V_f versus T_j) relation and (b) determination of junction temperature for different DC forward currents.

6.4 Measurement of junction temperature using forward voltage

This method consists of a V_f calibration measurement under pulsed-current injection, and a V_f measurement under DC-current injection. The two measurements are illustrated in Fig. 6.6. In the calibration measurement, the device under test is located in a temperature-controlled oven, so that the temperature of the device and junction is known. The temperature is varied from typically 20 °C to 120 °C. The calibration measurement is performed in a pulsed mode with a very small duty cycle (e.g. 0.1%), so that the heat generated by the injection current becomes negligibly small. The forward voltage is measured at each temperature for the current levels of interest. The calibration measurement establishes the relation between forward voltage and junction temperature for the I_f levels of interest.

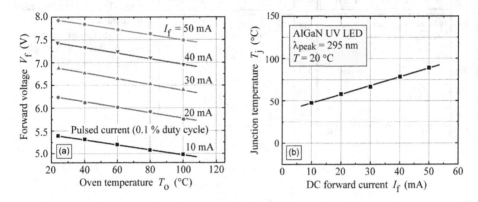

Fig. 6.7. (a) Pulsed calibration measurement (duty cycle 0.1%) and (b) junction temperature (T_j) versus DC current of an AlGaN UV LED (after Xi et al., 2005)

Subsequently the device is exposed to room-temperature ambient and subjected to a series of DC currents. Forward voltages are measured once thermal steady state has been reached. The measured DC forward voltages and the calibration measurement data are used to establish the junction temperature for different current levels. A calibration measurement and a junction temperature measurement for an AlGaN deep UV LED is shown in Fig. 6.7 (Xi et al., 2005).

Junction temperatures of several different devices are shown in Fig. 6.8 including red (AlGaInP, $\lambda = 625$ nm), green (GaInN, $\lambda = 525$ nm), blue (GaInN, $\lambda = 460$ nm), and UV (GaInN, $\lambda = 370$ nm) devices packaged in conventional 5 mm packages (Chhajed et al., 2005). The forward-voltage method is accurate to within a few degrees. The V_f method is more accurate than the peak-wavelength method. The latter method is limited by the uncertainty in the peak wavelength, which is difficult to determine accurately for broadened emission bands. Also shown in the figure is the carrier temperature derived from the high-energy slope of the spectrum. The accuracy of the carrier temperature suffers from alloy broadening, which decreases the high-energy slope (and thus increases the apparent carrier temperature).

6 Junction and carrier temperature

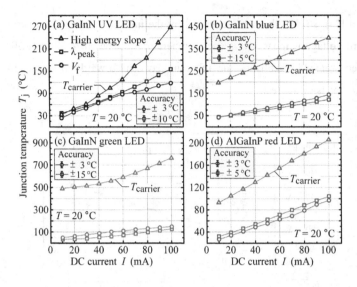

Fig. 6.8. Junction and carrier temperature of devices in conventional 5 mm packages as a function of DC injection current. The measured carrier temperature over-estimates the true carrier temperature due to alloy broadening (after Chhajed et al., 2005).

6.5 Constant-current and constant-voltage DC drive circuits

Different considerations play a role in designing the DC drive circuit of an LED operated under steady-state conditions. These considerations include the simplicity and cost of the drive circuit, the power efficiency, and the compensation of the temperature dependence of the light intensity emitted from the LED.

A simple drive circuit is a *constant-voltage supply* such as a battery or the rectified AC output of a transformer. There are, however, two drawbacks to constant-voltage drives of LEDs. Firstly, the diode current depends exponentially on the voltage, so a small variation in the drive voltage results in a large change in the current. Secondly, the threshold voltage of a diode depends on temperature, so any temperature change results in a significant change in current. A constant-voltage operating characteristic of a diode is shown in Fig. 6.9.

If a resistor is connected in series with the diode, the strong temperature dependence of the diode current is reduced. The series resistance together with the temperature dependence of the diode determines the temperature coefficient of the diode current.

When a diode is driven with a constant current, the emission intensity decreases with increasing temperature. A constant-voltage power supply with a series resistance can be used to reduce the temperature dependence of the emission intensity. The emission intensity of LEDs generally decreases with increasing temperature due to non-radiative recombination. In addition, the threshold voltage decreases with increasing temperature. However, for a constant-voltage supply, the diode current increases as the temperature increases, as shown in Fig. 6.9. Thus a series resistor can be used to compensate for the emission intensity decrease at elevated temperatures. It should be noted that the electrical-to-optical power-conversion efficiency drops due to the power consumed in the series resistor.

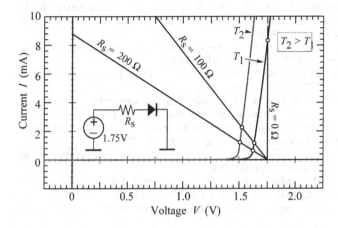

Fig. 6.9. LED drive circuit with series resistance R_S. The intersection between the diode I–V characteristics and the load lines are the points of operation. Small series resistances result in an increased diode current at high temperatures, thus allowing for compensation of a lower LED radiative efficiency.

The temperature dependence of LED intensity is an important factor for LEDs used in outdoor applications. On hot summer days, the temperature and the ambient light intensity are high. Owing to the high temperature, the LED intensity drops. In addition, high brightness is required due to high ambient light levels. This effect can be compensated for by driving the LEDs with a higher current as the temperature increases.

A constant-current drive circuit can consist of a transistor with the LED as a load. A constant-current drive circuit allows one to drive the LED irrespective of the diode threshold voltage and the diode temperature. However, a constant-current drive circuit does not compensate for the decrease of LED emission at elevated temperatures.

Exercise: *Compensation of the temperature dependence of an LED with a drive circuit.* Consider an LED with a characteristic temperature $T_1 = 100$ K, a turn-on voltage of 1.4 V at 20 °C, a temperature coefficient of the turn-on voltage of –2.1 mV/K, and a linear I–V characteristic with a differential resistance of 5 Ω for forward voltages larger than the turn-on voltage. Assume that the temperature dependence of the emission intensity is given by $I = I|_{300 \text{ K}} \exp[-(T - 300 \text{ K})/T_1]$.

Design a drive circuit consisting of a constant-voltage source and a resistor, which compensates for the temperature dependence of the emission intensity of the LED so that the LED emission intensity is the same at the water freezing-point temperature (0 °C) and 60 °C. The LED should draw 20 mA at the freezing-point temperature.

Solution: At 60 °C, the current needs to be 36.4 mA in order to keep the emission intensity independent of temperature. Constructing a load line that intersects the 0 °C and 60 °C diode I–V characteristic at 20 mA and 36.4 mA, respectively, yields the following values for the drive circuit: Constant-voltage source with $V = 1.6$ V and series resistance of 2.7 Ω.

References

Abdelkader H. I., Hausien H. H., and Martin J. D. "Temperature rise and thermal rise-time measurements of a semiconductor laser diode" *Rev. Sci. Instrum.* **63**, 2004 (1992)

Chhajed S., Xi Y., Li Y.-L., Gessmann Th., and Schubert E. F. "Influence of junction temperature on chromaticity and color rendering properties of trichromatic white light sources based on light emitting diodes" *J. Appl. Phys.* **97**, 054506 (2005)

Epperlein P. W. "Reflectance modulation – a novel approach to laser mirror characterization" in *Proceedings of 17th International Symposium of Gallium Arsenide and Related Compounds*, IOP Conference Series, IOP, London, **112**, 633 (1990)

Epperlein P. W. and Bona G. L. "Influence of the vertical structure on the mirror facet temperatures of visible GaInP quantum well lasers" *Appl. Phys. Lett.* **62**, 3074 (1993)

Gessmann Th., Schubert E. F., Graff J. W., Streubel K., and Karnutsch C. "Omni-directionally reflective contacts for light-emitting diodes" *IEEE Electr. Dev. Lett.* **24**, 683 (2003)

Gu Y. and Narendran N. "A non-contact method for determining junction temperature of phosphor-converted white LEDs" *Third International Conference on Solid State Lighting, Proceedings of SPIE*, San Diego, Calif., 2003 (to be published) see also J. Taylor "Non-intrusive techniques help to predict the lifetime of LED lighting systems" *Compound Semiconductors* October (2003)

Ioffe Physico-Technical Institute (Saint Petersburg, Russia) "Physical properties of semiconductors" www.ioffe.ru/SVA/NSM/Semicond (2004)

Hall D. C., Goldberg L., and Mehuys D. "Technique for lateral temperature profiling in optoelectronic devices using a photoluminescence microprobe" *Appl. Phys. Lett.* **61**, 384 (1992)

Millman J. and Halkias C. *Integrated Electronics: Analog and Digital Circuits and Systems* (McGraw-Hill, New York, 1972)

Murata S. and Nakada H. "Adding a heat bypass improves the thermal characteristics of a 50 μm spaced 8-beam laser diode array" *J. Appl. Phys.* **72**, 2514 (1992)

Rommel J. M., Gavrilovic P. and Dabkowski F. P. "Photoluminescence measurement of the facet temperature of 1 W gain-guided AlGaAs/GaAs laser diodes" *J. Appl. Phys.* **80**, 6547 (1996)

Schubert E. F., Göbel E. O., Horikoshi Y., Ploog K., and Queisser H. J. "Alloy broadening in photoluminescence spectra of AlGaAs" *Phys. Rev.* **B30**, 813 (1984)

Todoroki S., Sawai M., and Aiki K. "Temperature distribution along the striped active region in high-power GaAlAs visible lasers" *J. Appl. Phys.* **58**, 1124 (1985)

Varshni Y. P. "Temperature dependence of the energy gap in semiconductors" *Physica* **34**, 149 (1967)

Xi Y. and Schubert E. F. "Junction–temperature measurement in GaN ultraviolet light-emitting diodes using diode forward-voltage method" *Appl. Phys. Lett.* **85**, 2163 (2004)

Xi Y., Xi J.-Q., Gessmann T., Shah J. M., Kim J. K., Schubert E. F., Fischer A. J., Crawford M. H., Bogart K. H. A., and Allerman A. A. "Junction temperature measurements in deep-UV light-emitting diodes" *Appl. Phys. Lett.* **86**, 031907 (2005)

7

High internal efficiency designs

There are two general possibilities for attaining high internal quantum efficiency. The first possibility is to enhance the radiative recombination probability, and the second possibility is to decrease the non-radiative recombination probability. This can be accomplished in different ways which will be discussed below.

7.1 Double heterostructures

The lifetimes derived from the bimolecular rate equation show that the radiative rate increases with the free-carrier concentration for both the low excitation limit as well as the high excitation limit. It is therefore important that the region in which recombination occurs has a high carrier concentration. Double heterostructures are an excellent way to achieve such high carrier concentrations.

A double heterostructure (DH) consists of the active region in which recombination occurs and two confinement layers cladding the active region. A double heterostructure LED structure is shown schematically in Fig. 7.1. The two *cladding* or *confinement layers* have a larger bandgap than the active region. If the bandgap difference between the active and the confinement regions is ΔE_g, then the band discontinuities occurring in the conduction and valence bands follow the relation

$$E_g\big|_{\text{cladding}} - E_g\big|_{\text{active}} = \Delta E_g = \Delta E_C + \Delta E_V . \tag{7.1}$$

Both band discontinuities, ΔE_C and ΔE_V, should be much larger than kT in order to avoid carrier escape from the active region into the confinement regions.

The effect of a double heterostructure on the carrier concentration is illustrated schematically in Fig. 7.2. In the case of a homojunction, carriers diffuse to the adjoining side of the junction under forward bias conditions. Minority carriers are distributed over the electron and hole diffusion lengths as illustrated in Fig. 7.2 (a). In III–V semiconductors, diffusion lengths can be 10 µm or even longer.

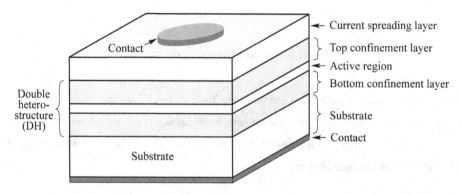

Fig. 7.1. Illustration of a double heterostructure consisting of a bulk or quantum well active region and two confinement layers. The *confinement* layers are frequently called *cladding* layers.

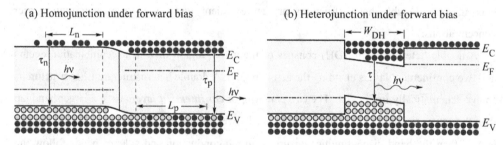

Fig. 7.2. Free carrier distribution in (a) a homojunction and (b) a heterojunction under forward bias conditions. In homojunctions, carriers are distributed over the diffusion length. In heterojunctions, carriers are confined to the well region.

The wide distribution of carriers and the correspondingly low carrier concentration (particularly towards the end of the diffusion tail) can be avoided by the employment of double heterostructures. Carriers are confined to the active region, as shown in Fig. 7.2 (b), as long as the barrier heights are much higher than the thermal energy kT. Today virtually all high-efficiency LEDs use double heterostructure designs.

Double heterostructures are used for *bulk* as well as *quantum well* active regions. Quantum well active regions provide additional carrier confinement to the narrow well regions, which can further improve the internal quantum efficiency. On the other hand, if a quantum well active region is used, the barriers between the wells will impede the flow of carriers between adjacent wells. Thus the barriers in a multi-quantum well active region need to be sufficiently "transparent" (low and/or thin barriers) in order to allow for efficient carrier transport between the wells and to avoid the inhomogeneous distribution of carriers within the active region.

The thickness of the active region in a DH has a strong influence on the internal quantum efficiency of an LED. Typical active region thicknesses are a few tenths of a micrometer for bulk active regions and even thinner for quantum well active regions. The dependence of device efficiency on the active region thickness is shown in Fig. 7.3 (Sugawara et al., 1992). Inspection of the figure shows that the optimum thickness for an AlGaInP active region is between 0.15 and 0.75 µm.

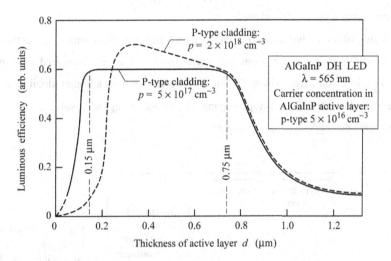

Fig. 7.3. Dependence of the luminous efficiency of an AlGaInP double heterostructure LED emitting at 565 nm on the active layer thickness. The figure reveals an optimum active region thickness of 0.15–0.75 µm (after Sugawara et al., 1992).

If the active region in a double heterostructure becomes too thick, e.g. larger than the diffusion length of carriers, the advantage of the double heterostructure is lost and carriers are distributed as they are in homojunctions. On the other hand, if the active region of a double heterostructure becomes too thin, the active region tends to overflow at high injection current levels.

7.2 Doping of active region

Doping of the active region and confinement layers plays a crucial role in the efficiency of double heterostructure (DH) LEDs. The influence of the doping on the internal efficiency is multifaceted. Firstly, we consider the doping of the active region.

The active region of III–V arsenide and phosphide DH LEDs must not be heavily doped. Heavy doping with either p-type or n-type dopants would place the p-n junction effectively at the edge of the DH well region, i.e. at the active/confinement interface, thereby promoting carrier spill-over into one of the confinement regions. Diffusion of carriers into the cladding region decreases the radiative efficiency. Thus heavy doping of the active region is rarely done in III–V arsenide and phosphide DH LEDs.

Therefore it is required that the active region is *either* doped at a level *lower* than the doping concentration in the confinement regions *or* left undoped. Typically the doping concentration in intentionally doped active regions is either p-type or n-type in the 10^{16} to low 10^{17} cm^{-3} range. Frequently the active layer is left undoped.

P-type doping of the active region is more common than n-type doping of the active region, due to the generally longer electron-minority-carrier diffusion length compared with the hole-minority-carrier diffusion length. (Note that electrons generally have a higher mobility than holes in III–V semiconductors.) Thus, p-type doping of the active region ensures a more uniform carrier distribution throughout the active region.

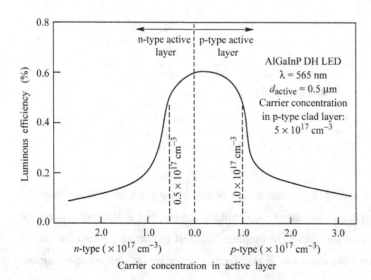

Fig. 7.4. Dependence of the luminous efficiency of an AlGaInP double heterostructure LED emitting at 565 nm on the active layer doping concentration (after Sugawara *et al.*, 1992).

116

The dependence of device efficiency on the doping concentration of the active region in an AlGaInP DH LED is shown in Fig. 7.4 (Sugawara *et al.*, 1992). Inspection of the figure shows that a high quantum efficiency is obtained only for n-type doping concentrations below 5×10^{16} cm^{-3}, and below 1×10^{17} cm^{-3} for p-type active layers.

Figure 7.4 also reveals that light p-type doping of the active region is preferable over light n-type doping. Therefore, most active layers of LEDs and lasers are lightly doped with acceptors. Electrons with their larger diffusion length are also more likely to diffuse into the p-type confinement layer if the active region is doped n-type.

Intentional doping of the active region can have advantages as well as disadvantages. The carrier lifetime depends on the concentration of majority carriers. In the low-excitation regime, the radiative carrier lifetime decreases with increasing free carrier concentration, i.e. doping concentration. As a result, the radiative efficiency increases. An example of a material whose luminescence efficiency increases with doping concentration is Be-doped GaAs. It is well known that the radiative efficiency of Be-doped GaAs increases with the Be doping level in the moderate doping concentration range.

On the other hand, dopants may, especially at high concentrations, introduce defects that act as recombination centers. High concentrations of intentional dopants lead to an increased concentration of native defects due to the dependence of the native and non-native defect concentrations on the Fermi level (see, for example, Longini and Greene, 1956; Baraff and Schluter, 1985; Walukiewicz, 1988, 1989, 1994; Neugebauer and Van de Walle, 1999).

The epitaxial growth process may also depend on doping. Doping atoms can act as **surfactants**, i.e. surface-active reagents. For example, a surfactant can increase the surface diffusion coefficient, thereby improving crystal quality. There are many other ways by which surfactants influence the growth process. These processes are generally not understood in great detail. However, the improvement of crystal quality has been found in a number of cases. During the growth of InGaN, for example, a marked improvement of crystal quality has been found upon doping with silicon (Nakamura *et al.*, 1996, 1998). The quantum barriers in III–V nitride multiple-quantum well structures are in fact often doped with silicon at a high level, e.g. 2×10^{18} cm^{-3}. The increase in device efficiency found for high doping may be due to the screening of internal polarization fields which reduces the overall potential drop within the active region.

7 High internal efficiency LED designs

7.3 p-n junction displacement

The displacement of the p-n junction from its intended location into the cladding layer can be a significant problem in DH LED structures. Usually, the lower confinement layer is n-type, the upper confinement layer is p-type, and the active region is undoped or lightly doped with n- or p-type dopants. However, if dopant redistribution occurs, the p-n junction can be displaced into one of the confinement layers. The diffusion of dopants can occur during growth and be caused by high growth temperature, a long growth time, or a strongly diffusing dopant. Dopants can redistribute due to diffusion, segregation, and drift.

Frequently, acceptors from the top confinement layer diffuse into the active region and also into the lower confinement layer. Impurities such as Zn and Be are small atoms that can easily diffuse through the crystal lattice. In addition, Zn and Be are known to have a strongly concentration-dependent diffusion coefficient. If a certain critical concentration is exceeded, Zn and Be acceptors diffuse very rapidly. As a result, the device will not work well and will not emit light at the intended emission wavelength.

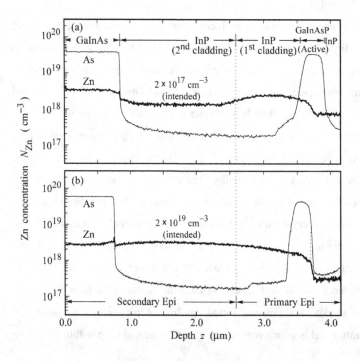

Fig. 7.5. Secondary ion mass spectrometry (SIMS) profile of Zn in a GaInAsP/InP double heterostructure. The structure uses Zn as a p-type dopant. Part (a) shows no p-n junction displacement. Part (b) shows p-n junction displacement caused by high Zn doping of the upper cladding region (after Schubert et al., 1995).

An example of a Zn acceptor profile in a GaInAsP/InP double heterostructure measured by secondary ion mass spectrometry (SIMS) is shown in Fig. 7.5. In Fig. 7.5 (a), the top confinement layer has a moderate intended doping concentration of 2×10^{17} cm^{-3}. The Zn profile indicates that the Zn is mostly confined to the upper confinement layer, even though some Zn has evidently diffused into the active region. In Fig. 7.5 (b), however, the top confinement layer has a high intended doping concentration of 2×10^{19} cm^{-3}. The profile indicates that Zn has strongly diffused into the active region. The p-n junction is displaced to the edge of the active region. As a result, the device shown in Fig. 7.5 (b) has a much lower quantum efficiency than the device shown in Fig. 7.5 (a).

A model explaining the p-n junction displacement in the GaInAsP/InP DH structure is illustrated in Fig. 7.6 (Schubert *et al.*, 1995). It is assumed in this model that the Zn diffusion coefficient increases rapidly above a *critical* concentration $N_{critical}$. If this concentration is exceeded during growth, Zn will redistribute until the concentration falls below the critical concentration. As a result, Zn can diffuse into and through the active region of the double heterostructure. It is remarkable that the p-n junction displacement can occur, even if the intended Zn concentration in the *confinement region in the vicinity* of the active region is quite low.

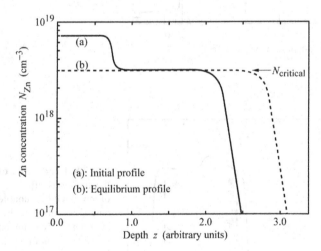

Fig. 7.6. P-n junction displacement process caused by excessive doping of the cladding region. If the acceptor dopant has a highly concentration-dependent diffusion constant and the diffusion constant increases strongly above a critical concentration, $N_{critical}$, p-n junction displacement occurs in the active region (after Schubert *et al.*, 1995).

7.4 Doping of the confinement regions

Doping of the confinement regions has a strong influence on the efficiency of double heterostructure LEDs. The resistivity of the confinement regions is one factor in determining the doping concentration in the confinement layers. The resistivity should be low to avoid resistive heating of the confinement regions.

Another factor is the residual doping concentration in the active region. The active region has a residual doping concentration, even if not intentionally doped. Typical doping concentrations in the active region are in the 10^{15}–10^{16} cm^{-3} range. The doping concentration in the confinement layers must be higher than the doping concentration of the active region to define the location of the p-n junction.

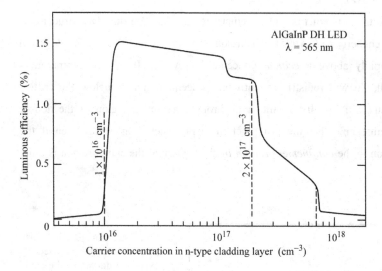

Fig. 7.7. Dependence of the luminous efficiency of an AlGaInP double heterostructure LED emitting at 565 nm on n-type confinement layer doping concentration (after Sugawara *et al.*, 1992).

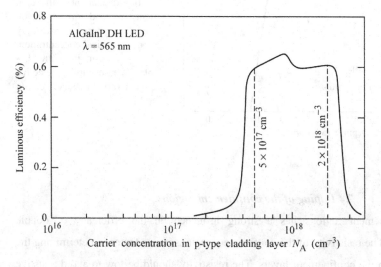

Fig. 7.8. Dependence of the luminous efficiency of an AlGaInP double heterostructure LED emitting at 565 nm on the p-type confinement layer doping concentration (after Sugawara *et al.*, 1992).

The influence of the confinement-layer doping concentration on the internal quantum efficiency was analyzed by Sugawara et al. (1992). The results are shown in Figs. 7.7 and 7.8. The figures reveal that there is an optimum doping range for the confinement regions. For n-type confinement regions, the optimum doping concentration ranges from 10^{16} to 2×10^{17} cm^{-3}. For p-type confinement regions, the optimum doping concentration ranges from 5×10^{17} to 2×10^{18} cm^{-3}, clearly higher than in the n-type cladding region. The reason for this marked difference could again lie in the larger diffusion length of electrons than that of holes. A high p-type concentration in the cladding region keeps electrons in the active region and prevents them from diffusing deep into the confinement region.

The carrier leakage out of the active region into the p-type cladding layer in double heterostructure lasers was investigated by Kazarinov and Pinto (1994). It was shown that the electron leakage out of the active region is more severe than the hole leakage. The difference is due to the generally higher diffusion constant of electrons compared with holes.

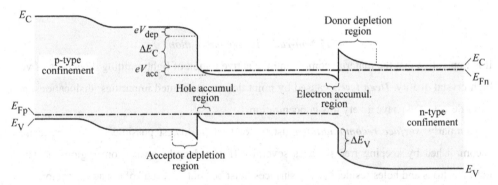

Fig. 7.9. Band diagram of a forward-biased double heterostructure. The p-type confinement layer consists of a lightly doped layer close to the active region and a higher doped layer further away from the active layer (adapted from Kazarinov and Pinto, 1994).

The band diagram of a double heterostructure under forward bias conditions is shown in Fig. 7.9. The figure illustrates that barriers are formed by the depletion layers at the two confinement–active-region interfaces. Compositional grading of these interfaces can be used to reduce these barriers.

The influence of the cladding layer doping concentration on the radiative efficiency of a double heterostructure laser at threshold is shown in Fig. 7.10 (Kazarinov and Pinto, 1994). Inspection of the figure reveals that the confinement layer doping has a severe influence on the luminous efficiency. Low doping concentrations in the p-type confinement layers facilitate electron escape from the active region, thereby lowering the internal quantum efficiency.

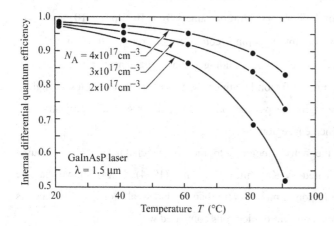

Fig. 7.10. Dependence of the internal differential quantum efficiency (emitted photons per injected electron) on temperature for different p-type doping levels in the cladding layer (after Kazarinov and Pinto, 1994).

7.5 Non-radiative recombination

It is imperative that the material comprising the active region of light-emitting devices is of very high crystal quality. **Deep levels** caused by point defects, unwanted impurities, dislocations, and other defects must have a very low concentration.

Similarly, **surface recombination** must be kept at the lowest possible levels. This can be accomplished by keeping free surfaces several diffusion lengths away from regions in which both electrons and holes reside, i.e. any surfaces must be "out of reach" of the active region.

Mesa-etched LEDs and lasers, in which the mesa etch exposes the active region to air, generally have low internal efficiencies due to surface recombination. Surface recombination also leads to a reduction of the lifetime of LEDs. Surface recombination generates heat at the semiconductor surface, which can lead to structural defects such as dark-line defects that further reduce the efficiency of LEDs.

Figure 7.11 shows the light intensity of two mesa-etched and two planar LEDs versus time. Inspection of Fig. 7.11 reveals that (*i*) the light intensity at $t = 0$ h of the mesa-etched LEDs is slightly lower than that of the planar structure and that (*ii*) the lifetime of the mesa-etched device is much shorter. In the planar device, electron–hole recombination occurs below the top metal contact far away from the sidewall surfaces of the device. Thus, no intensity reduction or degradation mediated by surface recombination is expected for the planar device.

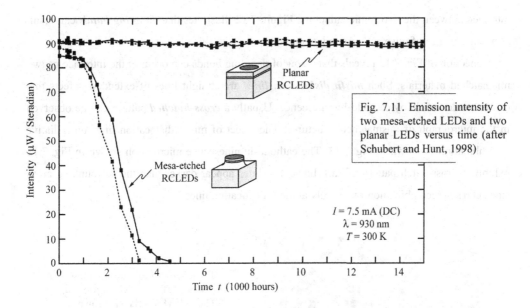

Fig. 7.11. Emission intensity of two mesa-etched LEDs and two planar LEDs versus time (after Schubert and Hunt, 1998)

Note that the presence of surfaces does *not* reduce the radiative efficiency if only *one type of carrier* is present, e.g. near the top contact of the device. Surfaces in such unipolar regions do not have any deleterious effects.

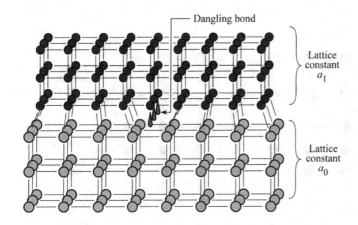

Fig. 7.12. Illustration of two crystals with mismatched lattice constant resulting in dislocations at or near the interface between the two semiconductors.

7.6 Lattice matching

In double heterostructures, the active region material is different from the confinement layer material. However, both materials should have the same crystal structure and lattice constant. If the semiconductors do not have the same lattice constant, defects will occur at or near the

7 High internal efficiency LED designs

interface between the two semiconductors. Figure 7.12 illustrates that ***dangling bonds*** can result as a consequence of mismatched semiconductors.

Inspection of Fig. 7.12 reveals that a *row* of dangling bonds can occur at the interface of two mismatched materials. Such ***misfit dislocation lines*** are straight lines of extended defects that can be made visible by cathodo-luminescence. Usually a ***cross-hatched*** pattern can be observed in the micrograph of mismatched structures. The effect of misfit dislocation lines on radiative recombination is shown in Fig. 7.13. The cathodo-luminescence micrograph shown in Fig. 7.13 exhibits a cross-hatch pattern of dark lines. The lines appear darker than the surrounding areas since carriers recombine non-radiatively at these dislocation lines.

Fig. 7.13. Cathodo-luminescence image of a 0.35 µm thick $Ga_{0.95}In_{0.05}As$ layer grown on a GaAs substrate. The dark lines forming a cross-hatch pattern are due to misfit dislocations (after Fitzgerald *et al.*, 1989).

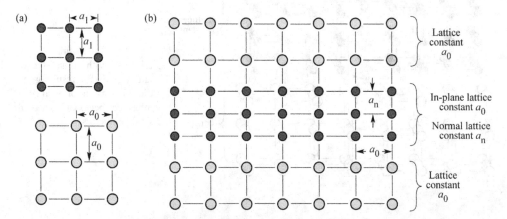

Fig. 7.14. (a) Cubic-symmetry crystals with equilibrium lattice constant a_1 and a_0. (b) Thin, coherently strained layer with equilibrium lattice constant a_1 sandwiched between two semiconductors with equilibrium lattice constant a_0. The coherently strained layer assumes an in-plane lattice constant a_0 and a normal lattice constant a_n.

Misfit dislocations may not occur directly *at* the interface between the mismatched materials but may start to form *near* the interface. This is because the mismatched crystal grown on top of the semiconductor will initially be strained elastically and assume the same in-plane lattice constant as the underlying substrate. This situation is shown in Fig. 7.14, where a thin layer is strained so that it has the same in-plane lattice constant as the underlying material. Once the energy needed to strain the lattice exceeds the energy required to form misfit dislocations, the thin film relaxes to its equilibrium lattice constant by forming misfit dislocations. The layer thickness at which misfit dislocations are formed is called the **critical thickness**. It has been calculated by Matthews and Blakeslee (1976). If the layer is thinner than the critical thickness given by the Matthews–Blakeslee law, a thin dislocation-free layer can be grown even if the layers have different lattice constants.

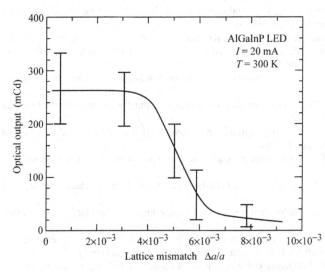

Fig. 7.15. Optical output intensity of an AlGaInP LED driven with an injection current of 20 mA versus the lattice mismatch between the AlInGaP active region and the GaAs substrate (after Watanabe and Usui, 1987).

The density of misfit dislocation lines per unit length is proportional to the lattice mismatch. Consequently, the efficiency of LEDs is expected to drop as the lattice mismatch is increased. Figure 7.15 shows the decrease of the optical intensity of an AlGaInP LED grown on a GaAs substrate. The AlGaInP material used for high-brightness red LEDs is matched to the GaAs substrate. Figure 7.15 reveals that the light output drops strongly as the mismatch, defined as $\Delta a / a$, exceeds 3×10^{-3}.

Red GaAsP LEDs are grown on GaAs substrates and the active layers are mismatched to the substrate. As a result, red GaAsP LEDs are low cost but also low efficiency. The homojunction

GaAsP LEDs grown on GaAs substrates are the lowest-cost red devices available.

Whereas the III–V arsenide and III–V phosphide material family is strongly affected by surface recombination and lattice mismatch, the III–V nitride material family is less so. One of the reasons for the insensitivity of the GaN material family to dislocation defects is the lower electrical activity that dislocations in these materials appear to exhibit. Another reason is the smaller diffusion length of carriers in the GaN material family. If the mean distance between dislocations is larger than the diffusion length, in particular the hole diffusion length, non-radiative recombination at these dislocations will not be severe. Another model explaining the high efficiency of GaInN is based on compositional fluctuations of the ternary alloy, which localize carriers preventing them from diffusing towards dislocation lines.

References

Baraff G. A. and Schluter M. "Electronic structure, total energies, and abundances of the elementary point defects in GaAs" *Phys. Rev. Lett.* **55**, 1327 (1985)

Fitzgerald E. A., Watson G. P., Proano R. E., Ast D. G., Kirchner P. D., Pettit G. D., and Woodall J. M. "Nucleation mechanism and the elimination of misfit dislocations at mismatched interfaces by reduction of growth area" *J. Appl. Phys.* **65**, 2220 (1989)

Kazarinov R. F. and Pinto M. R. "Carrier transport in laser heterostructures" *IEEE J. Quantum Electronics* **30**, 49 (1994)

Longini R. L. and Greene R. F. "Ionization interaction between impurities in semiconductors and insulators" *Phys. Rev.* **102**, 992 (1956)

Matthews J. W. and Blakeslee A. E. "Defects in epitaxial multilayers. III. Preparation of almost perfect multilayers" *J. Cryst. Growth* **32**, 265 (1976)

Nakamura S., Mukai T., and Iwasa N, "Light-emitting GaN-based compound semiconductor device" US Patent 5,578,839 (1996)

Nakamura S., Mukai T., and Iwasa N, "Light-emitting GaN-based compound semiconductor device" US Patent 5,747,832 (1998)

Neugebauer J. and Van de Walle C. G. "Chemical trends for acceptor impurities in GaN" *J. Appl. Phys.* **85**, 3003 (1999)

Schubert E. F., Downey S. W., Pinzone C., and Emerson A. B. "Evidence of very strong inter-epitaxial-layer diffusion in Zn doped GaInPAs/InP structures" *Appl. Phys. A* **60**, 525 (1995)

Schubert E. F. and Hunt N. E. J. "15,000 hours stable operation of resonant-cavity light-emitting diodes" *Appl. Phys. A* **66**, 319 (1998)

Sugawara H., Ishikawa M., Kokubun Y., Nishikawa Y., Naritsuka S., Itaya K., Hatakoshi G., Suzuki M., "Semiconductor light emitting device" US Patent 5,153,889, issued Oct. 6 (1992)

Walukiewicz W. "Fermi level dependent native defect formation: consequences for metal–semiconductor and semiconductor–semiconductor interfaces" *J. Vac. Sci. Technol. B*, **6**, 1257 (1988)

Walukiewicz W. "Amphoteric native defects in semiconductors" *Appl. Phys. Lett.* **54**, 2094 (1989)

Walukiewicz W. "Defect formation and diffusion in heavily doped semiconductors" *Phys. Rev. B* **50**, 5221 (1994)

Watanabe H. and Usui A. "Light emitting diode" US Patent 4,680,602, issued July 14 (1987)

8

Design of current flow

LEDs may be grown on conductive and insulating substrates. Whereas the current flow is mostly vertical (normal to the substrate plane) in structures grown on conductive substrates, it is mostly lateral (horizontal) in devices grown on insulating substrates. The location and size of ohmic contacts are relevant to light extraction, because metal contacts are opaque. The current chapter discusses the current flow patterns of different device structures aimed at high extraction efficiency.

8.1 Current-spreading layer

In LEDs with thin top confinement layers, the current is injected into the active region mostly under the top electrode. Thus, light is generated under an opaque metal electrode. This results in a low extraction efficiency. This problem can be avoided with a ***current-spreading layer*** that spreads the current under the top electrode to regions not covered by the opaque top electrode.

The *current-spreading layer* is synonymous with the ***window layer***. The term window layer is occasionally used to emphasize the *transparent character* of this layer and its ability to enhance the extraction efficiency.

The usefulness of current-spreading layers was realized during the infancy of LEDs. Nuese *et al.* (1969) demonstrated a substantial improvement of the optical output power in GaAsP LEDs by employing a current-spreading or window layer. The window layer is the top semiconductor layer located between the upper cladding layer and the top ohmic contact. The effect of a current-spreading layer is illustrated in Fig. 8.1. Light is emitted only around the perimeter of the top contact for LEDs without a current-spreading layer, as shown in Fig. 8.1 (a). The addition of a current-spreading layer results in more uniform and brighter surface emission as shown in Fig. 8.1 (b).

Nuese *et al.* (1969) demonstrated current-spreading layers composed of the ternary GaAsP and the binary GaP and discussed the requirements of the current-spreading layer. These

8 Design of current flow

requirements include low resistivity and large thickness for current spreading, and transparency to minimize absorption losses. To reduce absorption losses, Nuese et al. (1969) employed a high P mole fraction in the $GaAs_{1-x}P_x$ current-spreading layer, namely $0.45 < x \leq 1.0$, higher than the P mole fraction in the $GaAs_{1-x}P_x$ active region, where $x = 0.45$. Thus the bandgap energy of the current-spreading layer is higher than the bandgap of the active region. Although Nuese et al. discussed the properties of the current-spreading layer qualitatively, they did not provide a quantitative theoretical framework of current spreading. The theoretical foundation of current-spreading layers in devices with linear contact geometry was given by Thompson, as discussed below. The use of a current-spreading layer was adopted in most top-emitting LED designs, including AlGaAs LEDs (Nishizawa et al., 1983; Moyer 1988), GaP LEDs (Groves et al., 1977, 1978a, 1978b), and AlGaInP LEDs (Kuo et al., 1990; Sugawara et al., 1991, 1992a, 1992b).

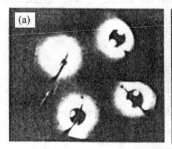 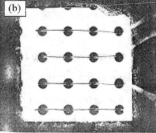

Fig. 8.1. Effect of the current-spreading layer on LED output. (a) Top view without a current-spreading layer. Emission occurs only near the perimeter of the contact. (b) Top view with a current-spreading layer (after Nuese et al., 1969).

The effect of the current-spreading layer is illustrated schematically in Fig. 8.2. Without a current-spreading layer, the current-injected area of the active region is limited to approximately the contact size, as indicated in Fig. 8.2 (a). The addition of the current-spreading layer results in a larger current-injected area, as shown in Fig. 8.2 (b).

Current-spreading layers are predominantly employed in top-emitting LEDs. Two different approaches for AlGaInP visible LEDs, grown on GaAs substrates, are shown in Figs. 8.2 (c) and (d). A GaP current-spreading layer was reported by Kuo et al. (1990) and Fletcher et al., (1991a, 1991b). GaP has a bandgap of $E_{g,GaP} = 2.26$ eV and is thus transparent for red, orange, yellow, and part of the green spectrum. AlGaInP LEDs with emission wavelengths as short as 550 nm have been fabricated. GaP, as a binary compound semiconductor, is very transparent for energies below the bandgap, i.e. the Urbach tail energy of GaP is small. Furthermore GaP is an indirect-gap semiconductor, which is inherently less absorbing compared with direct-gap semiconductors. Thus little light is absorbed even in thick GaP current-spreading layers.

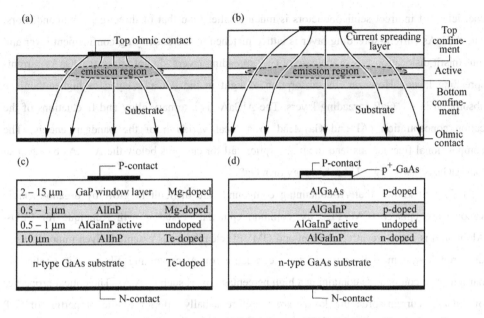

Fig. 8.2. Current-spreading structures in high-brightness AlGaInP LEDs. Illustration of the effect of a current-spreading layer for LEDs (a) without and (b) with a spreading layer on the light extraction efficiency. (c) GaP current-spreading structure (Fletcher et al., 1991a, 1991b). (d) AlGaAs currect-spreading structure (Sugawara et al., 1992a, 1992b).

However, GaP is lattice mismatched to the underlying epitaxial layers. The lower confinement, active, and upper confinement layers are lattice matched to the GaAs substrate. Since GaP has a lattice constant that is about 3.6% smaller than that of GaAs, a high density of threading dislocations and stacking faults is expected at the upper-cladding-layer-to-GaP interface. One could assume that these dislocations, which will act as non-radiative recombination centers, will not degrade the internal quantum efficiency of the LED because they are located at the confinement–window layer interface and in the window layer, far away from the active region. However, if the dislocations propagated downward towards the active region during device operation, the efficiency and the reliability of the LEDs would be affected. The issues associated with the confinement–window interface have apparently been resolved since AlGaInP/GaAs LEDs with GaP window layers have excellent reliability and efficiency.

An alternative approach for increasing the extraction efficiency in AlGaInP/GaAs LEDs uses AlGaAs current-spreading layers (Sugawara et al., 1991, 1992a, 1992b). $Al_xGa_{1-x}As$ is lattice-matched to GaAs for all chemical compositions $0 \leq x \leq 1$. AlAs has a bandgap energy of $E_{g,AlAs} = 2.9$ eV. For $x > 0.45$, $Al_xGa_{1-x}As$ becomes an indirect semiconductor. The absorption

8 Design of current flow

coefficient of indirect semiconductors is much smaller than that of direct-gap semiconductors. The AlGaAs current-spreading layer is lattice matched to the underlying confinement layer and thus misfit dislocations, as in the case of GaP spreading layers, do not arise for AlGaAs current-spreading layers. However, the absorption of light in the AlGaAs layers is higher than the absorption in the GaP spreading layers. The AlGaAs is a ternary alloy and fluctuations of the cation concentration (Al and Ga) lead to a local variation of the bandgap energy. The compositional fluctuations lead to an absorption tail for energies below the AlGaAs bandgap so that AlGaAs has a larger Urbach energy than GaP.

It is well known that Al-containing compounds are difficult to grow by organo-metallic vapor-phase epitaxy (OMVPE), the common epitaxial crystal growth technique for LEDs. Aluminum is a very reactive element and OMVPE cleanliness is essential. Even minor leaks in the growth system will result in the degradation of Al-containing films. This applies, in particular, to compounds containing a high percentage of Al such as AlAs. The optical properties of AlGaAs current-spreading layers are therefore usually inferior to the properties of GaP spreading layers. The electrical properties of AlAs or AlGaAs with high Al content are also inferior to GaP. Finally, AlAs tends to oxidize over time when exposed to water or humid air (Choquette *et al.*, 1997). Despite these difficulties, viable AlGaInP LEDs with AlGaAs current-spreading layers have been developed and are commercially available.

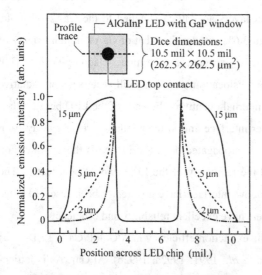

Fig. 8.3. The effect of GaP window thickness on current spreading is illustrated by surface light emission intensity profiles for three different AlGaInP LED chips with window layer thicknesses of 2, 5, and 15 μm. The profile is indicated by the dashed line in the inset. The dip in the middle of the profiles is due to the opaque ohmic contact pad. A microscope fitted with a video camera was used in the measurements (after Fletcher *et al.*, 1991a).

The increase in light extraction efficiency is illustrated in Fig. 8.3 for p-type GaP current-spreading layer with a resistivity of 0.05 Ω cm and thicknesses ranging from 2 to 15 μm

(Fletcher et al., 1991a). The data shown in Fig. 8.3 were obtained using a near-field microscope and a video analyzer. To obtain the light intensity profile, a single-line scan was taken across the chip, including the center p-type contact pad, as shown in the inset. Since the intensity is directly proportional to the p-n current density at any given point, the current-spreading characteristics are obtained by this measurement. For a window thickness of 2 µm, current spreading is limited. As the window layer thickness is increased to 15 µm, the current spreads well beyond the contact, reaching almost the edge of the chip. An even larger thickness of the window layer would spread the current to the edges of the chip. Such strong current spreading is not desirable due to surface recombination.

The effect of current spreading on the efficiency of an AlGaInP/GaAs LED with a GaP current-spreading layer is shown in Fig. 8.4. For as sufficiently thick window layer the extraction efficiency is increased by a factor of approximately 8. The comparison of pulsed with direct current (DC) measurements shown in Fig. 8.4 shows that the efficiency drop occurring at high currents is caused by heating of the device.

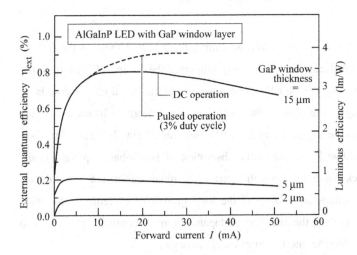

Fig. 8.4. Bare chip external quantum efficiency and luminous efficiency versus forward current for AlGaInP LEDs with GaP window layer thicknesses of 2, 5, and 15 µm. Solid curves are under DC conditions. Dashed curve is under pulsed condition using 400 ns pulses and a 3% duty cycle. Heating is essentially eliminated in this case (after Fletcher et al., 1991a).

The optimum thickness range of current-spreading layers in AlGaInP/GaAs LEDs with $Al_{0.70}Ga_{0.30}As$ current-spreading layers was investigated by Sugawara et al. (1991, 1992a, 1992b). The p-type doping concentration of the $Al_{0.70}Ga_{0.30}As$ spreading layer was 3×10^{18} cm^{-3}. The luminous efficiency of the LED versus current-spreading layer thickness is shown in Fig. 8.5. Inspection of the figure reveals that the optimum thickness of the current-spreading layer is between 5 and 30 µm. For a current-spreading layer thickness of 15 µm, the efficiency of the device increases by a factor of 30 compared with a device with no current-spreading layer at all. The optimum doping concentration in the p-type current-spreading layer was found in the low 10^{18} cm^{-3} range.

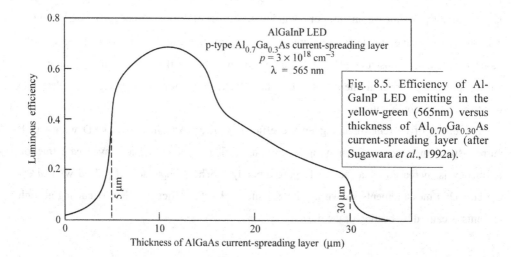

Fig. 8.5. Efficiency of AlGaInP LED emitting in the yellow-green (565nm) versus thickness of $Al_{0.70}Ga_{0.30}As$ current-spreading layer (after Sugawara et al., 1992a).

The disadvantage of no or a very thin current-spreading layer is that most of the light is generated under the opaque metal contact pad, thereby hindering the escape of light from the LED die. A very thick window layer is equally disadvantageous. Firstly, a thick window layer spreads the current all the way to the edge of the LED dies, which leads to increased surface recombination and thus lowers the efficiency of the LED. Secondly, light absorption increases with the thickness of the window layer due to the absorption of below-bandgap light in the window layer. Thirdly, a thick window layer will increase the ohmic resistance of the device thereby lowering the overall efficiency. Fourthly, the long growth times required for thick current-spreading layers may result in the diffusion of dopants from the confinement layers into the active region, thereby lowering the internal quantum efficiency.

Current spreading is an important issue in many LED materials, in particular in those materials that possess low conductivity. Current spreading in the top p-type layer of GaN/GaInN LEDs is very weak due to the high resistivity of the p-type top cladding layer. The hole mobility in III–V nitrides is typically 1–20 cm^2/(Vs) and the hole concentrations are in the 10^{17} cm^{-3} range, resulting in resistivities $> 1\,\Omega$ cm. To address this problem, Jeon et al. (2001) demonstrated an LED with a **tunnel junction** adjoining the p-type confinement region above the active region. An n-type layer on top of the tunnel junction allows for lateral current spreading under the top contact. Owing to the employment of the tunnel junction, the LED has *two* n-type but *no* p-type ohmic contacts.

8.2 Theory of current spreading

The theory of current spreading under a *linear stripe top contact* geometry has been reported by Thompson (1980). Such a stripe-like geometry is typical for semiconductor lasers. Figure 8.6 (a) shows the schematic cross section of a stripe-geometry semiconductor laser. The laser has a current-spreading layer located above the p-n junction. Because of the symmetry of the laser, only the right half of the laser is shown, so that the left edge of the contact shown in the diagram is actually the center of the laser stripe. The model assumes a constant potential and current density (J_0) under the metal ($x < r_c$). The potential throughout the substrate is assumed to be constant. The current density $J(x)$ extending away from the contact is given by

$$J(x) = \frac{2 J_0}{\left[(x - r_c)/L_s + \sqrt{2}\,\right]^2} \qquad (x \geq r_c) \qquad (8.1)$$

where L_s is the current spreading length given by

$$L_s = \sqrt{\frac{t\, n_{\text{ideal}}\, kT}{\rho\, J_0\, e}} \qquad (8.2)$$

where ρ is the resistivity of the current spreading-layer, t is the thickness of the current-spreading layer, and n_{ideal} is the diode ideality factor. The diode ideality factor has typical values of $1.05 < n_{\text{ideal}} < 1.35$.

We next develop a theoretical model that can be applied to *linear stripe* (see Fig. 8.6 (a)) as well as *circular contact* (see Fig. 8.6 (b)) shapes. We first consider the **linear stripe contact geometry**. We assume that the current at the edge of the spreading region ($x = r_c + L_s$) is a factor of e^{-1} lower than under the metal contact. Then the voltage drop across the junction at the edge of the current-spreading layer is $n_{\text{ideal}}\, kT/e$ lower than under the metal contact. This voltage drops within the current-spreading region. The resistance of the current-spreading region along the lateral direction per unit stripe length dy is given by

$$R = \rho\, \frac{L_s}{t\, dy} \,. \qquad (8.3)$$

8 Design of current flow

8.2 Theory of current spreading

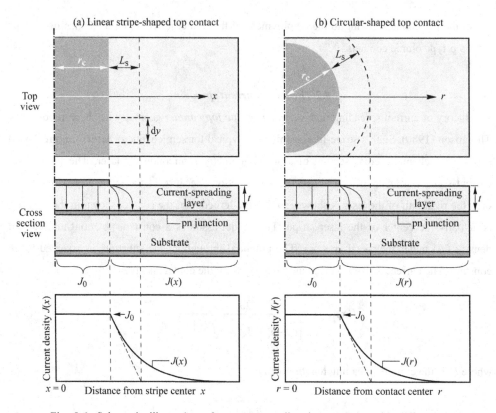

Fig. 8.6. Schematic illustration of current spreading in structures with different top contact geometries. (a) Linear stripe contact geometry. (b) Circular contact geometry.

The current flowing vertically through the junction in the current spreading region is given by

$$I = J_0 L_s \, dy . \quad (8.4)$$

Using Ohm's law, one obtains

$$\rho \frac{L_s}{t \, dy} J_0 L_s \, dy = \frac{n_{\text{ideal}} kT}{e} . \quad (8.5)$$

Solving this equation for t yields

$$\boxed{t = \rho L_s^2 J_0 \frac{e}{n_{\text{ideal}} kT}} . \quad (8.6)$$

Comparison of this equation with Eq. (8.2) yields that the two equations are identical.

Equation (8.6) allows one to calculate the required current-spreading layer thickness t for a given resistivity of this layer and the desired current-spreading length L_s.

We next consider the ***circular contact geometry*** shown in Fig. 8.6 (b). The circular geometry is relevant to LEDs with a circular top contact. Proceeding in a similar way, we write the lateral resistance from the edge of the contact to the edge of the current-spreading region. This resistance is given by

$$R = \int_{r_c}^{r_c+L_s} \rho \frac{1}{A} dr = \int_{r_c}^{r_c+L_s} \rho \frac{1}{t 2\pi r} dr = \frac{\rho}{2\pi t} \ln\left(1 + \frac{L_s}{r_c}\right). \tag{8.7}$$

The current flowing vertically through the junction in the current spreading region is given by

$$I = J_0 \left[\pi(L_s + r_c)^2 - \pi r_c^2\right] = J_0 \pi L_s (L_s + 2r_c). \tag{8.8}$$

Using Ohm's law, one obtains

$$\frac{\rho}{2\pi t} \ln\left(1 + \frac{L_s}{r_c}\right) J_0 \pi L_s (L_s + 2r_c) = \frac{n_{\text{ideal}} kT}{e}. \tag{8.9}$$

Solving this equation for t yields

$$\boxed{t = \rho L_s \left(r_c + \frac{L_s}{2}\right)\left(J_0 \frac{e}{n_{\text{ideal}} kT}\right) \ln\left(1 + \frac{L_s}{r_c}\right).} \tag{8.10}$$

Equation (8.10) allows one to calculate the required current-spreading layer thickness t for a given resistivity of this layer and the desired current-spreading length L_s. Note that for large values of r_c, we can simplify Eq. (8.10) using the approximation $\ln(1 + x) \approx x$, valid for $x \ll 1$. Thus, in the limit of large values of r_c (e.g. $r_c \to \infty$), Eq. (8.10) and Eq. (8.6) become identical, as expected.

Exercise: ***Current crowding occurring at very high current levels in devices with current-spreading layer***. In device structures with vertical current flow (current flowing from top to bottom of chip), the current-spreading layer ensures that the current spreads out over the entire p-n junction area. However, as the current increases to very high levels, the current tends to crowd under the top contact. This is illustrated in Fig. 8.7 (a) and (b). Explain the phenomenon of current crowding occurring at very high current densities.

Solution 1: The equation for the current-spreading length has the dependence $L_s \propto J_0^{-1/2}$. Thus, as the current density increases, L_s decreases, and the current "bunches" under the top contact.

8 Design of current flow

Solution 2: An intuitive explanation for current crowding can be obtained from the equivalent circuit shown in Fig. 8.7 (c). At very high current densities, the resistors that represent the p-n junction decrease (whereas the resistors representing the current-spreading layer remain constant), thereby causing the current to flow directly downward from the top contact.

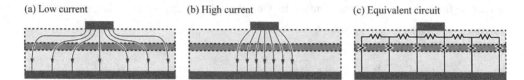

Fig. 8.7. Schematic current flow in device with current-spreading layer at (a) low and (b) high current. Current spreading decreases at very high current densities which results in current "bunching" under the top contact as shown in (b). (c) Equivalent circuit.

8.3 Current crowding in LEDs on insulating substrates

Current crowding also occurs in mesa-structure LEDs grown on insulating substrates. This type of LED includes GaInN/GaN LEDs grown on sapphire substrates. In these LEDs, the p-type contact is usually located on the top of the mesa, and the n-type contact is located on an n-type buffer layer at the bottom of the mesa. As a result, the current tends to *crowd* at the edge of the mesa contact adjoining the n-type contact.

A lateral p-side-up mesa LED grown on an insulating substrate is shown in Fig. 8.8 (a). It is intuitively clear that the p-n junction current *crowds* near the edge of the mesa as indicated in the figure. An equivalent circuit model is shown in Fig. 8.8 (b) and includes the p-type contact resistance and the resistances of the n-type and p-type cladding layers. The p-n junction is approximated by an ideal diode. The circuit model also shows several nodes separated by a distance dx. Assuming that V is the voltage in the n-type layer along the x-direction, then dV is the voltage drop across the n-layer resistance of length dx. The incremental current flowing downward through one of the diodes is given by d$I = J_0$ [exp (eV_j / kT) – 1] w dx, where J_0 is the saturation current density of the p-n junction. Calculating the *difference* in voltage drop between two adjacent resistors and applying Kirchhoff's current law to the node located between the two resistors yields the differential equation

$$\frac{d^2 V}{dx^2} = \frac{\rho_n}{t_n} J_0 \left[\exp\left(\frac{eV_j}{kT}\right) - 1 \right] . \qquad (8.11)$$

8.3 Current crowding in LEDs on insulating substrates

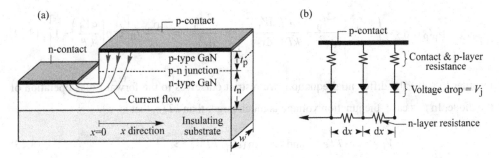

Fig. 8.8. (a) Current crowding in a mesa-structure GaN-based LED grown on an insulating substrate. (b) Equivalent circuit consisting of n-type and p-type layer resistances, p-type contact resistance, and ideal diodes representing the p-n junction.

In the case of zero or negligible resistance of the p-type layer, it is $dV = dV_j$. In this case, Eq. (8.11) can be easily solved and an analytic solution was given by Thompson (1980) who calculated the spreading length in p-n junction diodes grown on conductive substrates. In Thompson's study, the material resistivity of the top p-type cladding layer was considered, but the lower n-type cladding layer resistivity was neglected. However, in GaN/InGaN LEDs with the p-type layer on top and the n-type layer underneath, the resistive n-type layer causes current crowding and cannot be neglected. Furthermore, p-type resistances can be high, so that they should not be neglected either. As will be shown in the following calculation, both types of resistances play peculiar roles in the current-crowding problem.

Next, we take into account the resistance of the n-type layer, p-type layer, and the p-type contact resistance. The voltage drop across the p-n junction and the p-type resistors is given by

$$V = R_v I_0 \left[\exp(eV_j / kT) - 1 \right] + V_j \tag{8.12}$$

where R_v ("vertical resistance") is the sum of the p-type layer resistance and the p-type contact resistance of the area element $w\,dx$, that is

$$R_v = \rho_p \frac{t_p}{w\,dx} + \rho_c \frac{1}{w\,dx} \tag{8.13}$$

where ρ_p is the resistivity of the p-type layer and ρ_c is the p-type specific contact resistance. Forming the second derivative of V with respect to x in Eq. (8.12) and inserting the result into Eq. (8.11) yields the differential equation

8 Design of current flow

$$\frac{e}{kT}\left(\rho_c + \rho_p t_p\right) J_0 \exp\left(\frac{eV_j}{kT}\right) \left[\frac{d^2 V_j}{dx^2} + \frac{e}{kT}\left(\frac{dV_j}{dx}\right)^2\right] + \frac{d^2 V_j}{dx^2} = \frac{\rho_n}{t_n} J_0 \left[\exp\left(\frac{eV_j}{kT}\right) - 1\right]. \quad (8.14)$$

In order to solve the differential equation, we restrict ourselves to the forward-bias operation of the diode. In this case, the junction voltage is much larger than kT/e, that is

$$V_j \gg kT/e \quad \text{and} \quad \exp(eV_j/kT) \gg 1. \quad (8.15)$$

Furthermore, we assume that the voltage drop across the p-type series resistance and contact resistance is much larger than kT/e

$$(\rho_c + \rho_p t_p) J_0 \exp(eV_j/kT) \gg kT/e. \quad (8.16)$$

This condition applies to typical GaN/GaInN LEDs. Using the approximations of Eqs. (8.15) and (8.16), Eq. (8.14) can be simplified to

$$\frac{d^2 V_j}{dx^2} + \frac{e}{kT}\left(\frac{dV_j}{dx}\right)^2 = \frac{\rho_n}{(\rho_c + \rho_p t_p) t_n} \frac{kT}{e}. \quad (8.17)$$

Solving Eq. (8.17) for V_j yields $V_j(x) = V_j(0) - (kT/e)(x/L_s)$. Inserting V_j into the equation $J = J_0 \exp(eV_j/kT)$ yields the solution of the differential equation as

$$J(x) = J(0) \exp(-x/L_s) \quad (8.18)$$

where $J(0)$ is the current density at the p-type mesa edge and L_s is denoted as the **current spreading length**, that is, as the length where the current density has dropped to the $1/e$ value of the current density at the edge, so that $J(L_s)/J(0) = 1/e$. The current-spreading length is given by

$$L_s = \sqrt{(\rho_c + \rho_p t_p) t_n / \rho_n}. \quad (8.19)$$

Equation (8.19) shows that the current distribution depends on epitaxial layer thicknesses and material properties. A thick low-resistivity n-type buffer layer is needed to ensure that current crowding is minimized. Equation (8.19) also illustrates a somewhat surprising result; namely that the *decrease* of p-type specific contact resistance or p-type layer resistivity *enhances* the current-crowding effect. For *low* p-type contact and confinement resistances, strong current crowding results, unless the n-type buffer layer is very conductive so that t_n/ρ_n is very large. In GaN/GaInN devices, the sum of p-type contact and p-type layer resistances can be larger than the n-type cladding resistance, especially if t_n is small.

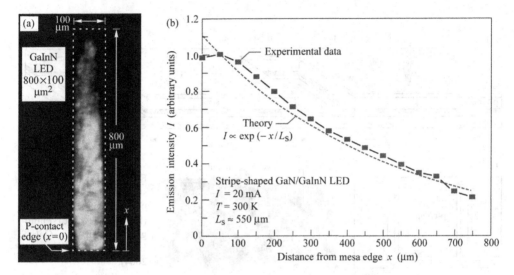

Fig. 8.9. (a) Micrograph of optical emission from mesa-structure GaInN/GaN LED grown on an insulating sapphire substrate. The LED has a stripe-shaped 800 μm × 100 μm p-type contact. (b) Theoretical and experimental emission intensity versus the distance from the mesa edge (after Guo and Schubert, 2001).

An experimental result on the current-crowding effect in GaInN/GaN LEDs grown on a sapphire substrate is shown in Fig. 8.9 (Guo and Schubert, 2001). A micrograph of the optical emission from a GaInN LED is shown in Fig. 8.9 (a). The picture was taken from the sapphire substrate side of the LED and shows the intensity of blue light emission. The micrograph clearly reveals that the emission intensity decreases with increasing distance from the mesa edge. Figure 8.9 (b) shows the experimental intensity as a function of the distance from the mesa edge. Also shown is a theoretical fit to the experimental data using the exponential decrease in current density derived above. The experimental and the theoretical data exhibit very good agreement if a current-spreading length of 550 μm is used in the calculation.

High contact resistances and high p-GaN resistivity are not desirable for high-power devices since these resistances generate heat. On the other hand, these resistances alleviate the current crowding effect. Note that with the expected future improvement of the contact and p-type doping in GaN devices, and larger device and contact sizes, current crowding will become increasingly severe unless novel contact geometries are introduced to alleviate the problem. Such novel contact geometries can include interdigitated structures (Guo et al., 2001; Steigerwald et al., 2001) with p-type finger widths of less than L_s. For device dimensions much smaller than L_s, the current-crowding effect becomes irrelevant.

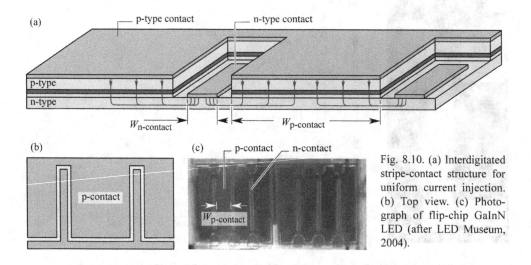

Fig. 8.10. (a) Interdigitated stripe-contact structure for uniform current injection. (b) Top view. (c) Photograph of flip-chip GaInN LED (after LED Museum, 2004).

The schematic structure and a photograph of an interdigitated stripe-contact geometry are shown in Fig. 8.10. Uniform current injection into the active region is achieved by the p-type contact width ($W_{p\text{-contact}}$) being smaller than the current spreading length. The width of the n-type contact ($W_{n\text{-contact}}$) must be at least equal to the contact transfer length to ensure low contact resistance. The contact transfer length follows from the transmission line model (TLM) used for characterization of ohmic contacts (see, for example, Schroder, 1998).

8.4 Lateral injection schemes

A device structure with a lateral current-injection scheme is shown in Fig. 8.11 (a). The current is transported laterally in both the n-type and p-type cladding layers. Ideally, the light would be generated in the region between the contacts where they would not hinder the extraction of light. If the n-type sheet resistance ρ_n / t_n (where ρ_n and t_n are the resistivity and layer thickness of the n-type material, respectively) is much lower than the p-type sheet resistance ρ_p / t_p, the current prefers to flow laterally in the low-resistance n-layer rather than the p-layer. As a result, the junction current crowds near the p-type contact.

8.4 Lateral injection schemes

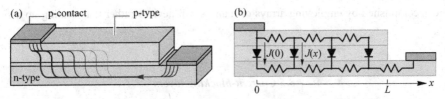

Fig. 8.11. (a) Lateral injection geometry and schematic current distribution for $\rho_n \ll \rho_p$. (b) Corresponding equivalent circuit.

A schematic equivalent circuit suitable for the quantitative analysis is shown in Fig. 8.11 (b) where a p-n junction current density of $J(0)$ is assumed at the edge of the p-type contact. The analytic solution of the equivalent circuit shown in Fig. 8.12 (Joyce and Wemple, 1970; Rattier et al., 2002) is an exponential given by

$$J(x) = J(0)\exp(-x/L_s) \qquad (8.20)$$

where

$$L_s = \sqrt{\frac{2V_a}{J(0)[(\rho_p/t_p) + (\rho_n/t_n)]}} . \qquad (8.21)$$

$J(x=0) = J(0)$ is the current density at the edge of the contact. Rattier et al. (2002) stated that the voltage V_a would be an activation voltage with magnitude of a few times kT/e, e.g. 50–75 mV.

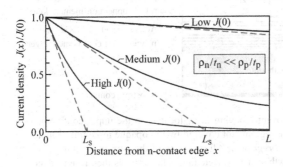

Fig. 8.12. Current density distribution at high, medium, and low current normalized to the initial current density for device with lateral injection geometry. The n-type sheet resistance is assumed to be lower than the p-type sheet resistance.

For uniform light generation across the gap between the contacts, it is desirable to have a long exponential decay length L_s. This can be achieved by high doping or thick confinement layers. To attain high powers, one may be tempted to scale the device structure in size. However, for large contact separations L, the device becomes generally more resistive unless very thick confinement layers are being used (which may be unpractical). Scaling such device structures

141

can be accomplished by employing arrays of many small devices rather than scaling up a single device.

8.5 Current-blocking layers

In conventional DH LEDs with small top contacts and large backside contacts, the current injected by the top contact enters the active region predominantly under the top contact. The extraction of light generated in the active region is thus strongly hindered by the opaque metal contact. One possibility to alleviate this problem is the use of a thick current-spreading layer. Another possibility is the use of a current-blocking layer. This layer blocks the current from entering the active region below the top contact. The current is deflected away from the top contact, thus allowing for higher extraction efficiency.

The schematic structure of an LED employing a current-blocking layer is shown in Fig. 8.13. The blocking layer is located on top of the upper confinement layer and has approximately the same size as the top metal contact. The current-blocking layer has n-type conductivity and is embedded in material with p-type conductivity. Owing to the p-n junction surrounding the current-blocking layer, the current flows around the current-blocking layer as indicated in Fig. 8.13.

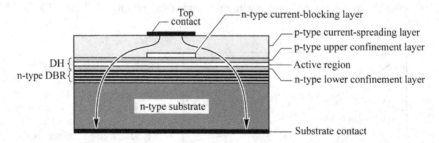

Fig. 8.13. LED with an n-type current-blocking layer located on the upper confinement layer. Light emission occurs in the regions not covered by the opaque top ohmic contact. The LED is fabricated by *epitaxial regrowth*. After growth of the current-blocking layer, the wafer is taken out of the growth system for etching. The wafer is then re-introduced into the epitaxial system for growth of the current-spreading layer.

Current-blocking layers can be fabricated by **epitaxial regrowth**. In this process, the DH and the thin n-type current-blocking layer are grown over the entire wafer surface. Subsequently, the wafer is taken out of the growth system for etching. The regions to be etched are defined by photolithography. The entire blocking layer is etched away *except* the region where the top ohmic contact is going to be located, as shown in Fig. 8.13. Frequently, the current-blocking-layer etch is *selective* so that it does not etch into the upper confinement layer. Subsequently, the wafer is returned to the growth system for resumption of epitaxial growth, i.e. for the growth of the current-spreading layer.

Regrowth processes are expensive due to the reduction of the device and wafer yield that usually accompanies regrowth processes. Cleaning of the wafer surface after the etching, directly before regrowth, is critical. Defects occurring at the regrowth interface can lead to a reduction in yield. Therefore, processes requiring epitaxial regrowth are more expensive and are not suited for low-cost devices. However, for more expensive devices, such as communication LEDs, epitaxial regrowth processes are used.

In AlGaInP LEDs, n-type GaAs has been used as the current-blocking layer. The n-type GaAs layer is located on top of the AlGaInP upper confinement layer. The GaAs current-blocking layer is lattice matched to the underlying AlGaInP confinement layer. Selective wet chemical etches are available that etch the GaAs but do not etch the AlGaInP (Adachi and Oe, 1983).

Vertical-cavity surface-emitting lasers (VCSELs) also employ current-blocking layers to funnel the current to the active region located between the mirrors of the laser. However, oxygen or hydrogen implantation rather than epitaxial regrowth is used in VCSELs to form current-blocking layers. The implantation depth is limited so that the lateral resistance can become substantial for devices with large-area contacts.

References

Adachi S. and Oe K. "Chemical etching characteristics of (001) GaAs" *J. Electrochem. Soc.* **130**, 2427 (1983)

Choquette K. D., Geib K. M., Ashby C. I. H., Twesten R. D., Blum O., Hou H. Q., Follstaedt D. M., Hammons B. E., Mathes D., and Hull R. "Advances in selective wet oxidation of AlGaAs alloys" *IEEE J. Sel. Top. Quantum Electron.* **3**, 916 (1997)

Fletcher R. M., Kuo C. P., Osentowski T. D., Huang K. H., and Craford M. G. "The growth and properties of high performance AlInGaP emitters using lattice mismatched GaP window layers" *J. Electron. Mater.* **20**, 1125 (1991a)

Fletcher R. M., Kuo C. P., Osentowski T. D., and Robbins V. M. "Light-emitting diode with an electrically conductive window" US Patent 5,008,718 (1991b)

Groves W. O. and Epstein A. S. "Epitaxial deposition of III–V compounds containing isoelectronic impurities" US Patent 4,001,056 (1977)

Groves W. O., Herzog A. H., and Craford M. G. "Process for the preparation of electroluminescent III–V materials containing isoelectronic impurities" US Patent Re. 29,648 (1978a)

Groves W. O., Herzog A. H., and Craford M. G. "GaAsP electroluminescent device doped with isoelectronic impurities" US Patent Re. 29,845 (1978b)

Guo X. and Schubert E. F. "Current crowding and optical saturation effects in GaInN/GaN light-emitting diodes grown on insulating substrates" *Appl. Phys. Lett.* **78**, 3337 (2001)

Guo X., Li Y.-L., and Schubert E. F. "Efficiency of GaN/GaInN light-emitting diodes with interdigitated mesa geometry" *Appl. Phys. Lett.* **79**, 1936 (2001)

Jeon S.-R., Song Y.-H., Jang H.-J., Yang G. M., Hwang S. W., and Son S. J. "Lateral current spreading in GaN-based light-emitting diodes utilizing tunnel contact junctions" *Appl. Phys. Lett.* **78**, 3265 (2001)

Joyce W. B. and Wemple S. H. "Steady-state junction-current distributions in thin resistive films on semiconductor junctions (solutions of $\nabla^2 v = \pm e^v$) *J. Appl. Phys.* **41**, 3818 (1970)

Kuo C. P., Fletcher R. M., Osentowski T. D., Lardizabal M. C., Craford M. G., and Robins V. M. "High performance AlGaInP visible light emitting diodes" *Appl. Phys. Lett.* **57**, 2937 (1990)

LED Museum, www.ledmuseum.org (2004)

Moyer C. D. "Photon recycling light emitting diode" US Patent 4,775,876 (1988)

Nishizawa J., Koike M., and Jin C. C. "Efficiency of GaAlAs heterostructure red light-emitting diodes" *J. Appl. Phys.* **54**, 2807 (1983)

Nuese C. J., Tietjen J. J., Gannon J. J., and Gossenberger H. F. "Optimization of electroluminescent efficiencies for vapor-grown GaAsP diodes" *J. Electrochem Soc.: Solid State Sci.* **116**, 248 (1969)

Rattier M., Bensity H., Stanley R. P., Carlin J.-F., Houdre R., Oesterle U., Smith C. J. M., Weisbuch C., and Krauss T. F. "Toward ultra-efficient aluminum oxide microcavity light-emitting diodes: Guided mode extraction by photonic crystals" *IEEE J. Selected Topics in Quant. Electron.* **8**, 238 (2002)

Schroder D. K. *Semiconductor Material and Device Characterization* (John Wiley and Sons, New York, 1998)

Sugawara H., Ishakawa M., and Hatakoshi G. "High-efficiency InGaAlP/GaAs visible light-emitting diodes" *Appl. Phys. Lett.* **58**, 1010 (1991)

Sugawara H., Ishakawa M., Kokubun Y., Nishikawa Y., Naritsuka S., Itaya K., Hatakoshi G., Suzuki M. "Semiconductor light-emitting device" US Patent 5,153,889, issued Oct. 6 (1992a)

Sugawara H., Itaya K., Nozaki H., and Hatakoshi G. "High-brightness InGaAlP green light-emitting diodes" *Appl. Phys. Lett.* **61**, 1775 (1992b)

Steigerwald D. A., Rudaz S. L., Thomas K. J., Lester S. D., Martin P. S., Imler W. R., Fletcher R. M., Kish Jr. F. A., Maranowski S. A. "Electrode structures for light-emitting devices" US Patent 6,307,218 (2001)

Thompson G. H. B. *Physics of Semiconductor Laser Devices* (John Wiley and Sons, New York, 1980)

9

High extraction efficiency structures

Owing to the high refractive index of semiconductors, light incident on a planar semiconductor–air interface is totally internally reflected, if the angle of incidence is sufficiently large. Snell's law gives the critical angle of total internal reflection. As a result of total internal reflection, light can be "trapped" inside the semiconductor. Light trapped in the semiconductor will eventually be reabsorbed, e.g. by the substrate, active region, cladding layer, or by a metallic contact.

If the light is absorbed by the substrate, the electron–hole pair will most likely recombine non-radiatively due to the inherently low efficiency of substrates. If the light is absorbed by the active region, the electron–hole pair may re-emit a photon or recombine non-radiatively. For active regions with internal quantum efficiencies of less than 100%, a reabsorption event by the active region reduces the efficiency of the LED.

The *external quantum efficiency* of an LED is the product of the *internal quantum efficiency*, η_{int}, and the *extraction efficiency*, $\eta_{extraction}$, i.e.

$$\eta_{ext} = \eta_{int}\, \eta_{extraction} \, . \qquad (9.1)$$

The extraction efficiency thus plays an important role in increasing the power efficiency of LEDs.

9.1 Absorption of below-bandgap light in semiconductors

To obtain high light-extraction efficiency and avoid absorption of light, all semiconductor layers other than the active region should have a bandgap energy larger than the photon energy. This can be done in different ways, for example by using double heterostructures, window layers, and other structures that will be discussed below. In this section, we discuss the absorption of light if the energy of the light is *below* the energy gap of the semiconductor.

Naively, one would assume that a semiconductor can absorb light only if the photon energy is higher than the bandgap energy and that the semiconductor is transparent for photon energies below the bandgap. However, semiconductors do absorb below-bandgap light, although with a much lower absorption coefficient.

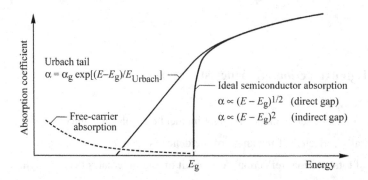

Fig. 9.1. Absorption coefficient of a semiconductor with bandgap E_g versus energy. The "Urbach tail" dominates absorption near but below the bandgap. Absorption further below the bandgap is dominated by free-carrier absorption.

The absorption coefficient versus energy of an idealized semiconductor and a real semiconductor is shown schematically in Fig. 9.1. In the idealized semiconductor at low temperatures, the absorption coefficient versus energy is given by (Pankove, 1971)

$$\alpha \propto (E - E_g)^{1/2} \quad \text{(direct gap)} \tag{9.2a}$$

$$\alpha \propto (E - E_g)^2 \quad \text{(indirect gap)}. \tag{9.2b}$$

The idealized semiconductor has a *zero* band-to-band absorption coefficient at the bandgap energy ($E = E_g$). The absorption strength in a *real* semiconductor, for below-bandgap light, can be expressed in terms of an exponentially decaying absorption strength. In this absorption tail, called the **Urbach tail**, the absorption coefficient versus energy is given by

$$\alpha = \alpha_g \exp\left[(E - E_g)/E_{\text{Urbach}}\right] \tag{9.3}$$

where α_g is the experimentally determined absorption coefficient at the bandgap energy and E_{Urbach} is the characteristic energy (here called the **Urbach energy**), which determines how rapidly the absorption coefficient decreases for below-bandgap energies.

Urbach (1953) measured the absorption tail for different temperatures and showed that the *Urbach energy* is approximately kT, the thermal energy. The temperature dependence of the Urbach tail led Knox (1963) to the conclusion that the below-bandgap transitions are phonon-assisted transitions. Thus, the Urbach energy is given by

$$E_{\text{Urbach}} = kT. \tag{9.4}$$

The Urbach tail can be caused by mechanisms other than phonon-assisted absorption. Any mechanism introducing a potential fluctuation will lead to local variations of the semiconductor band edges. As a result of these fluctuations, the bandgap energy fluctuates as well, and below-bandgap transitions can occur.

The most common potential fluctuations are fluctuations caused by random dopant distribution and local variations of the chemical composition of a ternary or quaternary alloy semiconductor.

Potential fluctuations caused by random dopant distribution can be calculated using Poisson statistics. The magnitude of these fluctuations is given by (see, for example, Schubert et al., 1997)

$$\Delta E_{\text{Urbach}} = \frac{2e^2}{3\varepsilon} \sqrt{\left(N_D^+ + N_A^-\right)\frac{r_s}{3\pi}} \, e^{-3/4} \tag{9.5}$$

where r_s is the screening radius.

Potential fluctuations caused by random compositional fluctuations can be calculated using binomial statistics. The magnitude of these fluctuations is given by (see, for example, Schubert et al., 1984)

$$\Delta E_{\text{alloy}} = \frac{dE_g}{dx} \left(\frac{x(1-x)}{4a_0^{-3} V_{\text{exc}}} \right)^{1/2} \tag{9.6}$$

where x is the alloy composition of the ternary semiconductor alloy, a_0 is the semiconductor lattice constant, and V_{exc} is the excitonic volume of the electron–hole pair.

It depends on the specific case at hand, which of the different physical effects dominates in the formation of the Urbach tail. Generally, binary semiconductors such as GaP or GaAs have a smaller tail than alloys such as AlGaAs or GaAsP. Furthermore, lightly doped semiconductors have a smaller Urbach tail than heavily doped semiconductors.

For energies sufficiently lower than the bandgap energy, absorption due to the Urbach tail is negligibly small and ***free-carrier absorption*** becomes the dominant absorption mechanism. As the name suggests, a free carrier is excited to a higher energy by absorption of a photon. The absorption transition must conserve momentum. Whereas photons have very small momenta, electrons must undergo a momentum change when excited higher within a parabolic band. This momentum change is provided by acoustic phonons, optical phonons, or by impurity scattering.

Free-carrier absorption is proportional to the free-carrier concentration, since a free carrier is required for an absorption event. Theoretical considerations in terms of the classical Drude free-electron model further show that free-carrier absorption increases as the square of the incident wavelength (Pankove, 1971). Thus the proportionalities

$$\alpha_{fc} \propto n \lambda^2 \quad \text{and} \quad \alpha_{fc} \propto p \lambda^2 \quad (9.7)$$

are valid for n-type and p-type semiconductors, respectively. Theoretical considerations in terms of a quantum mechanical treatment show that the absorption coefficient is proportional to $\lambda^{3/2}$, $\lambda^{5/2}$, and $\lambda^{7/2}$ depending upon whether acoustic phonon scattering, optical phonon scattering, or ionized impurity scattering is involved in the momentum conservation process, respectively (Swaminathan and Macrander, 1991).

In n-type and p-type GaAs, the room-temperature free-carrier absorption coefficient near the bandgap energy ($\lambda \approx 950$ nm) can be expressed as (Casey and Panish, 1978)

$$\alpha_{fc} = 3 \text{ cm}^{-1} \frac{n}{10^{18} \text{ cm}^{-3}} + 7 \text{ cm}^{-1} \frac{p}{10^{18} \text{ cm}^{-3}}. \quad (9.8)$$

Inspection of the equation indicates that the free-carrier absorption coefficient can be of the order of 10 cm^{-1} at high carrier concentrations. Approximate values for the free-carrier absorption coefficient in several compound semiconductors are given in Table 9.1.

Table 9.1. Free-carrier absorption coefficient (α_{fc}) of n-type semiconductors. (a) After Ioffe (2002). (b) After Wiley and DiDomenico (1970). (c) After Casey and Panish (1978). (d) After Kim and Bonner (1983) and Walukiewicz et al. (1980). (e) Data are extrapolated using the proportionality $\alpha_{fc} \propto n \lambda^2$.

Material	Wavelength	Electron concentration	α_{fc}
GaN	1.0 μm	1×10^{18} cm^{-3}	40 cm^{-1} (a, e)
GaP	1.0 μm	1×10^{18} cm^{-3}	22 cm^{-1} (b, e)
GaAs	1.0 μm	1×10^{18} cm^{-3}	3.0 cm^{-1} (c)
InP	1.0 μm	1.1×10^{18} cm^{-3}	2.5 cm^{-1} (d)

In LEDs, free-carrier absorption can affect the intensity of waveguided modes radiating out of the side of the chip. Free-carrier absorption also plays a role in LEDs with transparent semiconductor substrates. Such transparent substrates have a typical thickness greater than 100 μm. If the doping concentration of a transparent substrate is high, free-carrier absorption will reduce the light-output power. If it is low, the substrate becomes resistive. Thus a compromise needs to be made between the different doping requirements of a transparent substrate. For thin layers, such as confinement layers, free-carrier absorption effects are negligibly small if the optical path length within the layer is short.

9.2 Double heterostructures

Virtually all LED structures employ double heterostructures. They consist of two *confinement* layers and the *active* region. The band diagram of a double heterostructure is shown in Fig. 9.2. The active region has a smaller bandgap energy than the two confinement regions. As a result, the confinement regions are *transparent* to the light emitted by the active region. Since the confinement regions are relatively thin, they can be considered, for all practical purposes, to be totally transparent.

Reabsorption of light by the active region in the current-injected area below the top contact can also be neglected. The active region is, under normal injection conditions, injected with high current densities so that the electron and hole quasi-Fermi levels rise into the bands, as illustrated in Fig. 9.2. As a result, the active region is practically transparent for near-bandgap emission under high injection conditions.

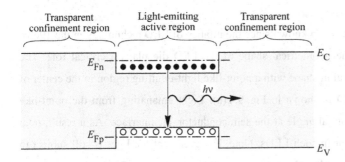

Fig. 9.2. Double hetero-structure with optically transparent confinement regions. Reabsorption in the active region is unlikely due to the high carrier concentration in the active region and the resulting Burstein–Moss shift of the absorption edge.

Note, however, that the active region is in equilibrium sufficiently far away from the current-injected region. These regions are not current injected and thus will absorb near-bandgap light emitted by the active region. To reduce optical losses by absorption, the active region should have a high internal quantum efficiency to make re-emission of absorbed photons likely.

9.3 Shaping of LED dies

One of the most important problems facing high-efficiency LEDs is the occurrence of *trapped light* within a high-index semiconductor. The occurrence of trapped light is illustrated in Fig. 9.3. A light ray emitted by the active region will be subject to total internal reflection, as predicted by Snell's law. In the high-index approximation, the angle of total internal reflection is given by

$$\alpha_c = \bar{n}_s^{-1} \tag{9.9}$$

where $\bar{n}_s$ is the semiconductor refractive index and the critical angle α_c is given in radians. For high-index semiconductors, the critical angle is quite small. For example, for a refractive index of 3.3, the critical angle for total internal reflection is only 17°. Thus most of the light emitted by the active region is trapped inside the semiconductor. The trapped light is most likely to be absorbed by the thick substrate. Once absorbed, the electron–hole pair is likely to recombine non-radiatively due to the comparatively low quality and efficiency of the substrate.

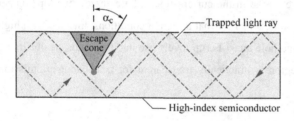

Fig. 9.3. "Trapped light" in a rectangular-parallelepiped-shaped semiconductor unable to escape for emission angles greater than α_c due to total internal reflection.

The light-escape problem has been known since the infancy of LED technology in the 1960s. It has also been known that the geometrical shape of the LED die plays a critical role. The optimum LED would be spherical in shape with a point-like light-emitting region in the center of the LED. Such a spherical LED is shown in Fig. 9.4 (a). Light emanating from the point-like active region is incident at a normal angle at the semiconductor–air interface. As a result, total internal reflection does not occur in such LEDs. Note, however, that the light is still subject to Fresnel reflection at the interface unless the sphere is coated with an anti-reflection coating.

LEDs with a hemispherical dome-like structure (Carr and Pittman, 1963) as well as other shapes, e.g. inverted, truncated cones (Franklin and Newman, 1964; Loebner, 1973) have been demonstrated to improve extraction efficiency over conventional designs, i.e. rectangular parallelepiped chips. However, the practical utility of such devices has not been realized primarily due to the high cost associated with shaping of individual LED dies.

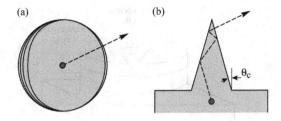

Fig. 9.4. Schematic illustration of different geometric shapes for LEDs with perfect extraction efficiency. (a) Spherical LED with a point-like light-emitting region at the center of the sphere. (b) A cone-shaped LED.

Unfortunately, spherical LEDs with a point-like light source in the center of the LED are somewhat impractical devices. Semiconductor fabrication technology is, in view of the flat substrates used in epitaxial growth, a *planar* technology. Thus spherical LEDs are difficult to fabricate using conventional planar technology.

Another interesting LED structure is a cone-shaped structure, as shown schematically in Fig. 9.4 (b). Light rays are emanating from the active region at or below the base of the cone. The light rays incident at the cone–air boundary are either transmitted through the semiconductor–air interface or guided by the cone. The guided rays undergo multiple reflections. As light rays undergo multiple reflections, they will form a progressively increasing angle of incidence at the semiconductor–air interface. As a result, light guided by the cone will eventually have near-normal incidence and escape from the cone. Although an interesting concept, cone-shaped LEDs are difficult to fabricate and manufacture.

The most common LED structure has the shape of a rectangular parallelepiped as shown in Fig. 9.5 (a). Such LED dies are fabricated by cleaving the wafer along its natural cleavage planes. The LEDs have a total of *six* escape cones, two of them perpendicular to the wafer surface, and four of them parallel to the wafer surface. The bottom escape cone will be absorbed by the substrate if the substrate has a lower bandgap than the active region. The four in-plane escape cones will be at least partially absorbed by the substrate. Light in the top escape cone will be obstructed by the top contact, unless a thick current-spreading layer is employed. Thus the simple rectangular parallelepiped LED is clearly a structure with low extraction efficiency. However, a substantial advantage of such LEDs is the low manufacturing cost.

An LED with a cylindrical shape is shown in Fig. 9.5 (b). The cylindrical LED has the advantage of higher extraction efficiency compared with a cube-shaped LED. An escape ring, as shown in Fig. 9.5 (b), replaces the four in-plane escape cones of the rectangular LED, which results in a substantial improvement of the extraction efficiency. Cylindrical-shaped LEDs require one more processing step (etching step) compared with rectangularly shaped LEDs.

9 High extraction efficiency structures

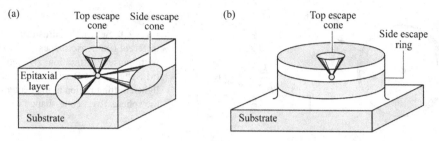

Fig. 9.5. Illustration of different geometric shapes of LEDs. (a) Rectangular parallelepipedal LED die with a total of six escape cones. (b) Cylindrical LED die with a top escape cone and a side escape ring.

Commercially available die-shaped devices include ***pedestal-shaped*** GaInN/SiC LEDs with trade name "Aton" (Osram, 2001) and ***truncated inverted pyramid*** (TIP) AlGaInP/GaP LEDs (Krames *et al.*, 1999). Photographs and schematic structures of the two devices are shown in Fig. 9.6. Ray traces indicated in the figure show that light rays entering the base of the pyramids escape from the semiconductor after undergoing one or multiple internal reflections. The pedestal and TIP geometries reduce the mean photon path length within the die, and thus reduce internal absorption losses. The increase in efficiency of pedestal-shaped devices is about a factor of 2 compared to rectangular-parallelepiped-shaped devices (Osram, 2001).

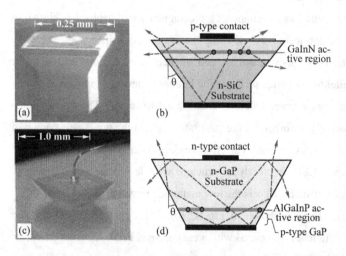

Fig. 9.6. Die-shaped devices: (a) Blue GaInN emitter on SiC substrate with trade name "Aton". (b) Schematic ray traces illustrating enhanced light extraction. (c) Micrograph of truncated inverted pyramid (TIP) AlGaInP/GaP LED. (d) Schematic diagram illustrating enhanced extraction (after Osram, 2001; Krames *et al.*, 1999).

The geometric shape of the LED, in particular the side-wall angle, is chosen in such a way that trapping of light is minimized. Ray tracing computer models are employed to maximize the escape probability from the semiconductor. The optimum angle of the TIP sidewall tilt is 35°

(Krames et al., 1999). The TIP LED is a high-power LED with a large p-n junction area of 500 μm × 500 μm. The luminous source efficiency of TIP LEDs exceeds 100 lm/W and is one of the highest ever achieved with LEDs.

The TIP LED performance versus injection current is shown in Fig. 9.7 (Krames et al., 1999). A peak luminous efficiency of 102 lm/W was measured for orange-spectrum ($\lambda \approx 610$ nm) devices at an injection current of 100 mA. This luminous efficiency exceeds that of most fluorescent (50–100 lm/W) and all metal-halide (68–95 lm/W) lamps. In the amber color regime, the TIP LED provides a photometric efficiency of 68 lm/W ($\lambda \approx 598$ nm). This efficiency is comparable to the source efficiency of 50 W high-pressure sodium discharge lamps. A peak external quantum efficiency of 55% was measured for red-emitting ($\lambda \approx 650$ nm) TIP LEDs. Under pulsed operation (1% duty cycle), an efficiency of 60.9% was achieved (data not shown), which sets a lower bound on the extraction efficiency of these devices.

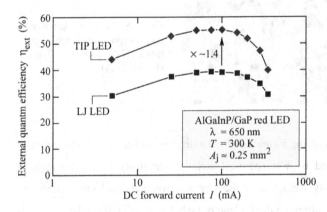

Fig. 9.7. External efficiency vs. forward current for red-emitting (650 nm) truncated inverted pyramid (TIP) LEDs and large junction (LJ) LEDs mounted in power-lamp packages. The TIP LED exhibits a 1.4 times improvement in extraction efficiency compared with the LJ device, resulting in a peak external quantum efficiency of 55 % at 100 mA (after Krames et al., 1999).

The rectangular parallelepiped and cylindrical LED structures can be fabricated with wafer-scale processing steps. Advanced structures, such as the TIP LED, are also fabricated by wafer-scale processes, namely a sawing process employing a beveled dicing blade (Haitz, 1992). The manufacturing cost of LEDs requiring die-level processing steps is much higher compared with LEDs manufactured with wafer-scale processes.

The use of tapered output couplers in LEDs has been reported by Schmid et al. (2000, 2001, 2002). For GaAs-based infrared devices, external quantum efficiencies near 50% have been demonstrated with devices having a tapered output coupler. However, due to the area requirements of the output coupler, the current-injected region of the device is much smaller than the overall device area and thus the total emission power is quite small.

9.4 Textured semiconductor surfaces

Other ways to increase the light extraction efficiency in GaAs-based devices include the use of **roughened** or **textured semiconductor surfaces** (see, for example, Schnitzer *et al.*, 1993; Windisch *et al.*, 1999, 2000, 2001, 2002). External quantum efficiencies near 50% have been reported for GaAs-based surface-textured devices. A detailed discussion of properties and fabrication of microstructured surfaces was given by Sinzinger and Jahns (1999).

After initial positive reports on enhancement of light extraction in devices having textured surfaces, the optimism of the technical community was dampened when it became clear that much of the increase in light extraction was lost during the polymer encapsulation process of the semiconductor chip. That is, although a strong increase of light extraction had been found in *unpackaged* chip measurements the gain practically vanished after chip encapsulation.

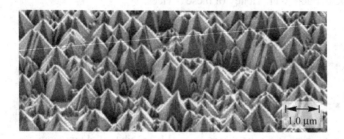

Fig. 9.8. Scanning electron micrograph of strongly textured GaN surface (after Haerle, 2004).

There are several recent reports on the increase of light extraction in GaN-based devices with substantial surface texture achieved by wet chemical etching. For many years, it had been assumed that GaN with Ga-polarity cannot be etched by any wet chemical etch. However, Stocker *et al.* (1998a) showed that, although the c-plane of GaN lacks etchability, the a- and m-planes of GaN *can* be etched by a number of wet chemical etches including hot KOH and H_3PO_4. Stocker *et al.* (1998b, 2000) also showed that these etches are crystallographic in nature, which allows one to create pyramid-like structures and atomically smooth surfaces suited for laser fabrication. Wet chemical etching and photoelectrochemical etching can be used to create a substantial amount of surface roughness on GaN. An example of a very rough GaN surface is shown in Fig. 9.8 (Haerle, 2004).

The increase in light extraction from a surface-textured GaInN device (with the device having the strongly textured surface shown in Fig. 9.8) is shown in Fig. 9.9 (Haerle, 2004). The author estimated the increase in output power to be 40–50%. Similar results were reported independently by Gao *et al.* (2004) and Fujii *et al.* (2004). Note that the interference fringes, observed for the GaN device having the smooth surface, completely vanish for the surface-textured film. The interference fringes are caused by the Fabry–Perot cavity formed by the GaN–air interface and the sapphire–GaN interface which form the two reflectors of the cavity (Billeb et al., 1997).

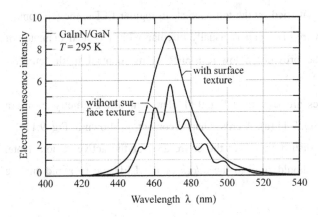

Fig. 9.9. Emission spectrum of GaInN blue LED with and without surface texture. The spectrum exhibits Fabry–Perot interference fringes for the device with a smooth surface (after Haerle, 2004).

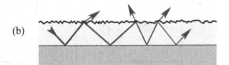

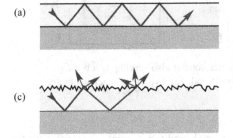

Fig. 9.10. Schematic illustration of waveguide with (a) no surface texture, (b) weak surface texture, and (c) strong surface texture, resulting in specular reflection, mixed reflection and scattering, and strong scattering, respectively.

The transition from a smooth surface to a strongly textured surface and the implication on ray propagation is shown in Fig. 9.10. A perfectly smooth surface, i.e. a specular surface, results in waveguided modes that cannot escape, as shown in Fig. 9.10(a). For a strongly scattering surface, i.e. a lambertian or diffuse surface, only one or a few scattering events are sufficient to out-couple the light as shown in Fig. 9.10(c). The intermediate case is shown in Fig. 9.10(b). Because the light-out-coupling efficiency depends on the degree of diffusivity, it would be desirable to quantify the degree of surface diffusivity (for a discussion of the quantification of the diffuseness of a surface, see the chapter entitled "Reflectors" in this book).

The optical microscopic inspection of chip surfaces of GaInN high-performance devices (Nichia, 2005) reveals that the chip surface is textured and appears as *white*, which is a clear indication that the surface was strongly textured to create a diffuse, strongly scattering surface.

9.5 Cross-shaped contacts and other contact geometries

Different requirements need to be satisfied by the top contact. In regular LEDs, the top contact provides a pad for the bonding wire. The pad is usually circular with a typical diameter of 100 µm. The top contact pad also provides a low-resistance ohmic contact to the current-spreading layer.

Typical top contact geometries are shown in Fig. 9.11. The simplest geometry is just a circular contact pad, as shown in Fig. 9.11 (a). A cross-shaped contact, as shown in Fig. 9.11 (b), provides a more uniform current distribution over the entire area of the active region. Note, however, that no or little current should flow to the *edges* of the LED die to avoid surface recombination.

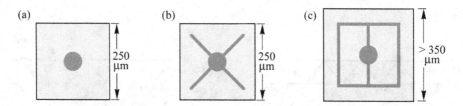

Fig. 9.11. Top view of an LED die with (a) a circular contact also serving as a bond pad and (b) a cross-shaped contact with a circular bond pad. (c) Typical contact geometry used for larger LED dies.

For large-area LEDs, a simple circular pad or a cross-shaped top contact are insufficient for uniform current distribution. In such large-area devices, patterns that include a ring, as shown in Fig. 9.11 (c), are better suited to providing uniform current distribution.

The area of the top contact is kept small so that the light emanating from the active region is not hindered by the opaque contact. However, the contact resistance scales with the contact area so that the top contact area cannot be scaled down arbitrarily.

9.6 Transparent substrate technology

Visible-spectrum $(Al_xGa_{1-x})_{0.5}In_{0.5}P$ LEDs with typical operating wavelengths of 560–660 nm are grown lattice matched on GaAs substrates. Since the energy gap of GaAs is $E_g = 1.424$ eV

(λ_g = 870 nm) at room temperature, GaAs substrates are absorbing at these emission wavelengths. As a result, the light emitted towards the substrate will be absorbed by the thick GaAs substrate. Thus, the extraction efficiency of AlGaInP/GaAs LEDs grown on GaAs substrates is low.

The extraction efficiency of AlGaInP LEDs can be increased substantially by removal of the GaAs substrate and bonding of the epitaxial layer to a GaP substrate (Kish et al., 1994). GaP is an indirect-gap semiconductor with E_g = 2.24 eV (λ_g = 553 nm). Thus, GaP does not absorb light with λ > 553 nm emitted by the AlGaInP active region.

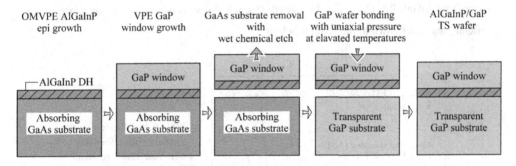

Fig. 9.12. Schematic fabrication process for wafer-bonded transparent substrate (TS) AlGaInP/GaP LEDs. After the selective removal of the original GaAs substrate, elevated temperature and uniaxial pressure are applied, resulting in the formation of a single TS LED wafer (after Kish et al., 1994).

The fabrication process of the AlGaInP LEDs wafer-bonded to a GaP substrate is illustrated schematically in Fig. 9.12. The AlGaInP double heterostructure (DH) is initially grown by OMVPE on a GaAs substrate. Subsequently, a thick GaP window layer (~ 50 µm) is grown on the DH by chloride VPE, a growth technique that allows for the low-cost growth of thick epitaxial layers. The GaAs substrate is then removed using a wet chemical selective etching process (Adachi and Oe, 1983; Kish et al., 1994). During the removal of the GaAs substrate, the thick GaP window layer serves as a temporary mechanical support for the DH. The DH with the GaP window is then wafer-bonded to a GaP substrate.

Wafer-bonding processes require a high degree of cleanliness, the lack of any particles between the wafers, and the absence of native surface oxides on both wafers. Frequently, the gap between two wafers is filled with a contact liquid. Rotating the wafers at a high rate spins out the contact liquid. Kish et al. (1995) and Hoefler et al. (1996) reported an AlGaInP-to-GaP wafer-bonding process, suitable for 50 mm (2 inch) GaP substrates. Uniaxial pressure and elevated temperatures (750–1000 °C) are used in this process (Hoefler et al., 1996). Kish et al. (1995) showed that the achievement of low-resistance ohmic conduction across wafer-bonded interfaces is critically dependent upon the crystallographic alignment of the bonded wafer surfaces, irrespective of the lattice mismatch between the surfaces. Furthermore, Kish et al. (1995) showed that the crystallographic surface orientation of the bonded surfaces must be nominally matched while simultaneously maintaining rotational alignment of the wafers. Low diode forward voltages of 2.2 V for AlGaInP/GaP LEDs are routinely achieved under high-volume manufacturing conditions with the process. The reliability of wafer-bonded LEDs is comparable to monolithic AlGaInP/GaAs LEDs. Usually the technical details of wafer-bonding processes are proprietary and not known to the general public.

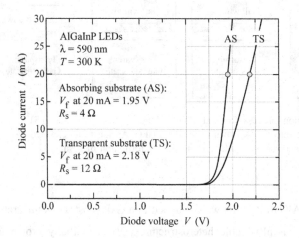

Fig. 9.13. Current–voltage characteristic, forward voltage, and series resistance of absorbing-substrate (GaAs) and transparent-substrate (GaP) LEDs with AlGaInP active regions.

The forward voltage is a critical figure of merit for wafer-bonded p-n junction devices. A low voltage indicates true semiconductor-to-semiconductor chemical bonding and an absence of interfacial oxide layers. The forward current–voltage characteristic of commercial absorbing-substrate (AS) and transparent-substrate (TS) AlGaInP LEDs is shown in Fig. 9.13. Inspection of the figure reveals that TS LEDs have a higher forward voltage and series resistance compared with AS devices.

The higher forward voltage in TS devices is probably due to either ohmic losses occurring at the wafer-bonded interface or in the GaP substrate. A detailed analysis by O'Shea et al. (2001) confirmed that a forward-voltage penalty results from the wafer-bonded interface, particularly if this interface is either contaminated with carbon or crystallographically misaligned. The n-type doping concentration in the GaP substrate is kept moderately low, to minimize free-carrier absorption.

Figure 9.14 compares AlGaInP/GaAs AS LEDs with AlGaInP/GaP TS LEDs emitting in the amber wavelength range (Kish and Fletcher, 1997). The substrate appears dark for the AS LED, in contrast to the TS LED. Transparent substrate AlGaInP/GaP LEDs have a factor of 1.5–3.0 higher external efficiency compared with AS AlGaInP/GaAs LEDs.

(a) AS LED (b) TS LED

Fig. 9.14. (a) Amber AlGaInP LED with a GaP window layer and absorbing GaAs substrate (AS). (b) Amber AlGaInP LED with a GaP window layer and a transparent GaP substrate (TS) fabricated by wafer bonding. Conductive Ag-loaded die-attach epoxy can be seen at bottom (after Kish and Fletcher, 1997).

9.7 Anti-reflection optical coatings

Anti-reflection (AR) coatings are frequently used in communication LEDs to reduce the Fresnel reflection at the semiconductor–air interface. For normal incidence, the intensity reflection coefficient for normal incidence is given by

$$R = \frac{(\bar{n}_s - \bar{n}_{air})^2}{(\bar{n}_s + \bar{n}_{air})^2} \tag{9.10}$$

where $\bar{n}_s$ and $\bar{n}_{air}$ are the refractive indices of the semiconductor and of air, respectively.

For normal incidence, Fresnel reflection at the semiconductor–air interface can be reduced to zero, if an AR coating cladding the semiconductor has the following parameters:

Thickness: $\lambda/4 = \lambda_0/(4\bar{n}_{AR})$ Refractive index: $\bar{n}_{AR} = \sqrt{\bar{n}_s \bar{n}_{air}}$. (9.11)

An AR coating with optimum thickness and refractive index is shown in Fig. 9.15. The refractive indices and the transparency ranges of different dielectric materials suitable for AR coatings are given in Table 9.2.

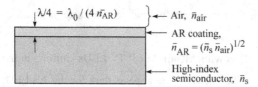

Fig. 9.15. Illustration of optimum thickness and refractive index of an anti-reflection (AR) coating.

Table 9.2. Refractive index and transparency range of common dielectrics suitable as anti-reflection coatings (after Palik, 1998).

Dielectric material	Refractive index	Transparency range
SiO_2 (silica)	1.45	> 0.15 μm
Al_2O_3 (alumina)	1.76	> 0.15 μm
TiO_2 (titania)	2.50	> 0.35 μm
Si_3N_4 (silicon nitride)	2.00	> 0.25 μm
ZnS (zinc sulphide)	2.29	> 0.34 μm
CaF_2 (calcium fluoride)	1.43	> 0.12 μm

9.8 Flip-chip packaging

For LEDs with two top contacts, such as GaInN/GaN LEDs grown on sapphire substrates, both regular packaging (epi-side up) and flip-chip packaging is used. Flip-chip packaging using **solder-bump bonding** is a more expensive packaging process compared with regular packaging where the LED top contact pads are contacted by wire bonding. An advantage of flip-chip packaging of GaInN/GaN LEDs is that the metal pads are not hindering the extraction of light radiating from the active region.

In packaged epi-side up GaInN LEDs, a large-area p-type top contact that covers the entire p-type mesa results in a uniform current distribution in the active region. However, such a large top contact will hinder the extraction of light. This fundamental trade-off can be avoided by flip-chip packaging, which is especially advantageous for high-power devices.

References

Adachi S. and Oe K. "Chemical etching characteristics of (001) GaAs" *J. Electrochem. Soc.* **130**, 2427 (1983)

Billeb A., Grieshaber W., Stocker D., Schubert E. F., and Karlicek R. F. Jr. "Microcavity effects in GaN epitaxial layers" *Appl. Phys. Lett.* **70**, 2790 (1997)

Carr W. N. and Pittman G. E. "One-Watt GaAs p-n junction infrared source" *Appl. Phys. Lett.* **3**, 173 (1963)

Casey Jr. H. C. and Panish M. B. *Heterostructure Lasers Part A: Fundamental Principles* pp. 46, 47, and 175 (Academic Press, San Diego, 1978)

Franklin A. R. and Newman R. "Shaped electroluminescent GaAs diodes" *J. Appl. Phys.* **35**, 1153 (1964)
Fujii T., Gao Y., Sharma R., Hu E. L., DenBaars S. P., and Nakamura S. "Increase in the extraction efficiency of GaN-based light-emitting diodes via surface roughening" *Appl. Phys. Lett.* **84**, 855 (2004)
Gao Y., Fujii T., Sharma R., Fujito K., DenBaars S. P., Nakamura S., and Hu E. L. "Roughening hexagonal surface morphology on laser lift-off (LLO) N-face GaN with simple photo-enhanced chemical wet etching" *Jpn. J. Appl. Phys.* **43**, L 637 (2004)
Haerle V. "Naturally textured GaN surface" *China Hi-Tech Fair* (CHTF) Shenzhen, China, October 12–17 (2004)
Haitz R. "Light-emitting diode with diagonal faces" US Patent 5,087,949 (1992)
Hoefler G. E., Vanderwater D. A., DeFevere D. C., Kish F. A., Camras M. D., Steranka F. M., and Tan I.-H. "Wafer bonding of 50-mm diameter GaP to AlGaInP–GaP light-emitting diode wafers" *Appl. Phys. Lett.* **69**, 803 (1996)
Ioffe Institute (Saint Petersburg, Russia) database on compound semiconductors available at www.ioffe.rssi.ru/SVA/NSM/Semicond/ (2002)
Kim O. K. and Bonner W. A. "Infrared reflectance and absorption of n-type InP" *J. Electron. Mater.* **12**, 827 (1983)
Kish F. A. and Fletcher R. M. "AlGaInP light-emitting diodes" in *High Brightness Light-Emitting Diodes* edited by G. B. Stringfellow and M. G. Craford, Semiconductors and Semimetals 48 (Academic, San Diego, 1997)
Kish F. A., Steranka F. M., DeFevere D. C., Vanderwater D. A., Park K. G., Kuo C. P., Osentowski T. D., Peanasky M. J., Yu J. G., Fletcher R. M., Steigerwald D. A., Craford M. G., and Robbins V. M. "Very high-efficiency semiconductor wafer-bonded transparent-substrate $(Al_xGa_{1-x})_{0.5}In_{0.5}P/GaP$ light-emitting diodes" *Appl. Phys. Lett.* **64**, 2839 (1994)
Kish F. A., Vanderwater D. A., Peanasky M. J., Ludowise M. J., Hummel S. G., and Rosner S. J. "Low-resistance ohmic conduction across compound semiconductor wafer-bonded interfaces" *Appl. Phys. Lett.* **67**, 2060 (1995)
Knox R. S. *Theory of Excitons* (Academic Press, New York, 1963)
Krames M. R., Ochiai-Holcomb M., Höfler G. E., Carter-Coman C., Chen E. I., Tan I.-H., Grillot P., Gardner N. F., Chui H. C., Huang J.-W., Stockman S. A., Kish F. A., Craford M. G., Tan T. S., Kocot C. P., Hueschen M., Posselt J., Loh B., Sasser G., and Collins D. "High-power truncated-inverted-pyramid $(Al_xGa_{1-x})_{0.5}In_{0.5}P/GaP$ light-emitting diodes exhibiting > 50% external quantum efficiency" *Appl. Phys. Lett.* **75**, 2365 (1999)
Loebner E. E. "The future of electroluminescent solids in display applications" *Proc. IEEE* **61**, 837 (1973)
Nichia Corporation. Visual inspection of the surface of a GaN LED chip manufactured by Nichia Corporation reveals that the surface is white, indicating that the chip surface is diffusive (2005)
O'Shea J. J., Camras M. D., Wynne D., and Hoefler G. E. "Evidence for voltage drops at misaligned wafer-bonded interfaces of AlGaInP light-emitting diodes by electrostatic force microscopy" *J. Appl. Phys.* **90**, 4791 (2001)
Osram Opto Semiconductors Corporation, Regensburg, Germany "Osram Opto enhances brightness of blue InGaN-LEDs" Press Release (January 2001)
Palik E. D. *Handbook of Optical Constants of Solids* (Academic Press, San Diego, 1998)
Pankove J. I. *Optical Processes in Semiconductors* p. 75 and section on Urbach tail (Dover, New York, 1971)
Schmid W., Eberhard F., Jager R., King R., Joos J., and Ebeling K. "45% quantum-efficiency light-emitting diodes with radial outcoupling taper" *Proc. SPIE* **3938**, 90 (2000)
Schmid W., Scherer M., Jager R., Strauss P., Streubel K., and Ebeling K. "Efficient light-emitting diodes with radial outcoupling taper at 980 and 630 nm emission wavelength" *Proc. SPIE* **4278**, 109 (2001)
Schmid W., Scherer M., Karnutsch C., Plobl A., Wegleiter W., Schad S., Neubert B., and Streubel K. "High-efficiency red and infrared light-emitting diodes using radial outcoupling taper" *IEEE J. Sel. Top. Quantum Electron.* **8**, 256 (2002)
Schnitzer I., Yablonovitch E., Caneau C., Gmitter T. J., and Scherer A. "30% external quantum efficiency from surface-textured, thin-film light-emitting diodes" *Appl. Phys. Lett.* **63**, 2174 (1993)

Schubert E. F., Goebel E. O., Horikoshi Y., Ploog K., and Queisser H. J. "Alloy broadening in photoluminescence spectra of AlGaAs" *Phys. Rev. B* **30**, 813 (1984)

Schubert E. F., Goepfert I. D., Grieshaber W., and Redwing J. M. "Optical properties of Si-doped GaN" *Appl. Phys. Lett.* **71**, 921 (1997)

Sinzinger S. and Jahns J. *Microoptics* (Wiley-VCH, New York, 1999)

Stocker D. A., Schubert E. F., and Redwing J. M. "Crystallographic wet chemical etching of GaN" *Appl. Phys. Lett.* **73**, 2654 (1998a)

Stocker D. A., Schubert E. F., Grieshaber W., Boutros K. S., and Redwing J. M. "Facet roughness analysis for InGaN/GaN lasers with cleaved facets" *Appl. Phys. Lett.* **73**, 1925 (1998b)

Stocker D. A., Schubert E. F., and Redwing J. M. "Optically pumped InGaN/GaN lasers with wet-etched facets" *Appl. Phys. Lett.* **77**, 4253 (2000)

Swaminathan V. and Macrander A. T. *Materials Aspects of GaAs and InP Based Structures* (Prentice Hall, Englewood Cliffs, 1991)

Urbach F. "The long-wavelength edge of photographic sensitivity of the electronic absorption of solids" *Phys. Rev.* **92**, 1324 (1953)

Walukiewicz W., Lagowski J., Jastrzebski L., Rava P., Lichtensteiger M., Gatos C. H., and Gatos H. C. "Electron mobility and free-carrier absorption in InP; determination of the compensation ratio" *J. Appl. Phys.* **51**, 2659 (1980)

Wiley J. D. and DiDomenico Jr. M. "Free-carrier absorption in n-type GaP" *Phys. Rev. B* **1**, 1655 (1970)

Windisch R., Schoberth S., Meinlschmidt S., Kiesel P., Knobloch A., Heremans P., Dutta B., Borghs G., and Doehler G. H. "Light propagation through textured surfaces" *J. Opt. A: Pure Appl. Opt.* **1**, 512 (1999)

Windisch R., Dutta B., Kuijk M., Knobloch A., Meinlschmidt S., Schoberth S., Kiesel P., Borghs G., Doehler G. H., and Heremans P. "40% efficient thin-film surface textured light-emitting diodes by optimization of natural lithography" *IEEE Trans. Electron Dev.* **47**, 1492 (2000)

Windisch R., Rooman C., Kuijk M., Borghs G., and Heremans P. "Impact of texture-enhanced transmission on high-efficiency surface-textured light-emitting diodes" *Appl. Phys. Lett.* **79**, 2315 (2001)

Windisch R., Rooman C., Dutta B., Knobloch A., Borghs G., Doehler G. H., and Heremans P. "Light-extraction mechanisms in high-efficiency surface-textured light-emitting diodes" *IEEE J. Sel. Top. Quantum Electron.* **8**, 248 (2002)

10

Reflectors

Ideal reflectors incorporated into a device structure should have (*i*) high reflectivity, (*ii*) a sufficiently broad spectral range of the high-reflectivity band, (*iii*) omnidirectional characteristics, and (*iv*) low resistivity (provided current flows through the reflector). It is a question of great practical and intellectual interest what type of reflector best meets these multiple requirements.

Different types of reflectors are shown in Fig. 10.1 including a metallic reflector, distributed Bragg reflector (DBR), hybrid metal–DBR, reflectors based on total internal reflection (TIR), and an omni-directional reflector. The characteristics of the different types of reflectors will be discussed below.

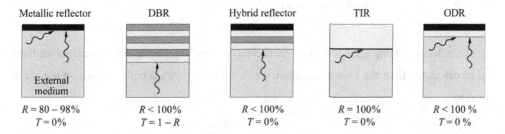

Fig. 10.1. Different types of reflectors including metallic reflector, distributed Bragg reflector (DBR), hybrid reflector, total internal reflector (TIR), and a triple-layer omni-directional reflector (ODR). Also given are angles of incidence for high reflectivity and typical reflectances and transmittances.

An ***external medium*** is indicated in the figure. In an LED structure with a reflector, the external medium is a semiconductor. The external medium has a significant influence on the reflector properties. For example, the reflectivity of a metal–semiconductor reflector is lower compared to a metal–air reflector.

10 Reflectors

10.1 Metallic reflectors, reflective contacts, and transparent contacts

Having been used by humankind for several millenniums, metal–air reflectors are the oldest type of reflector with high reflectivity. Metal reflectors are characterized by a broad spectral reflectivity band and a weak angular dependence of the reflectivity. The first high-quality metallic reflectors were used in reflection telescopes for astronomy applications (Bell, 1922).

An experimental reflectance spectrum of a silver–air reflector at normal incidence is shown in Fig. 10.2. The reflectance spectrum is characterized by a broad spectral band of high reflectivity and an average reflectance of 98.5%.

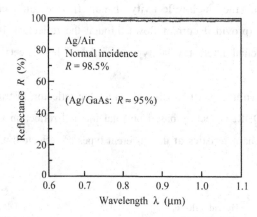

Fig. 10.2. Measured reflectance of a silver/air reflector for normal incidence. The average reflectivity in the visible spectrum is 98.5%.

The reflectance (amplitude reflection coefficient) of a metal reflector and an external medium can be calculated from the Fresnel equation, which for media with complex refractive indices is given by

$$r = \frac{\mathcal{E}_r}{\mathcal{E}_i} = \frac{\overline{N}_1 - \overline{N}_2}{\overline{N}_1 + \overline{N}_2} \quad (10.1)$$

where $\overline{N}_1$ and $\overline{N}_2$ are the complex refractive indices of the two media.

The reflected and transmitted *intensities* are proportional to the square of the electric field. The **reflectivity** or **reflectance** (power reflection coefficient) is given by

$$R = \frac{|\mathcal{E}_r|^2}{|\mathcal{E}_i|^2} = |r|^2 = \left|\frac{\overline{N}_1 - \overline{N}_2}{\overline{N}_1 + \overline{N}_2}\right|^2 = \frac{|\overline{N}_1 - \overline{N}_2|^2}{|\overline{N}_1 + \overline{N}_2|^2} \quad (10.2)$$

10.1 Metallic reflectors, reflective contacts, and transparent contacts

The energy conservation law requires that the power transmission coefficient in a lossless reflector is given by

$$T = 1 - R. \qquad (10.3)$$

Metals are lossy media so that the transmittance $T \approx 0$ for thick metal films. The Fresnel equation can be used to calculate the reflectance of a metal–dielectric interface. Assuming that the complex refractive indices of the dielectric and the metal are $\bar{N}_1 = \bar{n}_1$ and $\bar{N}_2 = \bar{n}_2 + i\bar{k}_2$, respectively, the reflectance is given by

$$r = \frac{\bar{n}_1 - (\bar{n}_2 + i\bar{k}_2)}{\bar{n}_1 + (\bar{n}_2 + i\bar{k}_2)} \quad \text{and} \quad R = \frac{(\bar{n}_1 - \bar{n}_2)^2 + \bar{k}_2^2}{(\bar{n}_1 + \bar{n}_2)^2 + \bar{k}_2^2}. \qquad (10.4)$$

An ideal metal has very high conductivity, i.e. $\sigma \to \infty$ and therefore $\bar{k} \to \infty$. (Note that conductivity σ and the imaginary part of the refractive index $\bar{k}$ are related by $\sigma = 2\bar{n}\omega\varepsilon_0 \bar{k}$. Also note that the absorption constant α and the imaginary part of the refractive index are related by $\alpha = 4\pi\bar{k}/\lambda_0$.) We thus obtain for an ideal metal

$$|r| \approx 1, \quad R \approx 1, \quad \text{and} \quad \phi_r = \pi. \qquad (10.5)$$

That is, the ideal metal has a unit reflection coefficient, unit reflectance, and a phase shift of π.

Real metals have high conductivity (but not an infinitely high conductivity) and the reflectivity of any real metal is therefore less than unity. Loss mechanisms in metals were first analyzed in terms of the Drude model (Drude, 1904). Table 10.1 gives the $\bar{n}$ and $\bar{k}$ values of several metals and semiconductors.

Table 10.1. Real and imaginary part of the refractive index for different semiconductors and metals at 0.5 and 1.0 μm.

Material	GaP	GaP	Si	Ag	Ag	Au	Au	Al	Al
λ =	0.5 μm	1.0 μm	1.0 μm	0.5 μm	1.0 μm	0.5 μm	1.0 μm	0.5 μm	1.0 μm
$\bar{n}$	3.5	3.1	3.6	0.05	0.04	0.86	0.26	0.77	1.35
$\bar{k}$	≈ 0	≈ 0	≈ 0	3.1	7.1	1.90	6.82	6.08	10.7

Using Eq. (10.4), the reflectance of a metal–air and metal–semiconductor boundary can be calculated; the results are given in Table 10.2. Inspection of the table reveals that the visible-spectrum reflectivity is generally lower for metal-semiconductor interfaces than that of metal-air interfaces. This is due to the lower refractive-index contrast of the metal-semiconductor boundary.

Table 10.2. Calculated reflectivity of metal–air and metal–semiconductor reflectors at 0.5 and 1.0 µm.

Material	R (%)	Material	R (%)	Material	R (%)
Ag/air (0.5 µm)	0.982	Al/air (0.5 µm)	0.923	Au/air (0.5 µm)	0.514
Ag/air (1.0 µm)	0.997	Al/air (1.0 µm)	0.955	Au/air (1.0 µm)	0.979
Ag/GaP (0.5 µm)	0.969	Al/GaP (0.5 µm)	0.805	Au/GaP (0.5 µm)	0.470
Ag/GaP (1.0 µm)	0.992	Al/GaP (1.0 µm)	0.876	Au/GaP (1.0 µm)	0.945
Ag/Si (1.0 µm)	0.991	Al/Si (1.0 µm)	0.861	Au/Si (1.0 µm)	0.939

Although metals are simple and viable reflectors, the reflection losses or ***mirror losses*** are quite high. The loss of one reflection event, $(1 - R)$, is about 5% for metal-semiconductor reflectors. Losses are particularly high for waveguided modes shown in Fig. 10.3. The intensity of waveguided modes decays according to

$$I/I_0 \;=\; R^N \;=\; (1-L)^N \;\approx\; 1 - NL \tag{10.6}$$

where N is the number of reflection events, and the mirror loss is $L = 1 - R$ ($L \ll 1.0$ and $R \approx 1.0$). The equation illustrates that a small difference (e.g. a few percent) in R can make a large difference (e.g. a factor of 2) in the intensity of waveguided modes after N reflection events.

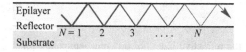

Fig. 10.3. Attenuation of waveguide mode due to lossy reflector.

Metal–semiconductor reflectors have been used in AlGaInP LEDs by Horng *et al.* (1999a, 1999b) to increase the light-extraction efficiency. The layer sequence of the finished device consisted of AlGaInP/AuBe/SiO$_2$/Si. The p-type AlGaInP–AuBe interface served as a reflector and as a broad-area ohmic contact. The AuBe layer also served as a supply layer for Be acceptors to attain low contact resistance. The contacts were annealed at 450 °C for 15 minutes. The LEDs were fabricated by a wafer-bonding process using a Si substrate. After the bonding process, the

GaAs substrate, on which the epitaxial layers had been grown, was removed. Si has a higher thermal conductivity than GaAs thus allowing for lower junction temperatures and reducing the joule-heating-induced emission-wavelength shift. The emission from the metal-reflector AlGaInP LEDs exceeded those of reference LEDs with DBRs fabricated on GaAs substrates.

Annealing and alloying forms low-resistance ohmic contacts. Typical annealing temperatures for alloyed contacts are between 375 and 450 °C for III–V arsenides and III–V phosphides and up to 600 °C for III–V nitrides. During the annealing process, the metal surface changes from smooth to rough and a concomitant decrease in the optical reflectivity results.

In contrast, non-alloyed contacts are just deposited on the semiconductor without annealing. Highly doped semiconductor surface layers are needed to obtain good ohmic I–V characteristics for such non-alloyed ohmic contacts. However, even for highly doped semiconductors, the contact resistance of non-alloyed contacts is usually higher than for alloyed contacts.

Thick metallic reflectors and hybrid reflectors are absorbing reflectors that should not be used as light-exit reflectors in LEDs and vertical-cavity surface emitting lasers. Metal contacts become practically opaque for thicknesses > 50 nm. That is, the transmittance of hybrid reflectors is near zero, unless the thickness of the metal is very thin (Tu et al., 1990).

Very thin metal contacts are semi-transparent. Most metal contacts have a transmittance of approximately 50% at a metal film thickness of 5–10 nm. The exact value of the transmittance needs to be calculated by taking into account the real and imaginary part of the refractive index (see, for example, Palik, 1998). However, very thin metallic contacts may form an islanded structure rather than a single continuous film. Furthermore, the electrical resistance of thin metal films can be large, in particular if an islanded structure is formed.

In LEDs with transparent substrates, e.g. AlGaInP LEDs on GaP, light emanating from the active region is incident on the substrate contact. To increase the reflectance of the backside, an ohmic contact geometry covering only a small fraction of the substrate surface can be used, such as a multiple-stripe or a ring-shaped contact. Using Ag-loaded conductive epoxy to attach the LED die to the package provides a high-reflectivity material in the regions not covered by the ohmic contact.

The die-attach epoxy can also serve as a reflector in LEDs grown on a transparent material, e.g. GaInN LEDs grown on sapphire substrates. Conductive Ag-loaded epoxy has a high conductivity as well as high reflectance. Such a highly reflective epoxy can increase the extraction efficiency in LEDs grown on transparent substrates.

There are ohmic contacts that are transparent to visible light. Such **transparent ohmic** *contacts* include indium tin oxide, frequently referred to as ITO (Ray *et al.*, 1983; Sheu *et al.*, 1998; Margalith *et al.*, 1999; Mergel *et al.*, 2000; Shin *et al.*, 2001). The material can be considered as a tin oxide semiconductor that is doped with indium. Indium substitutes for tin and therefore acts as an acceptor. Generally the specific contact resistance of ITO contacts is higher than the contact resistance of alloyed metal contacts.

10.2 Total internal reflectors

Total internal reflection is a fascinating phenomenon occurring at the boundary between two dielectric media with different refractive indices. Total internal reflection was first discovered by Johannes Kepler in the early 1600s (Kepler, 1611). Kepler attempted to explain the apparent bending of objects partially submersed in water. Kepler discovered that for rays near normal incidence, the ratio of the angles of incidence and refraction is proportional to the ratio of what are now known as the refractive indices of the media. Kepler's relationship can be expressed as

$$\bar{n}_1 \theta_1 = \bar{n}_2 \theta_2 \tag{10.7}$$

where the angles θ_1 and θ_2 are measured with respect to the surface normal. Comparison with the law discovered between 1621 and 1625 by Cornelius Willebrord Snell (***Snell's law***)

$$\bar{n}_1 \sin \theta_1 = \bar{n}_2 \sin \theta_2 \tag{10.8}$$

reveals that Kepler found the small-angle approximation of Snell's law. The angles used in Snell's law are shown in Fig. 10.4.

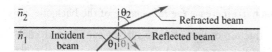

Fig. 10.4. Reflected and refracted light ray at the boundary between two media with refractive indices $\bar{n}_1$ and $\bar{n}_2$, where $\bar{n}_1 > \bar{n}_2$.

Kepler also discovered that, with suitable materials and sufficiently shallow angles of incidence, the refracted angle can be made to exceed 90°, resulting in total internal reflection. The critical angle for total internal reflection can be derived from Snell's law using the condition $\theta_2 = 90°$. Hence

$$\theta_{1,\text{crit}} = \arcsin(\bar{n}_2 / \bar{n}_1) . \tag{10.9}$$

10.2 Total internal reflectors

Because the ratio of the refractive indices on the right-hand side of the equation, $\bar{n}_2 / \bar{n}_1$, must be ≤ 1.0, total internal reflection can occur only in the optically denser material. Total internal reflection occurs for all angles of incidence $\theta_1 > \theta_{1,\,\text{crit}}$. For grazing angles of incidence and a sufficiently high index contrast, a light ray cannot leave a medium of high refractive index.

Isaac Newton later showed that, for most transparent media, the refractive index could be taken as unity plus a term proportional to the medium's mass density (measured in units of g/cm³). Media with high refractive index are therefore frequently called *optically dense* materials.

The most fascinating application of total internal reflection is fiber optic communication in which light rays, by undergoing total internal reflection events, are guided over thousands of kilometers in the core of a silica fiber. In 1841, the guiding of light was first demonstrated by Daniel Colladon using a jet of water as the high-index material (Hecht, 2001). This phenomenon is used to the present day to enhance the appearance of water fountains at night. The apparatus built by Daniel Colladon, who is considered the father of light guiding, is shown in Fig. 10.5.

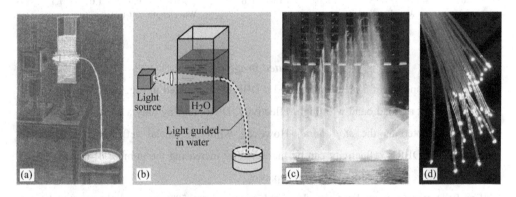

Fig. 10.5. (a) Historical drawing and (b) schematic illustration of apparatus used in 1841 by Swiss engineer Daniel Colladon to demonstrate the guiding of light by total internal reflection in a jet of water. (c) Light guiding used to beautify appearance of fountain in Las Vegas, Nevada. (d) Fiber-optic strands (after The Free Dictionary.com, 2005).

In light-emitting devices based on semiconductors, total internal reflection represents a major problem as it hinders the out-coupling of light out of the semiconductor die. As semiconductors have a large refractive index of typically 2.0–3.5, the critical angle for total internal reflection is small. Whereas the problem is severe for III–V arsenides and phosphides ($\bar{n} \approx 3.0$) it is less severe for III–V nitrides ($\bar{n} \approx 2.0$). Total internal reflection is of little concern in *organic* light-

10 Reflectors

emitting diodes due to the low refractive indices of organic materials.

One of the unique features of total internal reflection is that the magnitude of the reflection is $R = 1.0$ ("total"). This enables the demonstration of reflectors with zero mirror loss, a feature that has been employed advantageously in lasers. Edge-emitting lasers (Smith *et al.*, 1993) as well as microdisk lasers ("whispering gallery lasers") (McCall *et al.*, 1992) have employed total internal reflection to demonstrate cavities with very high cavity-quality factors.

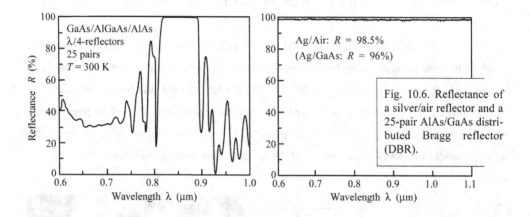

Fig. 10.6. Reflectance of a silver/air reflector and a 25-pair AlAs/GaAs distributed Bragg reflector (DBR).

10.3 Distributed Bragg reflectors

The reflectance spectra of a metal mirror and a DBR are compared in Fig. 10.6. Whereas metal reflectors exhibit a broad band with high reflectivity, DBRs display only a narrow band of high reflectivity denoted as the ***stop band***. However, while the metal reflector has a certain reflectivity, the DBR's reflectivity can be increased by increasing the number of reflector pairs so long as these reflector pairs are fully transparent.

For LED structures on light-absorbing substrates, about 50% of the light emitted by the active region is absorbed by the substrate. This represents a substantial loss. The absorption of light in the substrate can be avoided by placing a reflector between the substrate and the LED active layers. Light emanating from the active region towards the substrate will then be reflected and can escape from the semiconductor through the top surface.

Distributed Bragg reflectors (DBRs) are well suited for inclusion between the substrate and the active layers. The schematic LED structure with a DBR is shown in Fig. 10.7. LEDs with DBRs were first demonstrated by Kato *et al.* (1991) in the AlGaAs/GaAs material system. The LED had a 25-pair AlAs/GaAs or AlGaAs/GaAs DBR and emitted in the infrared at 870 nm.

10.3 Distributed Bragg reflectors

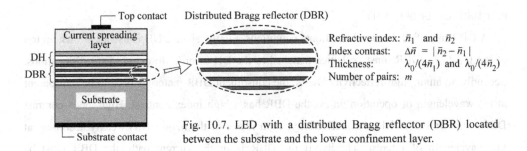

Fig. 10.7. LED with a distributed Bragg reflector (DBR) located between the substrate and the lower confinement layer.

A DBR is a multi-layer reflector consisting of typically 5–50 pairs of two materials with different refractive indices. As a result of the difference in refractive index, Fresnel reflection occurs at each of the interfaces. Usually the refractive index difference between the two materials is small so that the magnitude of the Fresnel reflection at one interface is also quite small. However, DBRs consist of many interfaces. More importantly, the thicknesses of the two materials are chosen in such a way that *all reflected waves* are in *constructive interference*. For normal incidence, this condition is fulfilled when both materials have a thickness of a *quarter wavelength* of the light, i.e.

$$t_{l,h} = \lambda_{l,h}/4 = \lambda_0/(4\bar{n}_{l,h}) \quad \text{(normal incidence)} \quad (10.10)$$

where λ_0 is the vacuum Bragg wavelength of the light, $t_{l,h}$ is the thickness of the low-index (l) and high-index (h) material, and $\bar{n}_{l,h}$ is the refractive index of the low-index (l) and high-index (h) material. The thickness of $t_{l,h}$ given in Eq. (10.10) can be not only $\lambda/4$, but also an odd-numbered integer multiple of $\lambda/4$, i.e. $3\lambda/4$, $5\lambda/4$, $7\lambda/4$, and so on. These thicknesses will also result in constructive interference of the reflected waves. For layer thicknesses greater than $\lambda/4$, e.g. $3\lambda/4$, the reflector will, however, have a narrower high-reflectivity stop band.

For an oblique angle of incidence, the wave vector can be separated into a parallel and a normal component. As in the normal-incident case, the thickness of the DBR layers must be a quarter wavelength for the wave vector component normal to the DBR layers. For an oblique angle of incidence $\Theta_{l,h}$, the optimum thicknesses for high reflectivity are given by

$$t_{l,h} = \lambda_{l,h}/(4\cos\Theta_{l,h}) = \lambda_0/(4\bar{n}_{l,h}\cos\Theta_{l,h}) \quad \text{(oblique incidence)}. \quad (10.11)$$

Again, the thickness $t_{l,h}$ given in Eq. (10.11) can also be an odd-numbered integer multiple of the value given by the equation. For DBRs with sufficiently many quarter-wave pairs, reflectivities near 100% can be obtained.

A DBR must fulfill several additional conditions. Firstly, since a DH is usually grown on top of the DBR, the DBR must be lattice matched to the DH in order to avoid misfit dislocations. Secondly, to attain high-reflectivity DBRs, the constituent DBR materials need to be transparent at the wavelength of operation unless the DBR has a high index contrast. High index-contrast DBRs (e.g. Si/SiO$_2$) yield high reflectivity, even if one of the materials is weakly absorbing at the wavelength of interest. Thirdly, if the DBR is in the current path, the DBR must be conductive.

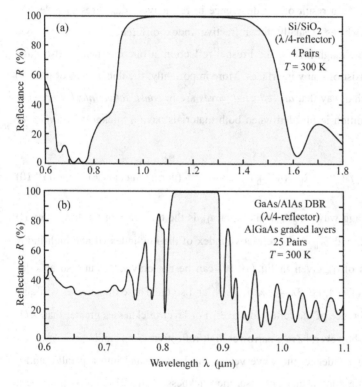

Fig. 10.8. Reflectance of two distributed Bragg reflectors (DBRs) versus wavelength. (a) Four-pair Si/SiO$_2$ reflector with high index contrast. (b) 25-pair AlAs/GaAs reflector. The high-index-contrast DBR only needs four pairs to attain high reflectivity. Note that the stop band of the high-index-contrast DBR is wider compared with the low-contrast DBR.

The reflectances of a Si/SiO$_2$ and an AlAs/GaAs DBR versus wavelength are shown in Fig. 10.8. The Bragg wavelength is located in the center of the high-reflectivity band or *stop band*. Inspection of the figure reveals that (*i*) the reflectivity of high-contrast DBRs (Si/SiO$_2$) is much higher than the reflectivity of low-index-contrast DBRs (AlAs/GaAs) for the same number of quarter-wave pairs and that (*ii*) the width of the stop band of the high-index-difference DBR is much wider than the stop band width of the low-contrast DBR.

10.3 Distributed Bragg reflectors

Properties of DBRs can be calculated by the matrix method (Born and Wolf, 1989). The properties of DBRs have been analyzed in detail by Coldren and Corzine (1995), Yariv (1989), and Björk *et al.* (1995). Here, only a brief summary will be given.

Consider a distributed Bragg reflector consisting of m pairs of two dielectric, lossless materials with refractive indices $\bar{n}_l$ and $\bar{n}_h$. The thicknesses of the layers are assumed to be a quarter wave, i.e. $L_l = \lambda_{\text{Bragg}} / (4\bar{n}_l)$, and $L_h = \lambda_{\text{Bragg}} / (4\bar{n}_h)$. The period of the DBR is $L_l + L_h$. The reflectivity of a single interface is given by Fresnel's equation for normal incidence

$$r = \frac{\bar{n}_h - \bar{n}_l}{\bar{n}_h + \bar{n}_l}. \tag{10.12}$$

Multiple reflections at the interfaces of the DBR and constructive interference of the multiple reflected waves increase the reflectivity with increasing numbers of pairs. The reflectivity has a maximum at the Bragg wavelength λ_{Bragg}. The reflectivity at the Bragg wavelength of a DBR with m quarter-wave pairs is given by (Coldren and Corzine, 1995)

$$R_{\text{DBR}} = |r_{\text{DBR}}|^2 = \left[\frac{1 - (\bar{n}_l / \bar{n}_h)^{2m}}{1 + (\bar{n}_l / \bar{n}_h)^{2m}} \right]^2. \tag{10.13}$$

The stop band of a DBR depends on the difference in refractive index of the two constituent materials, $\bar{n}_h - \bar{n}_l = \Delta\bar{n}$. The spectral width of the stop band is given by (Yariv, 1989)

$$\Delta\lambda_{\text{stopband}} = \frac{2\lambda_{\text{Bragg}} \Delta\bar{n}}{\pi \bar{n}_{\text{eff}}} \tag{10.14}$$

where $\bar{n}_{\text{eff}}$ is the effective refractive index of the DBR. For efficient operation of the LED, the stop band should be wider than the emission spectrum of the active region.

The effective refractive index of the DBR can be calculated by requiring the same optical path length normal to the layers for the DBR and an effective medium. The effective refractive index is then given by

$$\bar{n}_{\text{eff}} = 2 \left(\frac{1}{\bar{n}_l} + \frac{1}{\bar{n}_h} \right)^{-1}. \tag{10.15}$$

For small index differences, i.e. $\Delta\bar{n} \ll \Delta\bar{n}_l$, the effective refractive index can be approximated by

$$\bar{n}_{\text{eff}} = \tfrac{1}{2}\left(\bar{n}_{\text{l}} + \bar{n}_{\text{h}}\right). \tag{10.16}$$

The optical wave penetrates into the DBR only by a finite number of quarter-wave pairs. That is, a finite number out of the total number of quarter-wave pairs are effectively reflecting the wave. The effective number of pairs "seen" by the wave electric field is given by (Coldren and Corzine, 1995)

$$m_{\text{eff}} \approx \frac{1}{2} \frac{\bar{n}_{\text{h}} + \bar{n}_{\text{l}}}{\bar{n}_{\text{h}} - \bar{n}_{\text{l}}} \tanh\left(2m \frac{\bar{n}_{\text{h}} - \bar{n}_{\text{l}}}{\bar{n}_{\text{h}} + \bar{n}_{\text{l}}}\right). \tag{10.17}$$

For thick DBRs ($m \to \infty$), the tanh function approaches unity and one obtains

$$m_{\text{eff}} \approx \frac{1}{2} \frac{\bar{n}_{\text{h}} + \bar{n}_{\text{l}}}{\bar{n}_{\text{h}} - \bar{n}_{\text{l}}}. \tag{10.18}$$

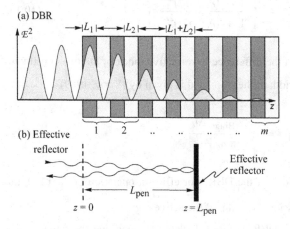

Fig. 10.9. Illustration of the DBR penetration depth. (a) DBR consisting of two materials with thickness L_1 and L_2. (b) Ideal (metallic) reflector displaced from the DBR surface by the penetration depth.

At the Bragg wavelength ($\lambda = \lambda_{\text{Bragg}}$), the phase change of the reflected wave is zero. In the vicinity of the Bragg wavelength ($\lambda \approx \lambda_{\text{Bragg}}$), the phase of the reflected wave changes *linearly* with wavelength. It is therefore possible to approximate a DBR with a metal-like mirror located a distance L_{pen} behind the first dielectric interface, as shown in Fig. 10.9. The reflection of the DBR can thus be expressed as

10.3 Distributed Bragg reflectors

$$r_{DBR} \approx |r_{DBR}|\, e^{-2i(\beta - \beta_{Bragg})L_{pen}} \tag{10.19}$$

where $\beta = 2\pi/\lambda$ is the average phase constant of the wave. The phase change at $z = 0$ (see Fig. 10.9) of the wave reflected by the metal mirror is given by

$$r_{metal}|_{z=0} = |r_{metal}|\, e^{2i(2\pi/\lambda)L_{pen}}. \tag{10.20}$$

Equating the phase changes given by Eqs. (10.19) and (10.20) and using the phase changes of a DBR (Coldren and Corzine, 1995), the penetration depth is given by

$$L_{pen} = \frac{L_1 + L_2}{4r}\tanh(2mr). \tag{10.21}$$

For a large number of pairs ($m \to \infty$), the penetration depth is given by

$$L_{pen} \approx \frac{L_1 + L_2}{4r} = \frac{(L_1 + L_2)}{4}\frac{\bar{n}_h + \bar{n}_l}{\bar{n}_h - \bar{n}_l}. \tag{10.22}$$

Comparison of Eqs. (10.22) and (10.18) yields that

$$L_{pen} = (1/2)\, m_{eff}\,(L_1 + L_2). \tag{10.23}$$

The factor of (1/2) in Eq. (10.23) is due to the fact that m_{eff} applies to the effective number of periods seen by the *electric field*, whereas L_{pen} applies to the optical power. The optical power is equal to the square of the electric field and hence it penetrates half as far into the mirror. The effective length of a cavity consisting of two DBRs is thus given by the sum of the thickness of the center region plus the two penetration depths into the DBRs. The effective length of a cavity with DBRs is thus longer than the effective length of a cavity with metal mirrors.

The reflectivity of the DBR depends on the polar angle of incidence and on the wavelength. Although an analytic result can be obtained for the reflectivity of a DBR at the Bragg wavelength for normal incidence, the reflectivity of a DBR at an arbitrary wavelength and an arbitrary angle of incidence can only be calculated numerically. If light from an isotropic source is reflected by a DBR, the reflected intensity can be obtained by integration over all angles. The angle-integrated reflectance at a certain wavelength λ is then given by

$$R_{\text{int}}(\lambda) = \frac{\int_0^{\pi/2} R(\lambda,\Theta)\, 2\pi \sin\Theta\, d\Theta}{\int_0^{\pi/2} 2\pi \sin\Theta\, d\Theta} = \int_0^{\pi/2} R(\lambda,\Theta) \sin\Theta\, d\Theta\, . \tag{10.24}$$

The total light intensity reflected by the DBR is calculated by

$$I_r = \int_\lambda I_i(\lambda)\, R_{\text{int}}(\lambda)\, d\lambda \tag{10.25}$$

where $I_i(\lambda)$ is the emission intensity spectrum of the active region incident on the DBR and I_r is the intensity reflected by the DBR. For an isotropic emitter such as the active region of an LED, it can be assumed that the emission spectrum incident on the DBR is independent of the emission angle.

An efficient DBR will be optimized in such a way that it maximizes the intensity of the reflected light. In addition, the escape of the light reflected by the DBR from the LED die must be taken into account. Maximizing the LED extraction efficiency using a DBR is a complicated problem that cannot be solved analytically. *Ray tracing computer software* is used to maximize the extraction efficiency in LED structures with DBRs.

A priori, it is not clear that the use of conventional DBRs provides the best extraction efficiency. DBRs with layer thicknesses thinner or thicker than a quarter-wave should also be considered for maximization of the extraction efficiency. Such variable-thickness DBRs have a *lower reflectivity* but a *wider stop-band width* compared with quarter-wave DBRs. For active regions with broad emission spectra, such variable-period DBRs can be advantageous.

Ideally, the layers comprising the DBR are transparent. Layers of **transparent DBRs** have negligible absorption losses. However, transparent materials may not always be available so that *absorbing* materials must be used. Such **absorbing DBRs** have a maximum reflectivity of less than 100%, even if an infinite number of pairs are used.

An example of a partially absorbing Si/SiO$_2$ DBR is shown in Fig. 10.8. Silicon absorbs light for $\lambda < 1.1$ μm, i.e. for $h\nu > E_g$. However, the results shown in the figure demonstrate that very high reflectivities can be attained at 1.0 μm, where Si is absorbing. This is due to the high index contrast between Si and SiO$_2$.

Transparent and absorbing DBRs lattice matched to GaAs are used in the AlGaInP/GaAs material system. This material system is used for high-efficiency visible LEDs emitting at $\lambda > 550$ nm (green, yellow, amber, orange, and red). The properties of transparent and absorbing

DBRs used in this material system have been compiled by Kish and Fletcher (1997) and are summarized in Table 10.3. Inspection of the table reveals that the absorbing $Al_{0.5}In_{0.5}P$/GaAs DBRs have the advantage of a high refractive index contrast. However, the absorbing nature of the DBR imposes an upper limit on the maximum reflectivity. High-contrast DBRs have a *wider* stop-band width. The transparent $Al_{0.5}In_{0.5}P/(AlGa)_{0.5}In_{0.5}P$ DBRs have the advantage of negligible optical losses. However, many pairs are needed to attain high reflectivity and the stop-band width is narrower than in high index-contrast DBRs.

Table 10.3. Properties of distributed Bragg reflector (DBR) materials used for visible and infrared LED applications. The DBRs marked as "lossy" are absorbing at the Bragg wavelength (data after Adachi, 1990; Adachi et al., 1994; Kish and Fletcher, 1997; Babic et al., 1999; Palik, 1998).

Material system	Bragg wavelength	$\bar{n}_{low}$	$\bar{n}_{high}$	$\Delta\bar{n}$	Transparency range
$Al_{0.5}In_{0.5}P$/GaAs	590 nm	3.13	3.90	0.87	> 870 nm (lossy)
$Al_{0.5}In_{0.5}P/Ga_{0.5}In_{0.5}P$	590 nm	3.13	3.74	0.61	> 649 nm (lossy)
$Al_{0.5}In_{0.5}P/(Al_{0.3}Ga_{0.7})_{0.5}In_{0.5}P$	615 nm	3.08	3.45	0.37	> 592 nm
$Al_{0.5}In_{0.5}P/(Al_{0.4}Ga_{0.6})_{0.5}In_{0.5}P$	590 nm	3.13	3.47	0.34	> 576 nm
$Al_{0.5}In_{0.5}P/(Al_{0.5}Ga_{0.5})_{0.5}In_{0.5}P$	570 nm	3.15	3.46	0.31	> 560 nm
AlAs/GaAs	900 nm	2.97	3.54	0.57	> 870 nm
SiO_2/Si	1300 nm	1.46	3.51	2.05	> 1106 nm

In practice, transparent layers are used at and near the top (epitaxial side) of the DBR whereas absorbing layers are used towards the bottom (substrate side) of the DBR. For DBRs used in manufactured AlGaInP/GaAs LEDs, each of the layers is different and optimized to keep the number of pairs low, the absorption low, and the reflectivity spectrum broad (Streubel, 2000).

Table 10.3 also shows properties of the AlAs/GaAs and the SiO_2/Si material systems. The SiO_2/Si material system is an example of a high index-contrast system. However, it cannot be used for current conduction due to the insulating nature of the SiO_2. The AlAs/GaAs material system is used in resonant-cavity LEDs and vertical-cavity surface-emitting lasers emitting in the range 880–980 nm.

A DBR that is resonant with the peak-emission wavelength, is not necessarily the optimum reflector for an absorbing substrate (AS) LED. Although a DBR has a high reflectivity for normal incidence, it rapidly decreases for off-normal angles of incidence. Assume that the angle of incidence is $\theta = 0°$ at normal incidence. Since the solid angle (per angle interval $d\theta$) increases with angle θ according to a *sine* function, it is advantageous to shift the normal-incidence resonance wavelength of the DBR to wavelengths *longer* than the peak emission wavelength.

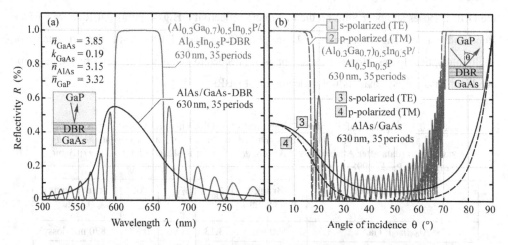

Fig. 10.10. Calculated reflectivity (inside the cladding GaP) versus (a) wavelength and (b) polar angle of a transparent AlGaInP/AlInP DBR and an absorbing AlAs/GaAs DBR.

The calculated reflectivities of a transparent and an absorbing DBR versus wavelength and polar angle of incidence are shown in Fig. 10.10. Whereas the DBR made of the transparent materials has a reflectivity close to 100% at the Bragg wavelength, the DBR that includes the absorbing GaAs layers has a maximum reflectivity of about 55%. This value cannot be increased by adding additional pairs to the DBR, because the limitation of the value lies in the absorptive nature of the DBR. The figure also shows the reflectivity of the DBRs versus the angle of incidence. Inspection of the figure reveals a major drawback of DBRs, namely the limitation of the high reflectivity band to small angles of incidence. For angles of incidence greater 20°, the reflectivity strongly decreases to assume values close to zero. Thus, for oblique angles of incidence (20° < θ < 70°), the DBR becomes non-reflective. The lack of reflectivity is a major loss mechanism in AlGaInP LEDs whose active layers are grown on top of a DBR that in turn is located on the absorbing GaAs substrate.

Next a formula will be derived that gives the critical angle, Θ_c, at which the reflectivity of a DBR strongly decreases. The structure of a DBR and the critical angle are illustrated in Fig. 10.11 (a) and (b), respectively. Note that the outside medium is a semiconductor with refractive index $\bar{n}_0$. For normal incidence ($\Theta = 0°$), the Bragg condition is fulfilled at the Bragg wavelength, λ_{Bragg}, which is located at the center of the high-reflectivity stop band. The Bragg wavelength shifts with angle of incidence, as shown in Fig. 10.11 (c). However, the *width* of the stop band does not depend on the angle of incidence as long as it is small. We can thus write the following condition for the critical angle Θ_c

$$\Delta\lambda_{\text{Bragg}} = \lambda_{\text{Bragg}}(\Theta = 0°) - \lambda_{\text{Bragg}}(\Theta_c) = \tfrac{1}{2}\Delta\lambda_{\text{stop band}}. \tag{10.26}$$

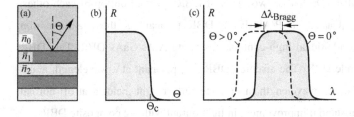

Fig. 10.11. (a) DBR stucture used in calculation. (b) Reflectivity versus angle of incidence and critical angle at which reflectivity decreases. (c) DBR reflectivity versus wavelength for two angles of incidence.

Using the expression for the angle-dependent Bragg wavelength and the expression for the width of the stop band given earlier in this chapter, one can write

$$\lambda_{\text{Bragg}}(\Theta = 0°)\left[1 - \cos\left(\frac{\bar{n}_0}{\bar{n}_1}\Theta_c\right)\right] = 2\lambda_{\text{Bragg}}(\Theta = 0°)\frac{\Delta\bar{n}}{\bar{n}_1 + \bar{n}_2}\frac{1}{\bar{n}_0}. \tag{10.27}$$

Dividing both sides of the equation by $\lambda_{\text{Bragg}}(\Theta = 0°)$ and solving the equation for Θ_c yields

$$\boxed{\Theta_c = \frac{\bar{n}_1}{\bar{n}_0}\arccos\left[1 - \left(\frac{2\Delta\bar{n}}{\bar{n}_1 + \bar{n}_2}\frac{1}{\bar{n}_0}\right)\right].} \tag{10.28}$$

Using the approximation $\cos x \approx 1 - (1/2)x^2$ (valid near $x = 0$) and $\arccos x \approx [2(1-x)]^{1/2}$ (valid near $x = 1$), the following approximation is obtained

$$\boxed{\Theta_c \approx \frac{\bar{n}_1}{\bar{n}_0}\sqrt{\frac{2}{\bar{n}_0}\frac{2\Delta\bar{n}}{\bar{n}_1 + \bar{n}_2}}.} \tag{10.29}$$

The equation shows that the critical angle strongly depends on the refractive index of the outside medium, $\bar{n}_0$, with $\Theta_c \propto (\bar{n}_0)^{-3/2}$. Thus the critical angle is small for outside media with a high refractive index. For this reason, it is difficult if not impossible to attain omnidirectional reflection characteristics with a DBR if the outside medium is a high-index semiconductor.

As a numerical example, we consider an AlAs/GaAs DBR ($\bar{n}_{AlAs} = 3.0$; $\bar{n}_{GaAs} = 3.5$) with GaP as the outside medium ($\bar{n}_{GaP} = 3.1$). Insertion of these values in the equation for the critical angle yields $\Theta_c = 20.5°$, which is relatively close to normal incidence. Even high-contrast DBRs, such as SiO_2/Si DBRs, do not have omnidirectional characteristics if the outside medium is a high-index semiconductor.

Different strategies have been employed to optimize DBRs. Chiou et al. (2000) used a composite DBR in an AlGaInP LED, that is, two types of DBRs stacked on top of each other, namely a non-absorbing $(Al_{0.4}Ga_{0.6})_{0.5}In_{0.5}P/Al_{0.5}In_{0.5}P$ DBR resonant at the peak emission wavelength of 590 nm and an additional high-contrast absorbing AlAs/GaAs DBR. This DBR was located below the phosphide DBR. The arsenide DBR was resonant at a wavelength about 10% longer than the peak emission wavelength in order to reflect light incident at off-normal angles. The authors found a substantial improvement in light output with the composite DBR.

Non-periodic DBRs have a wider stop band and thereby a high reflectivity over a wider range of angles. To find optimum non-periodic DBR structures, numerical simulations and optimization procedures have been conducted (see, for example, Li et al., 1999).

As shown by the calculation above, an increase in index contrast will provide a wider range of angles within which a high reflectivity is maintained. The use of such high-contrast DBRs in LEDs has been proposed by Chiou et al. (2003) who disclosed the use of an AlGaAs/Al_xO_y DBR in an LED. Al_2O_3 has a refractive index of about 1.75 whereas Al-rich AlGaAs has a refractive index of about 3.25 thus providing a large index contrast of $\Delta\bar{n} = 1.5$. The Al_xO_y layers of the DBR were fabricated from epitaxially grown AlAs by using an oxidation process that is performed in water vapor at temperatures of 400–450 °C. The Al_xO_y layers of the DBR are not conductive thus necessitating non-oxidized AlAs openings in the Al_xO_y layers, which provide a current path between substrate and active layers.

DBRs can have a high electrical resistance for current transport perpendicular to the layers. The resistance can pose a substantial problem in LED and laser structures, which manifests itself in a high forward voltage. Early experiments on VCSELs or vertical-cavity surface-emitting lasers (Jewell et al., 1989; Koyama et al., 1989) revealed forward voltages as high as 30.0 V (Jewell, 1992) and prevented these first devices from lasing in continuous-wave mode. The high resistance is caused by abrupt heterojunctions which pose barriers for carrier transport. Fortunately, heterojunction barriers can be completely eliminated by parabolic compositional grading (Schubert et al., 1992a, 1992b). Such compositional grading is now routinely used in

DBRs so that the resistance issue of DBR heterojunctions no longer is a concern.

10.4 Omnidirectional reflectors

Electrically conductive high-reflectivity omnidirectional reflectors are highly desirable. With air as the outside medium, omnidirectional reflection characteristics can be demonstrated by using high-contrast DBRs. The high-contrast materials Si ($\bar{n} \approx 3.5$ at $\lambda = 1$ μm) and SiO$_2$ ($\bar{n} \approx 1.46$) are natural candidates for such ODRs. The optical properties of Si/SiO$_2$ distributed Bragg omnidirectional reflectors (DB-ODRs) and other material systems have been investigated (Chen et al., 1999; Bruyant et al., 2003).

Whereas omnidirectional TE reflectivity is readily obtained in DB-ODRs, the Brewster angle, at which the TM reflectivity decreases to zero, is an impediment in achieving omnidirectional characteristics for TM waves. An outside angle range of $0° \leq \Theta \leq 90°$ (in a low-index outside material such as air) results in an angle range $0° \leq \theta \ll 90°$ inside the material (i.e. inside the DBR). The inside range may not include the Brewster angle under which circumstances a DBR becomes omnidirectionally reflective.

Highly omnidirectional reflection characteristics were also obtained with DB-ODRs using polystyrene and Te layers (Fink et al., 1998). Owing to the very large difference of the refractive indices, $\bar{n}_{\text{polystyrene}} = 1.8$ and $\bar{n}_{\text{Te}} = 5$, the interfacial Brewster angle θ_B is not accessible from light incident from air, resulting in a complete photonic bandgap in the wavelength range from 10 to 15 μm.

Another intriguing approach uses birefringent polymers with two different refractive indices parallel and vertical to the layer planes (Weber et al., 2000). By adjusting the differences between the vertical and in-plane indices, the value of the Brewster angle can be controlled. Brewster angles up to 90° (grazing incidence) and even imaginary values are possible, resulting in a high reflectivity for TM-polarized light at virtually all angles of incidence.

Unfortunately the applicability of the above-mentioned DB-ODRs to LEDs is limited due to the insulating electrical characteristics of the constituent materials.

Metallic layers are capable of reflecting light over a wide range of wavelengths and incident angles with the high-reflectivity band being limited to frequencies below the plasma frequency of the free-electron gas (for historical references, see Drude, 1904; Lorentz 1909). However, electron oscillations induced by the incident light waves not only result in reflection but also in absorption caused by electron–phonon scattering. Thus, pure metal reflectors have significant

10 Reflectors

reflection losses, particularly when used on high-index materials.

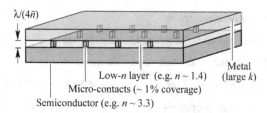

Fig. 10.12. Structure of omnidirectional reflector consisting of semiconductor, low-refractive index dielectric layer, and metal layer. The dielectric is perforated by an array of microcontacts providing electrical conductivity (after Gessmann *et al.*, 2003).

An electrically conductive reflector that has high-reflectivity and omnidirectional characteristics is shown in Fig. 10.12 (Schubert, 2001, 2004). It consists of three layers, a semiconductor, a dielectric layer, and a metal layer (triple-layer ODR). The dielectric layer is perforated by an array of microcontacts that provide electrical conductivity. The dielectric layer should have a refractive index as low as possible to provide a high index contrast to the semiconductor and the metal. The metal has a complex refractive index with a large extinction coefficient.

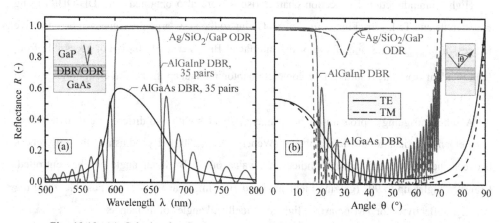

Fig. 10.13. (a) Calculated reflectance at normal-incidence versus wavelength and (b) reflectance versus angle of incidence for an omnidirectional reflector (ODR), a transparent AlGaInP/AlInP DBR, and an absorbing AlGaAs/GaAs DBR (after Gessmann *et al.*, 2003).

The reflectivity of the triple-layer ODR calculated by the matrix method is shown in Fig. 10.13 versus wavelength and versus angle along with the reflectivity of two DBRs (Gessmann *et al.*, 2003). The triple-layer ODR provides a broad reflectivity and omnidirectional characteristics. A small dip in the reflectivity is found for the TM wave at an incidence angle of about 30° due to the reduced reflectivity of the semiconductor/dielectric interface at the Brewster angle. The angle-integrated reflectivity of the triple-layer ODR is very high and can exceed 99%.

At normal incidence ($\theta = 0$), the reflectivity of the triple-layer ODR can be calculated analytically and is given by

$$R_{ODR} = \frac{[(\bar{n}_s - \bar{n}_{li})(\bar{n}_{li} + \bar{n}_m) + (\bar{n}_s + \bar{n}_{li})k_m]^2 + [(\bar{n}_s - \bar{n}_{li})k_m + (\bar{n}_s + \bar{n}_{li})(\bar{n}_{li} - \bar{n}_m)]^2}{[(\bar{n}_s + \bar{n}_{li})(\bar{n}_{li} + \bar{n}_m) + (\bar{n}_s - \bar{n}_{li})k_m]^2 + [(\bar{n}_s + \bar{n}_{li})k_m + (\bar{n}_s - \bar{n}_{li})(\bar{n}_{li} - \bar{n}_m)]^2} \quad (10.30)$$

where $\bar{n}_{li}$ and $\bar{n}_s$ are the refractive indices of the dielectric and semiconductor, respectively, and $N_m = \bar{n}_m + ik_m$ is the complex refractive index of the metal. The equation applies to a thickness of $\lambda_0/(4n_{li})$ for the low-index dielectric layer, i.e. a quarter-wave layer. For an AlGaInP/SiO$_2$/Ag structure emitting at 630 nm, the equation yields a normal-incidence reflectivity $R_{ODR}(\theta = 0)$ of 98.8% compared to a value of 96.1% for a structure without a dielectric layer.

650 nm AlGaInP LEDs using a triple-layer ODR have been demonstrated. The surface coverage of the microcontact array was 1%. The light-output power versus current characteristics of the ODR-LED and several reference devices are shown in Fig. 10.14 (Gessmann et al., 2003). Comparison of the ODR-LED output characteristics with those of DBR-LEDs yields that the ODR-LED provides higher output powers.

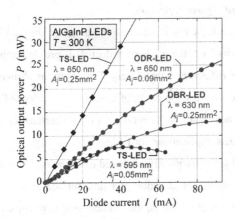

Fig. 10.14. Light-output power versus injection current of different types of LEDs. The ODR device has a higher output power than the DBR device (after Gessmann et al., 2003).

ODR-LEDs have also been demonstrated in the GaInN material system (Kim et al., 2004). The ODR consisted of a RuO$_2$ ohmic contact to p-type GaN, a quarter-wave-thick SiO$_2$ low-index layer perforated by an array of microcontacts, and an Ag layer. Calculations predict a 98% angle-averaged reflectivity at λ = 450 nm for an GaN/SiO$_2$/Ag ODR, much higher than that for

10 Reflectors

an $Al_{0.25}Ga_{0.75}N$/GaN distributed Bragg reflector (49%) and Ag (94%). It was shown that the RuO_2/SiO_2/Ag ODR has a higher reflectivity than Ni/Au and even Ag reflectors, leading to a higher light extraction efficiency of the GaInN ODR-LED. The electrical properties of the ODR-LED were found to be comparable to those LEDs using conventional Ni/Au contacts. A comparison of the electrical and optical characteristics of the GaInN ODR device is shown in Fig. 10.15.

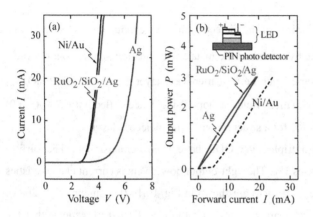

Fig. 10.15. Current–voltage and light-output-versus-current characteristics of a GaInN LED with a GaInN/RuO_2/SiO_2/Ag omni-directional reflector (after Kim et al., 2004).

10.5 Specular and diffuse reflectors

In *specular reflectors*, the angle of the reflected light ray is equal to the angle of the incident light ray. Specular reflectors are thus deterministic, i.e. the angle of reflection is pre-determined by the angle of the incident light ray. *Diffuse reflectors* behave very differently, as shown in Fig. 10.16. The reflected intensity is distributed over a wide range of angles, independent of the incidence angle of the incoming ray. We will next discuss the properties of lambertian sources and lambertian reflectors and subsequently discuss implications for LEDs.

Lambertian surface sources are motivated by the frequently observed *experimental fact that the source radiance* (i.e. the optical power emitted per steradian per unit surface area of the source) *is a constant*, independent of the viewing angle. That is, the radiance and luminance of a lambertian source are independent of the viewing angle. The sun is the prime example of a lambertian source. As shown in Fig. 10.17 (a), the sun's surface has the same brightness (luminance), irrespective of the viewing angle. That is, the brightness (luminance) is the same for, e.g., the normal-incidence viewing angle and an oblique-incidence viewing angle with respect to the sun's surface. For the same reason, the moon, shown in Fig. 10.17 (b), can be considered a good example of a lambertian reflector. Light-diffusing reflectors randomize the propagation direction of incoming photons and are referred to as **diffuse or lambertian reflectors**.

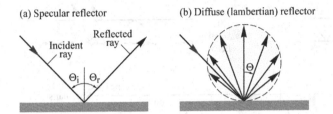

Fig. 10.16. Schematic of a specular and a diffuse (lambertian) reflector. The reflected power distribution of a lambertian reflector follows a $\cos \Theta$ dependence.

We now assume that a lambertian source has an intensity (i.e. optical power emitted per steradian) along the direction given by the angle Θ is given by

$$I = I_n \cos \Theta \tag{10.31}$$

where I_n is the intensity emitted normal to the reflector surface. The angular $\cos \Theta$ dependence of the equation is known as **Lambert's cosine law**.

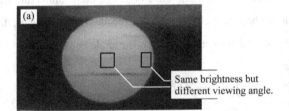

Fig. 10.17. (a) The sun's surface brightness is independent of viewing angle with respect to the sun's surface. It is a good example of a lambertian source. (b) The moon is a good example of a lambertian reflector.

Next, we will show that Lambert's cosine law results in a source luminance (or radiance, in radiometric units) that is *independent* of the viewing angle with respect to the source surface. Assume a lambertian surface source with area A. The projected area visible to an observer positioned at angle Θ is given by $A \cos \Theta$. Thus the luminance found by the observer is given by

$$\text{Luminance} = \frac{I_n A \cos \Theta}{A \cos \Theta} = I_n \tag{10.32}$$

where $A \cos\Theta$ is the surface area seen by the observer. Thus the luminance is constant, independent of the viewing angle. This fact is corroborated by the photographs shown in Fig. 10.17.

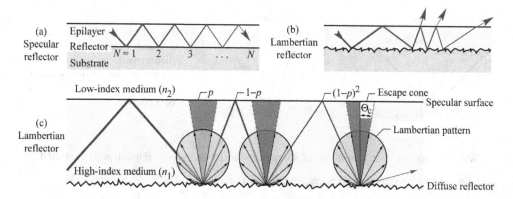

Fig. 10.18. (a) Optical ray, propagating in epilayer, guided by specular epilayer/substrate interface and epilayer/air interface. (b) and (c) Optical ray guided by lambertian reflector at epilayer/substrate interface and specular epilayer/air interface.

Light propagating in a layer clad by specular reflectors will be waveguided within the layer. However, light propagating in a layer clad by a lambertian reflector can be out-coupled into free space as shown in Fig. 10.18. As a lambertian reflector follows the $\cos\Theta$ dependence, the probability of reflected light falling into the escape cone defined by the angle Θ_c is given by

$$p = \frac{\int_0^{\Theta_c} I_n \cos\Theta \, 2\pi \sin\Theta \, d\Theta}{\int_0^{90°} I_n \cos\Theta \, 2\pi \sin\Theta \, d\Theta} = \frac{\int_0^{\Theta_c} \sin 2\Theta \, d\Theta}{\int_0^{90°} \sin 2\Theta \, d\Theta} = \frac{1 - \cos 2\Theta_c}{2}. \quad (10.33)$$

Using Snell's law ($\bar{n}_1 \sin\Theta_c = \bar{n}_2$, where $\bar{n}_1$ is the refractive index of the waveguide and $\bar{n}_1 > \bar{n}_2$), one obtains

$$p = \frac{1 - \cos[2\arcsin(\bar{n}_2/\bar{n}_1)]}{2} = \left(\frac{\bar{n}_2}{\bar{n}_1}\right)^2. \quad (10.34)$$

Assuming a lambertian reflector with unit reflectivity ($R = 1.0$), the light intensity inside the semiconductor decreases according to a geometric series. After N reflection events, the intensity of the light ray will have fallen to $(1-p)^N$. Defining N as the number of reflection events after which the light intensity has decreased to 1/e, we can write the equation

$$(1-p)^N = 1/e. \tag{10.35}$$

Solving the equation for N allows one to calculate the average number of reflection events before the light ray will escape into the free space surrounding the semiconductor. One obtains

$$N = -\left[\ln\left(1 - \bar{n}_2^2/\bar{n}_1^2\right)\right]^{-1}. \tag{10.36}$$

As an example, we consider $\bar{n}_1 = 2.5$ (GaN) and $\bar{n}_2 = 1$ (air) and obtain $N = 5.7$. That is, light escapes from the waveguide after about six diffuse reflection events. Thus the introduction of diffuse reflectors into LED structures is a fruitful strategy in attaining laterally scalable emitters that do not suffer from an efficiency penalty usually associated with the up-scaling of the LED chip dimension (Kim *et al.*, 2006).

Mechanical roughening of reflective surfaces generally results in a change from specular to diffuse reflection characteristics. Such reflective surfaces can be, e.g., metallic surfaces or dielectric surfaces. In addition, porous silica (optionally coated with a metal) is known to have diffuse reflection characteristics as multiple refraction, reflection, and scattering events at pore/silica interfaces randomize the photon propagation direction.

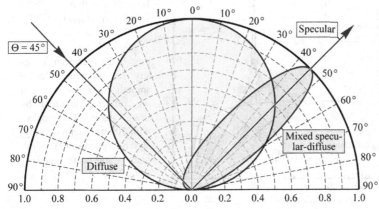

Fig. 10.19. Schematic reflection contour plots of a specular reflector, diffuse reflector, and a mixed specular-diffuse reflector for an angle of incidence of 45°.

The fabrication of *ideal* lambertian reflectors requires substantial roughness (greater than the wavelength of light, λ). Such ideal diffuse reflectors would have a reflection characteristic that is independent of the angle of incidence. Many real surface-textured reflectors have a mixed

10 Reflectors

specular-diffuse reflection characteristic with a *preferential* direction along the specular-reflection direction. Figure 10.19 shows the schematic reflection contour plots of a diffuse reflector, a mixed specular-diffuse reflector, and a specular reflector for an angle of incidence of 45°.

Figure 10.20 shows the measured reflected intensity versus reflection angle for a planar smooth Ag reflector and a surface-textured Ag reflector. The surface roughening was achieved by natural lithography using 700 nm diameter polystyrene spheres on the substrate, followed by ion-beam etching and Ag deposition. Inspection of the figure reveals a strong diffuse background for the textured reflector that is about two orders of magnitude higher than that of the planar Ag reflector. However, the textured reflector still exhibits a specular component. The relative strength of the diffuse and specular components of mixed specular-diffuse reflectors can be assessed quantitatively using the model developed by Xi et al. (2006). Using this model, it is found that the mixed specular-diffuse reflector, shown in Fig. 10.20, has a diffuse power ratio of $P_{\text{diff}}/(P_{\text{spec}} + P_{\text{diff}}) = 42.8\%$. The root-mean-square (rms) roughness of the partially diffuse reflector shown in Fig. 10.20 is 21.2 nm. By further increasing the rms roughness, diffusive power ratios of 100% can be attained.

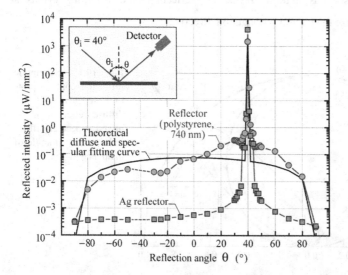

Fig. 10.20. Reflected intensity versus angle for a smooth Ag reflector and a Ag reflector textured by using "natural lithography" with 700 nm diameter polystyrene spheres and subsequent ion-beam etching (after Xi et al., 2006).

Smooth Ag reflector: $P_{\text{diff}}/(P_{\text{diff}} + P_{\text{spec}}) < 1\%$

Textured Ag reflector: $P_{\text{diff}}/(P_{\text{diff}} + P_{\text{spec}}) = 42.8\%$

Exercise: *Lambertian reflectors in LEDs*. Assume a lambertian reflector with reflectivity 1.0 that is incorporated in a lossless GaAs LED structure with refractive index of 3.5. Assume that the outside medium is air. Calculate the critical angle of the escape cone, the probability that a reflected light ray falls within the escape cone and the average number of reflection events before a photon escapes from the high-index GaAs layer.

Solution: $\Theta_c = 16.6°$; $p = 8.2\%$; $N = 11.7$.

Would a hypothetical planar reflector that reflected light coming from any incoming direction towards the surface normal be useful? Is there a physical principle that prevents a reflector from reflecting light in such a way?

Answer: Although such a reflector would be very useful, such a reflector would unfortunately violate the ***conservation of radiance theorem*** (previously called the conservation of brightness theorem), which states that it is impossible to increase the radiance of light by a passive optical system beyond a value of L/n^2 where L is the radiance in vacuum and n is the refractive index of the medium in which the light propagates.

References

Adachi S. in *Properties of Gallium Arsenide* EMIS Datareview Ser. **2**, 513 INSPEC (IEE, New York, 1990)

Adachi S., Kato H., Moki A., and Ohtsuka K. "Refractive index of $(Al_xGa_{1-x})_{0.5}In_{0.5}P$ quaternary alloys" *J. Appl. Phys.* **75**, 478 (1994)

Babic D. I., Piprek L., and Bowers J. E. in *Vertical-Cavity Surface-Emitting Lasers* edited by C. W. Wilmsen, H. Temkin, and L. A. Coldren (Cambridge University Press, Cambridge, 1999)

Bell L. *The Telescope* this book discusses metal reflectors in a historical context; see pp. 220–227 (McGraw Hill, New York, 1922)

Björk G., Yamamoto Y., and Heitmann H. "Spontaneous emission control in semiconductor microcavities" in *Confined Electrons and Photons* edited by E. Burstein and C. Weisbuch (Plenum Press, New York, 1995)

Born M. and Wolf E. *Principles of Optics* 6th edition (Pergamon Press, New York, 1989)

Bruyant A., Lérondel G., Reece P. J., and Gal M. "All-silicon omnidirectional mirrors based on one-dimensional photonic crystals" *Appl. Phys. Lett.* **82**, 3227 (2003)

Chen K. M., Sparks A. W., Luan H.-C., Lim D. R., Wada K., and Kimerling L. C. "SiO_2/TiO_2 omnidirectional reflector and microcavity resonator via the sol-gel method" *Appl. Phys. Lett.* **75**, 3805 (1999)

Chiou S.-W., Lee C. P., Huang C. K., and Chen C. W. "Wide angle distributed Bragg reflectors for 590 nm amber AlGaInP light-emitting diodes" *J. Appl. Phys.* **87**, 2052 (2000)

Chiou S.-W., Chang H., Chen T.-P., and Chang C.-S. "Light emitting diodes and fabrication method thereof" US Patent 6,552,369 (2003)

Coldren L. A. and Corzine S. W. *Diode Lasers and Photonics Integrated Circuits* (John Wiley and Sons, New York, 1995)

Drude P. "Optische Eigenschaften und Elektronen Theorie I" ("Optical properties and electron theory I") *Annalen der Physik* **14**, 677 (1904); see also "Optische Eigenschaften und Elektronen Theorie II" ("Optical properties and electron theory II") *Annalen der Physik* **14**, 936 (1904)

Gessmann Th., Schubert E. F., Graff J. W., Streubel K., and Karnutsch C. "Omni-directionally reflective contacts for light-emitting diodes" *IEEE Electron. Dev. Lett.* **24**, 683 (2003)

Fink Y., Winn J. N., Fan S., Chen C., Michel J., Joannopoulos J. D., Thomas E. L. "A dielectric omnidirectional reflector" *Science* **282**, 1679 (1998)

Hecht J. *Understanding Fiber Optics* (Pearson Education, Upper Saddle River NJ, 2001)

Horng R. H., Wuu D. S., Wei S. C., Huang M. F., Chang K. H., Liu P. H., and Lin K. C. "AlGaInP/AuBe/glass light-emitting diodes fabricated by wafer bonding technology" *Appl. Phys. Lett.* **75**, 154 (1999a)

Horng R. H., Wuu D. S., Wei S. C., Tseng C. Y., Huang M. F., Chang K. H., Liu P. H., and Lin K. C. "AlGaInP light-emitting diodes with mirror substrates fabricated by wafer bonding" *Appl. Phys. Lett.* **75**, 3054 (1999b)

Jewell J. L., Huang K. F., Tai K., Lee Y. H., Fischer R. J., McCall S. L., and Cho A. Y. "Vertical cavity single quantum well laser" *Appl. Phys. Lett.* **55**, 424 (1989)

Jewell J. L. personal communication (1992)

Kato T., Susawa H., Hirotani M., Saka T., Ohashi Y., Shichi E., and Shibata S. "GaAs/GaAlAs surface emitting IR LED with Bragg reflector grown by MOCVD" *J. Cryst. Growth* **107**, 832 (1991)

Kepler J. *Dioptrice* (1611)

Kim J. K., Gessmann T., Luo H., and Schubert E. F. "GaInN light-emitting diodes with $RuO_2/SiO_2/Ag$ omni-directional reflector" *Appl. Phys. Lett.* **84**, 4508 (2004)

Kim J. K., Luo H., Xi Y., Shah J. M., Gessmann T., and Schubert E. F. "Light extraction in GaInN light-emitting diodes using diffuse omnidirectional reflectors" *Journal of the Electrochemical Society* **153**, G105 (2006)

Kish F. A. and Fletcher R. M. "AlGaInP light-emitting diodes" in *High Brightness Light-Emitting Diodes* edited by G. B. Stringfellow and M. G. Craford, Semiconductors and Semimetals 48 (Academic, San Diego, 1997)

Koyama F., Kinoshita S., and Iga K. "Room temperature continuous wave lasing characteristics of a GaAs vertical-cavity surface-emitting laser" *Appl. Phys. Lett.* **55**, 221 (1989)

Li H., Gu G., Chen H., and Zhu S. "Disordered dielectric high reflectors with broadband from visible to infrared" *Appl. Phys. Lett.* **74**, 3260 (1999)

Lorentz H. A. *The Theory of Electrons and its Applications to the Phenomena of Light and Radiant Heat* (Teubner, Leipzig, Germany, 1909)

Margalith T., Buchinsky O., Cohen D. A., Abare A. C., Hansen M., DenBaars S. P., and Coldren L. A. "Indium tin oxide contacts to gallium nitride optoelectronic devices" *Appl. Phys. Lett.* **74**, 3930 (1999)

McCall S. L., Levi A. F. J., Slusher R. E., Pearton S. J., and Logan R. A. "Whispering-gallery mode microdisk lasers" *Appl. Phys. Lett.* **60**, 289 (1992)

Mergel D., Stass W., Ehl G., and Barthel D. "Oxygen incorporation in thin films of In_2O_3:Sn prepared by radio frequency sputtering" *J. Appl. Phys.* **88**, 2437 (2000)

Palik E. D. *Handbook of Optical Constants of Solids* (Academic Press, San Diego, 1998)

Ray S., Banerjee R., Basu N., Batabyal A. K., and Barua A. K. "Properties of tin doped indium oxide thin films prepared by magnetron sputtering" *J. Appl. Phys.* **54**, 3497 (1983)

Schubert E. F., Tu L.-W., Zydzik G. J., Kopf R. F., Benvenuti A., and Pinto M. R. "Elimination of heterojunction band discontinuities by modulation doping" *Appl. Phys. Lett.* **60**, 466 (1992a)

Schubert E. F., Tu L.-W., and Zydzik G. J. "Elimination of heterojunction band discontinuities" US Patent No. 5,170,407 (1992b)

Schubert E. F. "Light-emitting diode with omni-directional reflector" US Patent application 60/339,335 (2001)

Schubert E. F. "Light-emitting diode with omni-directional reflector" US Patent 6,784,462; issued Aug. 31 (2004)

Sheu J. K., Su Y. K., Chi G. C., Jou M. J., and Chang C. M. " Effects of thermal annealing on the indium tin oxide Schottky contacts of n-GaN" *Appl. Phys. Lett.* **72**, 3317 (1998)

Shin J. H., Shin S. H., Park J. I., and Kim H. H. "Properties of dc magnetron sputtered indium tin oxide films on polymeric substrates at room temperature" *J. Appl. Phys.* **89**, 5199 (2001)

Smith G. M., Forbes D. V., Coleman J. J., and Verdeyen J. T. "Optical properties of reactive ion etched corner reflector strained-layer InGaAs–GaAs–AlGaAs quantum-well lasers" *IEEE Photonics Technol. Lett.* **5**, 873 (1993)

Streubel K., personal communication (2000)

Tu L. W., Schubert E. F., Kopf R. F., Zydzik G. J., Hong M., Chu S. N. G., and Mannaerts J. P. "Vertical cavity surface emitting lasers with semitransparent metallic mirrors and high quantum efficiencies" *Appl. Phys. Lett.* **57**, 2045 (1990)

Weber M. F., Stover C. A., Gilbert L. R., Nevitt T. J., and Ouderkirk A. J. "Giant birefringent optics in multilayer polymer mirrors" *Science* **287**, 2451 (2000)

Xi Y., Kim J. K., Mont F., Gessmann Th., Luo H., and Schubert E. F. "Quantitative assessment of diffusivity and specularity of surface-textured reflectors for light extraction in light-emitting diodes" manuscript in preparation (2006)

Yariv A. *Quantum Electronics* 3rd edition (John Wiley and Sons, New York, 1989)

11

Packaging

11.1 Low-power and high-power packages

Virtually all LEDs are mounted in a package that provides two electrical leads, a transparent optical window for the light to escape, and, in power packages, a thermal path for heat dissipation. The chip-encapsulating material advantageously possesses high optical transparency, a high refractive index, chemical inertness, high-temperature stability, and hermeticity. The refractive index contrast between the semiconductor and air is reduced by including an encapsulant thereby increasing the light extraction efficiency. Virtually all encapsulants are polymers with a typical refractive index of 1.5 to 1.8. A reduced index contrast at the semiconductor surface increases the angle of total internal reflection thereby enlarging the light escape cone and the extraction efficiency.

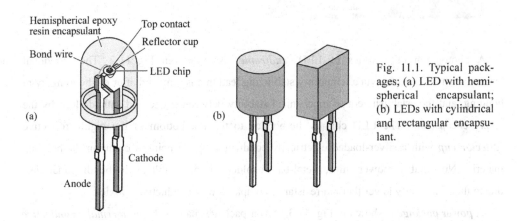

Fig. 11.1. Typical packages; (a) LED with hemispherical encapsulant; (b) LEDs with cylindrical and rectangular encapsulant.

A *low-power package* is shown in Fig. 11.1 (a). The active device is die-bonded or soldered to the bottom of a cup-like depression ("reflector cup") in one of the lead wires (usually the cathode lead). A bond wire connects the LED top contact to the other lead wire (usually the

11 Packaging

anode lead). The LED package shown in the figure is frequently referred to as a "5 mm" or "T1-3/4" package.

In low-power LEDs, the encapsulant has the shape of a hemisphere, as shown in Fig. 11.1 (a), so that the angle of incidence at the encapsulant–air interface is always normal. As a result, total internal reflection does not occur at the encapsulant–air interface. There are types of LEDs that do *not* have a hemispherical shape for the encapsulant. Some LEDs have a rectangular or cylindrical shape with a planar front surface. Examples of such shapes are shown in Fig. 11.1 (b). Planar-surface LEDs are frequently used under circumstances where the intended viewing angle is close to normal incidence or where the LED is intended to blend in with a planar surface. Encapsulants provide protection against unwanted mechanical shock, humidity, and chemicals. The encapsulant also stabilizes the anode and cathode lead, the LED chip and bonding wire.

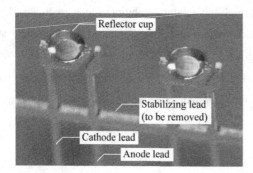

Fig. 11.2. Leadframe of a conventional 5 mm package for mounting and connecting LED chips. The stabilizing lead is cut off once mechanical stability between the anode and cathode lead has been established by the epoxy encapsulant.

A photograph of a series of LED *leadframes* is shown in Fig. 11.2. The individual leadframes are connected via a temporary stabilizing lead that is removed after die bonding, wire bonding, and the establishment of mechanical stability between anode and cathode lead by the epoxy encapsulant. The LED chip is die-bonded to the flat bottom of the highly reflective *reflector cup* with a silver-loaded electrically conductive epoxy being a common die-bonding material. Note that for power chips, metal-based solders are preferable as die-bonding materials due to their inherently lower thermal resistance compared with conductive epoxies.

A *power package* is shown in Fig. 11.3. Power packages have a *direct, thermally conductive path* from the LED chip, through the package, to a heat sink, e.g. a printed circuit board. The power package shown in the figure has several advanced features. Firstly, the package contains an Al or Cu heatsink slug with low thermal resistivity to which the LED submount is soldered by a metal-based solder. Secondly, the chip is encapsulated with silicone. Because standard silicone retains mechanical softness in its cured state, the silicone encapsulant is covered with a plastic cover that also serves as lens. Thirdly, the chip is directly mounted on a Si submount that includes electrostatic discharge protection (ESD).

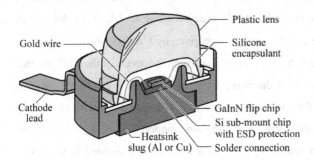

Fig. 11.3. Cross section through high-power package. The heatsink slug can be soldered to a printed circuit board for efficient heat removal. This package, called the *Barracuda package*, was introduced by Lumileds Corp. (adapted from Krames, 2003).

Photographs of the high-power package are shown in Fig. 11.4 including a high-magnification micrograph, Fig. 11.4 (b), revealing interdigitated contacts of a GaN-based LED. Fig. 11.4 (c) shows the package soldered to a printed-circuit board that possesses high thermal conductivity for efficient cooling (LED Museum, 2003). Several light-emitting diodes located on a single chip may be interconnected in series to increase operating voltage and decrease operating current of a device (Krames *et al.*, 2002, 2003).

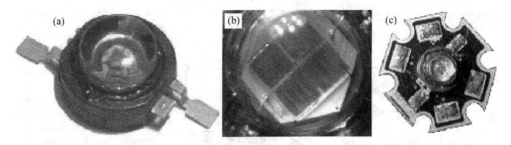

Fig. 11.4. (a) High-power package; (b) LED die in package; (c) package on printed circuit board with high thermal conductivity ((a) after Krames, 2003; (b), (c) after after LED Museum, 2003).

11.2 Protection against electrostatic discharge (ESD)

Electrostatic discharge (ESD) can be a major failure mechanism for electronic and optoelectronic components (Voldman, 2004). Consider that a charge $+Q$ is brought into contact with one of the diode electrodes. Consider further that the charge $+Q$ is discharged uniformly over a time Δt, so that a current of $I = +Q/\Delta t$ flows through the device.

Let us first assume that the charge is brought into contact with the *cathode* of an LED and that the anode is grounded. The current will discharge with the diode in the reverse polarity. Taking the equivalent circuit of the reverse-biased diode as a capacitor C and a parallel resistor R_p, the voltage across the p-n junction in the steady state will rise to IR_p. Thus the energy dissipated in the device during the reverse discharge is then given by $I^2 R_p \Delta t$.

Let us next assume that the charge is brought into contact with the *anode* of the LED and that the cathode is grounded. The current will discharge with the diode in the forward polarity. Taking the equivalent circuit of the forward-biased diode as a voltage source with voltage V_{th} with a series resistor R_s, the steady-state voltage across the p-n junction is $V_{th} + IR_s$, which, in the limit of a high current, can be approximated by IR_s. Thus the energy dissipated in the device during the forward discharge is approximately $I^2 R_s \Delta t$.

Because $I^2 R_p \Delta t \gg I^2 R_s \Delta t$, it is evident that the energy dissipated per reverse discharge event is much higher than the energy dissipated per forward discharge event, suggesting that *reverse* discharges are more damaging than *forward* discharges. This contention has indeed been confirmed by experiments (Wen *et al.*, 2004).

Wide-bandgap diodes (such as GaN-based diodes) are particularly prone to ESD failures, due to inherently high values of R_p (low reverse saturation currents and high breakdown voltages). This has spurred the development of ESD protection circuits for III–V nitride diodes (Steigerwald *et al.*, 2002; Sheu, 2003).

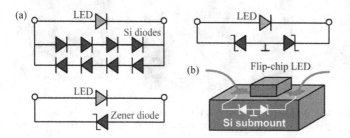

Fig. 11.5. (a) Electrostatic discharge (ESD) protection circuits using multiple Si p-n junctions, one Zener diode, and two Zener diodes. (b) ESD protection integrated into a Si submount (two-Zener diode circuit after Lumileds, 2004).

Electrostatic discharge-protection circuits can consist of a series of Si diodes, one Si Zener diode, or two Si Zener diodes, as shown in Fig. 11.5, with two Zener diodes being a common configuration (Steigerwald *et al.*, 2002; Lumileds, 2004). The current caused by an electrostatic discharge will bypass the LED and flow through the ESD protection circuit, particularly for reverse discharges. ESD protection circuits have been integrated with Si submounts (Steigerwald *et al.*, 2002; Lumileds, 2004). Using one or two Zener diodes or placing several Si diodes in series increases the threshold voltage of the ESD circuit to values beyond the turn-on voltage of the LED. Thus, under normal operating conditions, the current through the ESD circuit is negligibly small.

Sheu (2003) proposed a Schottky diode integrated with the LED on the same chip. The structure, shown in Fig. 11.6 (a), consists of a large-area p-n junction diode and, separated by a deep trench, a small-area Schottky diode. The Schottky diode, fabricated on the n-type buffer layer of a GaInN LED, is forward biased when the LED is biased in the reverse polarity. For reverse electrostatic discharges, the current flows mostly through the Schottky diode, thereby bypassing the p-n junction and preventing damage to the p-n junction. For forward electrostatic discharges, the current flows through the p-n junction. In an alternative structure, shown in Fig. 11.6 (b), the Schottky diode is replaced by a p-n junction diode (Cho, 2005).

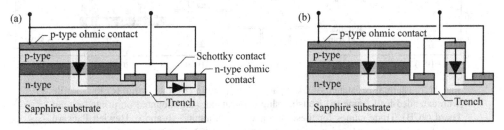

Fig. 11.6. On-chip ESD protection using (a) a small-area Schottky diode on the n-type buffer layer of a GaInN device and (b) a small-area p-n junction diode (Schottky diode circuit after Sheu, 2003).

11.3 Thermal resistance of packages

The thermal resistance of LED packages together with the maximum temperature of operation determines the maximum thermal power that can be dissipated in the package. The maximum temperature of operation may be determined by reliability considerations, by the degradation of the encapsulant, and by internal-quantum-efficiency considerations. Several types of LED packages and their thermal resistance are shown in Fig. 11.7 (Arik *et al.*, 2002). Early LED packages, introduced in the late 1960s and still used for low-power packages at the present time, have a high thermal resistance of about 250 K/W. Packages using **heatsink slugs** made of Al or Cu that transfer heat from the chip directly to a printed circuit board (PCB), which in turn spreads the heat, have thermal resistances of 6–12 K/W. It is expected that thermal resistances of

< 5 K/W will be achieved for advanced passively cooled power packages.

Note that the packages shown in Fig. 11.7 do not use *active cooling*, i.e. fan cooling. Heatsinks with cooling fins and fan are commonly used to cool electronic microchips including Si CMOS microprocessors. They have thermal resistances < 0.5 K/W. The use of active cooling devices would reduce the power efficiency of LED-based lighting systems.

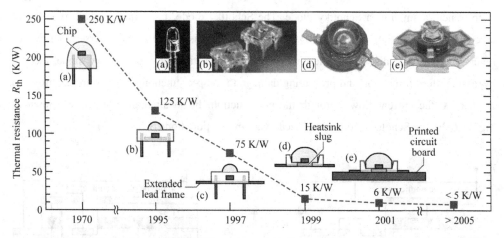

Fig. 11.7. Thermal resistance of LED packages: (a) 5mm (b) low-profile (c) low-profile with extended lead frame (d) heatsink slug (e) heatsink slug mounted on printed circuit board (PCB). Trade names for these packages are "Piranha" (b and c, Hewlett Packard Corp.), "Barracuda" (d and e, Lumileds Corp.), and "Dragon" (d and e, Osram Opto Semiconductors Corp.) (adapted from Arik *et al.*, 2002).

11.4 Chemistry of encapsulants

Encapsulants have several requirements including high transparency, high refractive index, chemical stability, high-temperature stability, and hermeticity. All encapsulants are based on polymers, several of which are shown in Fig. 11.8. A simple polymer molecule consisting of a hydrocarbon chain is shown in Fig. 11.8 (a). Branching and cross linking the polymer molecule results in rubber compounds as shown in Fig. 11.8 (b). Such rubber compounds lack transparency and cannot be used as LED encapsulants. However, it is well known that oxides are frequently transparent. In fact, all encapsulants used for LEDs contain oxygen.

A common encapsulant is *epoxy resin* (also called epoxy), which remains transparent and does not show degradation over many years for long-wavelength visible-spectrum and IR LEDs. However, it has been reported that epoxy resins lose transparency in LEDs emitting at shorter

11.4 Chemistry of encapsulants

wavelengths, i.e. in the blue, violet, and UV (Barton et al., 1998). Epoxy resins are chemically stable up to temperatures of about 120 °C. Prolonged exposure to temperatures greater than 120 °C leads to yellowing (loss of transparency).

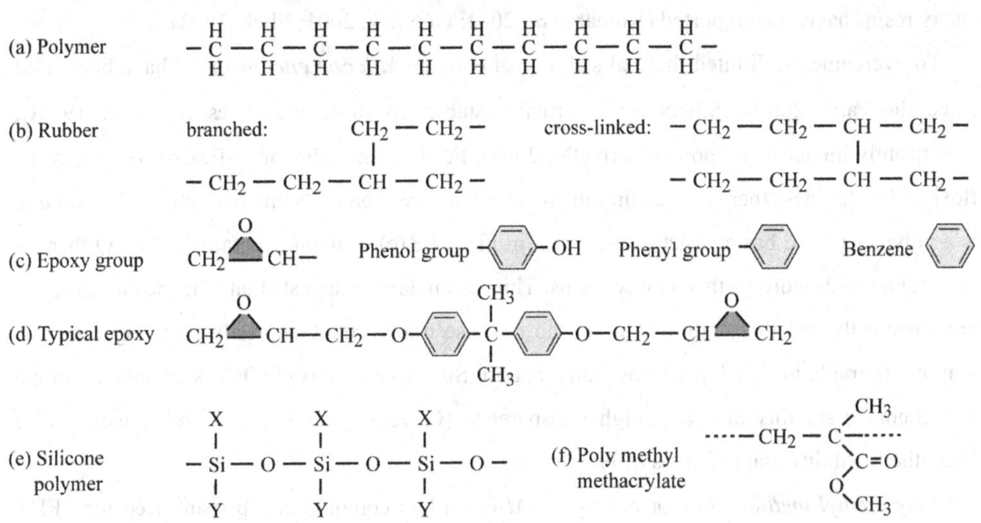

Fig. 11.8. Chemical structures of polymers. Epoxy resins, silicone polymers, and poly methyl methacrylate (PMMA) are used as LED encapsulants. In the silicone structure, X and Y represent atoms or molecules such as H, CH_3 (methyl), C_6H_5 (phenyl).

The *epoxy group* shown in Fig. 11.8 (c) contains an O atom attached to two C atoms already bonded to each other. Such a three-membered ring consisting of one oxygen and two carbon atoms is part of the epoxy resin structure shown in Fig. 11.8 (d). Solid epoxy resins are formed by stoichiometrically mixing two liquid compounds, an epoxide with another compound, a resin, having two hydroxyl groups. *Resins* are oil-like substances that frequently have phenol groups. The phenol group, $-C_6H_4-OH$, is derived from the phenyl group, $-C_6H_5$ by removing one H atom and replacing it with the hydroxyl group, $-OH$. The phenyl group is derived from benzene, the well-known six-atom carbon ring, C_6H_6, by removing one H atom. Under a thermal-setting process, the epoxide groups co-polymerize with hydroxyl groups of the resin.

Typical epoxy resins for encapsulation are thermally curable two-part liquid systems consisting of bisphenol-A based or cycloaliphatic epoxide and anhydride (Kumar et al., 2001). The formation of the epoxy resin requires a short high-temperature cure (120 °C). The two-part system has to be in stoichiometric proportions. Resin-rich compositions lead to lower glass transition temperatures while hardener-rich compositions may lead to discoloration of the encapsulant. The refractive index of epoxy resin is near 1.6. Besides being transparent, epoxy resin is noted for its good mechanical properties and good thermal stability. However, prolonged exposure of the epoxy to temperatures exceeding 120 °C will lead to discoloration and loss of transparency. In addition to thermo-setting epoxy resins, UV-curable and microwave-curable epoxy resins have been reported (Kumar *et al.* 2001; Gorczyk, 2001; Flick, 1993).

To overcome the limited thermal stability of epoxies, **silicone encapsulants** have been used since the early 2000s. Silicone is thermally stable up to temperatures of about 190 °C, significantly higher than epoxies (Crivello, 2004). Furthermore silicone is flexible (and remains flexible for decades) thereby reducing the mechanical stress on the semiconductor chip. **Silicone** is a polymer whose basic structure is shown in Fig. 11.8 (e). Silicone contains Si and O thereby resembling SiO_2 more so than epoxy resins. This resemblance suggests that silicone encapsulants are chemically and thermally stable and do not lose transparency as easily as epoxy resins. It may be desirable to develop encapsulants that are SiO_2-like because SiO_2 has excellent thermal and chemical stability and very high transparency (Crivello, 2004). On the other hand, silica lacks the flexibility that silicones offer.

Poly methyl methacrylate or briefly ***PMMA*** is a less common encapsulant used for LEDs. The chemical structure of methyl methacrylate, the elementary cell of PMMA, is shown in Fig. 11.8 (f). PMMA is also known under the name of acrylic glass and under the product name Plexiglas. The relatively low refractive index of PMMA ($\bar{n} = 1.49$ in the wavelength range 500–650 nm) results in a limited extraction efficiency when used with high-index semiconductors.

11.5 Advanced encapsulant structures

Graded-index encapsulants consisting of several layers with different refractive indices were demonstrated by Lee *et al.* (2004). The layer of highest refractive index is in contact with the semiconductor chip. The outer layers of the encapsulant have lower refractive indices. Extraction efficiencies exceeding those with a constant refractive index can be attained by using such refractive-index graded encapsulants.

Encapsulants containing **mineral diffusers** cause light to reflect, refract, and scatter, thereby randomizing the propagation direction and isotropizing the far-field distribution. For multi-color devices (e.g. multi-chip white LEDs), mineral diffusers uniformize the color distribution. Mineral diffusers are optically transparent substances, such as TiO_2, CaF_2, SiO_2, $CaCO_3$, and $BaSO_4$, with a refractive index different from that of the encapsulant (Reeh *et al.*, 2003).

Encapsulants containing **nanoparticles** with a high refractive index (e.g. titania, magnesia, yttria, zirconia, alumina, GaN, AlN, ZnO, ZnSe) have been proposed by Lester *et al.* (1998). Nanoparticles embedded into a host (usually a polymer) do not scatter light if they are uniformly distributed and if their size is much smaller than the wavelength. The refractive index of the of the nanoparticle-loaded encapsulant is given by

$$\bar{n} = \frac{\bar{n}_{host} V_{host} + \bar{n}_{nano} V_{nano}}{V_{host} + V_{nano}} \qquad (11.1)$$

where V_{host} and V_{nano} refer to the volume of the host and nanoparticles, respectively. For high loading factors, the refractive index of the encapsulant can significantly exceed that of the host, thereby enlarging the semiconductor escape cone and increasing the extraction efficiency.

References

Arik M., Petroski J., and Weaver S. "Thermal challenges in the future generation solid state lighting applications: light emitting diodes" *Eighth Intersociety Conference on Thermal and Thermomechanical Phenomena in Electronic Systems* (Cat. No.02CH37258) May 30–June 1 2002, p. 113 (IEEE, Piscataway NJ, 2002)

Barton D. L., Osinski M., Perlin P., Helms C. J., and Berg N. H. "Life tests and failure mechanisms of GaN/AlGaN/InGaN light-emitting diodes" *Proc. SPIE* **3279**, 17 (1998)

Cho J., Samsung Advanced Institute of Technology, Suwon, Korea, personal communication (2005)

Crivello J. V., Rensselaer Polytechnic Institute, personal communication (2004)

Flick E. W. *Epoxy Resins, Curing Agents, Compounds, and Modifiers: An Industrial Guide* (Noyes Data Corporation/Noyes Publications, Park Ridge NJ, 1993)

Gorczyk J., Bogdal D., Pielichowski J., and Penczek P. "Synthesis of high molecular weight epoxy resins under microwave irradiation" *Fifth International Electronic Conference on Synthetic Organic Chemistry* (ECSOC-5), http://www.mdpi.org/ecsoc-5.htm, 1–30 (September 2001)

Krames M. R., Steigerwald D. A., Kish Jr. F. A., Rajkomar P., Wierer Jr. J. J., and Tan T. S. "III-nitride light-emitting device with increased light generating capability" US Patent 6,486,499 B1 (2002)

Krames M. R. "Overview of current status and recent progress of LED technology" *US Department of Energy Workshop "Solid State Lighting – Illuminating the Challenges"* Crystal City, VA, Nov. 13–14, 2003

Kumar R. N., Keem L. Y., Mang N. C., and Abubakar A. "Ultraviolet radiation curable epoxy resin encapsulant for light-emiting diodes" *4th International Conference on Mid-Infrared Optoelectronics Materials and Devices* (MIOMD) (2001)

LED Museum, http://ledmuseum.home.att.net/agilent.htm (2003)

Lee B. K., Goh K. S., Chin Y. L., and Tan C. W. "Light emitting diode with gradient index layering" US Patent 6,717,362 B1 (2004)

Lester S. D., Miller J. N., and Roitman D. B. "High refractive index package material and light emitting device encapsulated with such material" US Patent 5,777,433 (1998)

Lumileds Corporation *Luxeon reliability* Application Brief **AB25**, 11 (2004)

Reeh U., Höhn K., Stath N., Waitl G., Schlotter P., Schneider J., and Schmidt R. "Light-radiating semiconductor component with luminescence conversion element" US Patent 6,576,930 B2 (2003)

Sheu J.-K. "Group III–V element-based LED having ESD protection capacity" US Patent 6,593,597 B2 (2003)

Steigerwald D. A., Bhat J. C., Collins D., Fletcher R. M., Holcomb M. O., Ludowise M. J., Martin P. S., and Rudaz S. L. "Illumination with solid state lighting technology" *IEEE J. Sel. Top. Quantum Electron.* **8**, 310 (2002)

Voldman S. H. *ESD: Physics and Devices* (John Wiley and Sons, New York, 2004)
Wen T. C., Chang S. J., Lee C. T., Lai W. C., and Sheu J. K. "Nitride-based LEDs with modulation-doped AlGaN-GaN superlattice structures" *IEEE Trans. Electron Dev.* **51**, 1743 (2004)

12

Visible-spectrum LEDs

Originally, LEDs were exclusively used for low-brightness applications such as indicator lamps. In these applications, the efficiency and the overall optical power of the LED are *not* of primary importance. However, in more recent applications, for example traffic light applications, the light emitted by LEDs must be seen even in bright sunlight and from a considerable distance. LEDs with high efficiency and brightness are required for such applications.

In this chapter, low-brightness as well as high-brightness LEDs are discussed. GaAsP and nitrogen-doped GaAsP LEDs are suitable only for low-brightness applications. AlGaAs LEDs are suitable for low- as well as high-brightness applications. AlGaInP and GaInN LEDs are used in high-brightness applications.

12.1 The GaAsP, GaP, GaAsP:N, and GaP:N material systems

The $GaAs_{1-x}P_x$ and $GaAs_{1-x}P_x$:N material system is used for emission in the red, orange, yellow, and green wavelength range. The GaAsP system is lattice mismatched to GaAs substrates, resulting in a relatively low internal quantum efficiency. As a result these LEDs are suitable for low-brightness applications only.

$GaAs_{1-x}P_x$ was one of the first material systems used for visible-spectrum LEDs (Holonyak and Bevacqua 1962; Holonyak *et al.* 1963, 1966; Pilkuhn and Rupprecht 1965; Wolfe *et al.*, 1965; Nuese *et al.* 1966). In the early 1960s, GaAs substrates were already available. Bulk growth of GaAs substrates was initiated in the 1950s and epitaxial growth by LPE and VPE started in the 1960s. As phosphorus is added to GaAs, the ternary alloy $GaAs_{1-x}P_x$, or briefly GaAsP, is formed. The addition of phosphorus increases the bandgap of GaAs, which emits in the infrared at 870 nm. The visible wavelength range starts at about 750 nm, so that a small amount of phosphorus is sufficient to attain visible-spectrum light emitters. Note, however, that the sensitivity of the human eye is low at the edges of the visible spectrum.

A significant problem with GaAsP LEDs is the lattice mismatch between the GaAs substrate

12 Visible-spectrum LEDs

and the GaAsP epitaxial layer. A large mismatch exists between GaAs and GaP (about 3.6%) so that many misfit dislocations occur when the critical thickness of GaAsP on GaAs is exceeded. As a result, the luminescence efficiency decreases substantially in GaAsP with increasing phosphorus content. GaAsP LEDs are therefore useful for low-brightness applications only.

It was realized early in the GaAsP work that the lattice mismatch between the GaAs substrate and the GaAsP epilayer reduces the radiative efficiency. It was also found that the radiative efficiency of the active p-n junction layer strongly depends on the growth conditions and, in particular, on the thickness of the GaAsP buffer layer (Nuese et al., 1969). A thick buffer layer reduces the dislocation density by annihilation of misfit dislocations. However, the dislocation density does not approach the low dislocation density of GaAs substrates, so that even with thick GaAsP buffer layers the dislocation density is substantial.

The band structure of GaAs, GaAsP, and GaP is shown schematically in Fig. 12.1. The figure shows that GaAsP is a direct-gap semiconductor for low phosphorus mole fractions. Beyond the direct–indirect crossover occurring at phosphorus mole fractions of about 45–50%, the semiconductor becomes indirect and the radiative efficiency drops rapidly (Holonyak et al., 1963, 1966). GaP is an indirect-gap semiconductor and therefore is unsuitable as an efficient LED material.

GaAsP and GaP LEDs are frequently doped with isoelectronic impurities such as nitrogen (Grimmeiss and Scholz, 1964; Logan et al., 1967a, 1967b, 1971; Craford et al., 1972; Groves and Epstein 1977; Groves et al. 1978a, 1978b). The isoelectronic impurities form an optically active level within the forbidden gap of the semiconductor so that carriers recombine radiatively via the nitrogen levels, as indicated in Fig. 12.1.

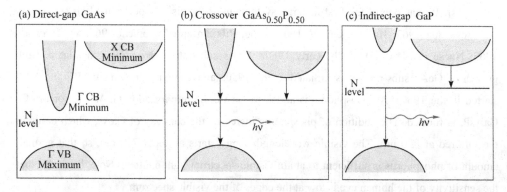

Fig. 12.1. Schematic band structure of GaAs, GaAsP, and GaP. Also shown is the nitrogen level. At a P mole fraction of 45–50%, the direct–indirect crossover occurs.

Isoelectronically doped LEDs are also interesting from a fundamental point of view. They are one of the first practical applications of Heisenberg's uncertainty principle. Isoelectronic impurities have an electronic wave function that is strongly localized in position space (small Δx). Therefore, the wave function is delocalized in momentum space (large Δp). Since the level is delocalized in momentum space, two vertical transitions can occur via the isoelectronic trap, with one of them being radiative. Physically speaking, the change in momentum, occurring when an electron makes a transition from the indirect X valley of the conduction band to the central Γ valley of the valence band, is absorbed by the isoelectronic impurity atom.

The emission wavelength of undoped and nitrogen-doped GaAsP is shown in Fig. 12.2 (Craford et al., 1972). The emission energy of GaAsP and GaP doped with the isoelectronic impurity nitrogen is below the bandgap of the semiconductor. Figure 12.2 illustrates that the emission energy is about 50–150 meV below the bandgap of the semiconductor. As a result, reabsorption effects are much less likely in nitrogen-doped structures compared with LEDs based on band-edge emission. This is a substantial advantage of LEDs doped with isoelectronic impurities.

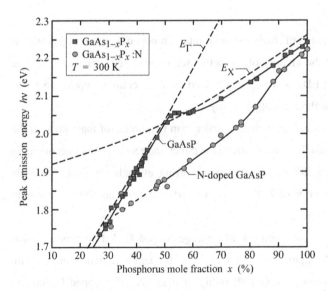

Fig. 12.2. Room-temperature peak emission energy versus alloy composition for undoped and nitrogen-doped GaAsP LEDs injected with a current density of 5 A/cm^2. Also shown is the energy gap of the direct-to-indirect (E_Γ-to-E_X) transition. The direct–indirect crossover occurs at $x \approx 50\%$ (after Craford et al., 1972).

Groves et al. (1978a, 1978b) showed that this advantage is particularly pronounced if only the active region is doped with nitrogen. In this case, the region of the p-n junction plane and the regions located within the carrier diffusion lengths from the junction plane are doped with nitrogen. Other regions, such as the confinement and window layers, are not doped with the isoelectronic impurity, so that reabsorption of light by the isoelectronic impurities is limited to the narrow active region. Quantum efficiencies of several percent can be attained with GaP:N LEDs in which the nitrogen doping is limited to the active region.

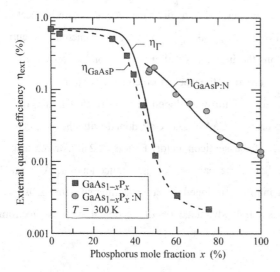

Fig. 12.3. Experimental external quantum efficiency of undoped and N-doped GaAsP versus the P mole fraction. Also shown is the calculated direct-gap (Γ) transition efficiency, η_Γ, and the calculated nitrogen (N) related transition efficiency, η_N (solid lines). Note that the nitrogen-related efficiency is higher than the direct-gap efficiency in the indirect bandgap ($x > 50\%$) regime (after Campbell et al., 1974).

The external quantum efficiency of undoped and nitrogen-doped GaAsP is shown in Fig. 12.3 (Campbell, 1974). Only the vicinity of the active region is doped with nitrogen. The efficiency of the nitrogen-doped LEDs is strongly enhanced over the entire composition range compared with the GaAsP LEDs without nitrogen doping.

Note that the GaAsP LED efficiency decreases by more than two orders of magnitude in the composition range $x = 40$–60%. This decrease is due to the direct–indirect crossover occurring in GaAsP and due to the increasing dislocation density occurring at higher phosphorus mole fractions. At a phosphorus mole fraction of 75%, the GaAsP external quantum efficiency is only 0.002%.

The external quantum efficiency of undoped and nitrogen-doped GaAsP versus emission wavelength is shown in Fig. 12.4. Again, only the vicinity of the active region is doped with nitrogen. The efficiency of nitrogen-doped GaAsP is higher than that of undoped GaAsP, in particular in the orange, yellow, and green wavelength range where the improvement is a factor of 2–5. In the red wavelength range, the undoped and nitrogen-doped GaAsP LEDs have similar efficiencies.

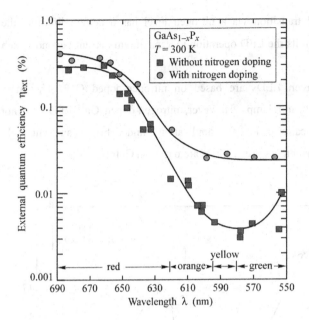

Fig. 12.4. External quantum efficiency versus emission wavelength in undoped and nitrogen-doped $GaAs_{1-x}P_x$ (after Groves *et al.*, 1978a, 1978b).

The ratio of external quantum efficiencies of undoped and nitrogen-doped GaAsP LEDs is shown in Fig. 12.5. It is inferred from the figure that nitrogen-doped devices have a higher efficiency over the entire composition range.

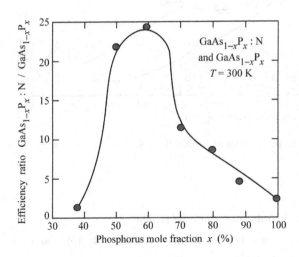

Fig. 12.5. Efficiency ratio between nitrogen-doped and undoped $GaAs_{1-x}P_x$ at 300 K (after Groves *et al.*, 1978a, 1978b).

The brightness of LEDs based on isoelectronic impurity transitions is limited by the *finite solubility* of nitrogen. For example, nitrogen is soluble in GaP up to nitrogen concentrations of

12 Visible-spectrum LEDs

about 10^{20} cm^{-3}. Since an optical transition via a nitrogen level has a certain lifetime, the maximum nitrogen concentration limits the LED operation to a maximum current beyond which the LED efficiency decreases.

Commercial low-brightness green LEDs are based on nitrogen-doped GaP. The main application of GaP:N LEDs is indicator lamps. However, nitrogen-doped GaP LEDs are not suitable for high-brightness applications, i.e. for applications under bright ambient light conditions such as sunlight. High-brightness green LEDs are based on GaInN.

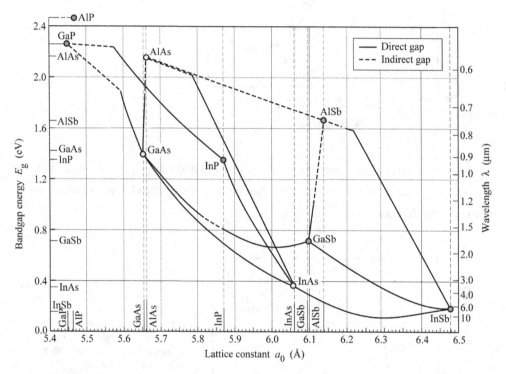

Fig. 12.6. Bandgap energy and lattice constant of various III–V semiconductors at room temperature (after Tien, 1988).

12.2 The AlGaAs/GaAs material system

The Al$_x$Ga$_{1-x}$As/GaAs material system was developed in the 1970s and early 1980s and it was the first material system suitable for high-brightness LED applications (for a review, see Steranka, 1997). Owing to the very similar atomic radii of Al (1.82 Å) and Ga (1.81 Å), the material system Al$_x$Ga$_{1-x}$As (or briefly AlGaAs) is lattice matched to GaAs for all Al mole fractions. The lack of dependence of the lattice constant on the Al mole fraction can be inferred from Fig. 12.6, which shows the energy gap and lattice constant of several III–V semiconductors and of its ternary and quaternary alloys as a function of the lattice constant (adopted from Tien, 1988).

GaAs and $Al_xGa_{1-x}As$ for Al mole fractions $x < 0.45$ are direct-gap semiconductors. The energy gap of $Al_xGa_{1-x}As$ versus the Al mole fraction is shown in Fig. 12.7 (Casey and Panish, 1978). For Al mole fractions $x < 45\%$, the Γ conduction-band valley is the *lowest* minimum and the semiconductor has a *direct* gap. For $x > 45\%$, the X valleys are the *lowest* conduction-band minimum and the semiconductor becomes *indirect*.

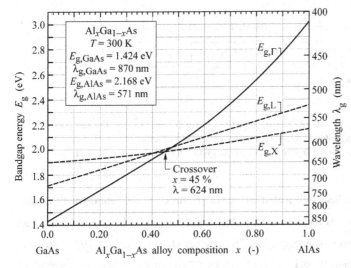

Fig. 12.7. Bandgap energy and emission wavelength of AlGaAs at room temperature. E_Γ denotes the direct gap at the Γ point and E_L and E_X denote the indirect gap at the L and X point of the Brillouin zone, respectively (adapted from Casey and Panish, 1978).

$E_{g,\Gamma}/eV = 1.424 + 1.247\,x$ $(0 \le x \le 0.45)$

$E_{g,\Gamma}/eV = 1.424 + 1.247x +$ $1.147(x - 0.45)$ $(0.45 \le x \le 1.0)$

$E_{g,L}/eV = 1.708 + 0.642\,x$ $(0 \le x \le 1.0)$

$E_{g,X}/eV = 1.900 + 0.125\,x + 0.143\,x^2$ $(0 \le x \le 1.0)$

The AlGaAs material system is suited for high-brightness visible-spectrum LEDs emitting in the red wavelength range. The direct–indirect crossover occurs at a wavelength of 621 nm. At that wavelength, the radiative efficiency of the AlGaAs system becomes quite low due to the direct–indirect transition. To maintain high efficiency, the emission energy must be several kT lower than the bandgap energy at the direct–indirect crossover point.

There are several possible strategies for AlGaAs-based red LEDs, including $Al_xGa_{1-x}As$ bulk active regions, $Al_xGa_{1-x}As/GaAs$ quantum well active regions, and $Al_xGa_{1-x}As/Al_yGa_{1-y}As$ ($x > y$) double heterostructure active regions. The first possibility, $Al_xGa_{1-x}As$ bulk active regions, lacks the advantages of a heterostructure and this approach is therefore not used in high-brightness LEDs. The two other possibilities are more attractive due to the employment of heterostructures. The quantum well and double heterostructure active region is used in high-efficiency red LEDs and the two structures are shown schematically in Fig. 12.8. In the $Al_xGa_{1-x}As/GaAs$ quantum well case shown in Fig. 12.8 (a), size quantization is used to increase the emission energy. In the case of the $Al_xGa_{1-x}As/Al_yGa_{1-y}As$ double heterostructures shown in Fig. 12.8 (b), AlGaAs is used for both the barrier region and the well region. A drawback of the $Al_xGa_{1-x}As/GaAs$ quantum well active regions is the requirement of very thin GaAs quantum wells clad by $Al_xGa_{1-x}As$ barriers. Vertical transport in multi-quantum well (MQW) structures can lead to non-uniform carrier distribution in the MQW active region unless the barriers are very thin. Consequently, the $Al_xGa_{1-x}As/Al_yGa_{1-y}As$ double heterostructure approach is usually preferred.

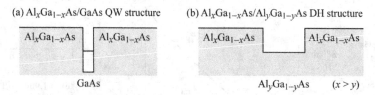

Fig. 12.8. Band diagrams of AlGaAs/GaAs structures suited for emission in the red part of the visible spectrum. (a) AlGaAs/GaAs quantum well (QW) structure with a thin GaAs well. (b) AlGaAs/AlGaAs double heterostructure (DH) with an AlGaAs active region.

AlGaAs/GaAs LEDs have been fabricated as homostructures, single heterostructures, and double heterostructures (Nishizawa et al., 1983). The most efficient AlGaAs red LEDs are double-heterostructure transparent-substrate (DH-TS) devices (Ishiguro et al., 1983; Steranka et al. 1988; Ishimatsu and Okuno, 1989). *AlGaAs DH-TS LEDs* are grown on temporary GaAs substrates and consist of a thick (e.g. 125 µm) $Al_xGa_{1-x}As$ lower confinement layer with an Al mole fraction $x > 60\%$, an $Al_xGa_{1-x}As$ active layer ($x = 35\%$, for red devices), and a thick (e.g. 125 µm) $Al_xGa_{1-x}As$ upper confinement layer, also with an Al mole fraction $x > 60\%$. For devices emitting in the IR, the Al mole fractions of the active and confinement layers can be lower. After epitaxial growth, the absorbing GaAs substrate is removed by polishing and selective wet chemical etching. AlGaAs DH-TS LEDs are more than a factor of 2 brighter than double-heterostructure absorbing-substrate (DH-AS) devices (Steranka et al., 1988).

In the 1980s the growth method of choice for AlGaAs DH-TS LEDs was liquid-phase epitaxy (LPE). This growth method is capable of growing, at a high growth rate, very thick high-quality AlGaAs layers with high Al content. LPE can be scaled up for high-volume production

(Ishiguro et al., 1983; Steranka et al., 1988; Ishimatsu and Okuno, 1989). AlGaAs/GaAs DH-AS LEDs have also been grown by OMVPE (Bradley et al., 1986). However, the OMVPE growth rate is lower than the LPE growth rate. OMVPE growth of thick layers, as required for DH-TS devices, is therefore difficult. Historically AlGaAs DH-TS LEDs were the first high-brightness LEDs suitable for demanding applications such as automotive brake lights and traffic lights, which must be clearly visible under bright ambient conditions.

The reliability of AlGaAs devices is known to be lower than that of AlGaInP devices that do not contain any AlGaAs. High-Al-content AlGaAs layers are subject to oxidation and corrosion, thereby lowering the device lifetime. Dallesasse et al. (1990) reported the deterioration of AlGaAs/GaAs heterostructures by hydrolysis. Cracks, fissures, and pinholes were found after long-term exposure to room environmental conditions, especially for thick AlGaAs layers (> 0.1 μm) with a high Al content such as 85%. The authors found very thin AlGaAs layers (e.g. 20 nm) to be stable, even for Al contents of 100%. Hermetic packaging is required to avoid oxidation and hydrolysis of AlGaAs layers. Steranka et al. (1988) stated that some AlGaAs devices on the market have exhibited severe degradation. However, accelerated aging data taken at 55 °C with an injection current of 30 mA showed no degradation at all after 1000 h of stress. Such a performance requires excellent understanding and control of the device fabrication and packaging process.

12.3 The AlGaInP/GaAs material system

The AlGaInP material system was developed in the late 1980s and early 1990s and today is the primary material system for high-brightness LEDs emitting in the long-wavelength part of the visible spectrum, i.e. in the red, orange, amber, and yellow wavelength range. The AlGaInP material system and AlGaInP LEDs have been reviewed by Stringfellow and Craford (1997), Chen et al. (1997), and Kish and Fletcher (1997). Further reviews and recent developments were published by Mueller (1999, 2000) and Krames et al. (2002).

Figure 12.9 shows the energy gap and the corresponding wavelength versus the lattice constant of AlGaInP (Chen et al., 1997). AlGaInP can be lattice matched to GaAs. Replacing all As atoms in the GaAs lattice by *smaller* P atoms and some of the Ga atoms in the GaAs lattice by *larger* In atoms, forms GaInP, which at the particular composition $Ga_{0.5}In_{0.5}P$, is lattice matched to GaAs. Since Al and Ga have very similar atomic radii, the material $(Al_xGa_{1-x})_{0.5}In_{0.5}P$ is also lattice matched to GaAs.

12 Visible-spectrum LEDs

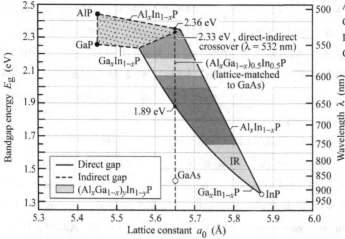

Fig. 12.9. Bandgap energy and corresponding wavelength versus lattice constant of $(Al_xGa_{1-x})_yIn_{1-y}P$ at 300 K. The vertical dashed line shows $(Al_xGa_{1-x})_{0.5}In_{0.5}P$ lattice matched to GaAs (adapted from Chen et al., 1997).

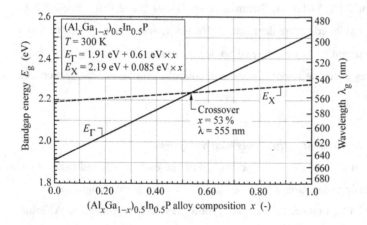

Fig. 12.10. Bandgap energy and emission wavelength of unordered AlGaInP lattice matched to GaAs at room temperature. E_Γ denotes the direct gap at the Γ point and E_X denotes the indirect gap at the X point of the Brillouin zone (after Prins et al., 1995 and Kish and Fletcher, 1997).

According to Chen et al. (1997), $(Al_xGa_{1-x})_{0.5}In_{0.5}P$ has a direct bandgap for $x < 0.5$ and an indirect bandgap for $x > 0.5$. At the crossover point ($x = 0.5$), the bandgap energy is 2.33 eV, corresponding to a wavelength of 532 nm. Kish and Fletcher (1997) compiled data from Prins et al. (1995) and concluded that $(Al_xGa_{1-x})_{0.5}In_{0.5}P$ is a direct-gap semiconductor for Al mole fractions $x < 0.53$. The energy gap versus Al mole fraction is shown in Fig. 12.10 (Prins et al., 1995; Kish and Fletcher, 1997). At Al mole fractions $x < 53\%$, the Γ conduction-band valley is the lowest minimum and the semiconductor has a direct gap. For $x > 53\%$, the X valleys are the lowest conduction-band minimum and the semiconductor becomes indirect. The emission wavelength at the direct–indirect crossover point is approximately 555 nm. The exact wavelength of the crossover point may depend on the degree of atomic ordering present in a particular material (Kish and Fletcher, 1997).

A contour plot of the lattice constant and the energy gap of the AlGaInP materials system is shown in Fig. 12.11 (Chen et al., 1997). The bandgap energy values and the composition of the direct–indirect crossover shown in Fig. 12.11 are slightly different from the data shown in Fig. 12.10, which can be attributed to atomic ordering in AlGaInP. Atomic ordering lowers the bandgap energy by values up to 190 meV (Kish and Fletcher, 1997).

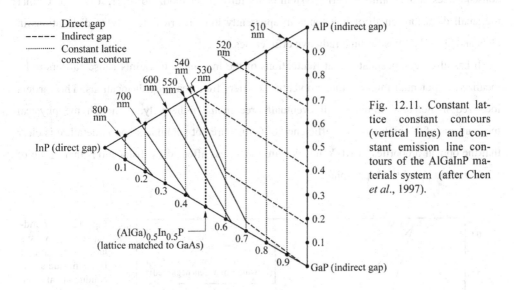

Fig. 12.11. Constant lattice constant contours (vertical lines) and constant emission line contours of the AlGaInP materials system (after Chen et al., 1997).

The AlGaInP material system is suited for high-brightness visible-spectrum LEDs emitting in the red, orange, amber, and yellow wavelength range. At the direct–indirect crossover the radiative efficiency of the AlGaInP system is quite low due to the direct–indirect transition. To maintain high efficiency, the emission energy should be several kT lower than the bandgap energy at the direct–indirect crossover point.

12.4 The GaInN material system

The GaInN material system was developed in the early 1990s. GaInN LEDs emitting in the blue and green wavelength range became commercially available in the late 1990s. To date GaInN is the primary material system for high-brightness blue and green LEDs. The GaInN material system and GaInN LEDs have been reviewed by Nakamura and Fasol (1997) and by Strite and Morkoc (1992).

12 Visible-spectrum LEDs

One of the greatest surprises of the GaInN materials system is its high radiative efficiency despite the presence of a very high concentration of threading dislocations in GaInN/GaN epitaxial films. These threading dislocations are due to the lattice mismatch between the commonly used sapphire and SiC substrates and the GaN and GaInN epitaxial films. Typical densities of the threading dislocations are in the 10^7–10^9 cm^{-2} range.

In the III–V arsenide and III–V phosphide material systems, misfit dislocations have disastrous consequences for the radiative efficiency. The lack of such strongly detrimental consequences in the GaInN material system is not fully understood. However, it is believed that the small diffusion length of holes and an apparently low electrical activity of dislocations in GaN and GaInN allows for high radiative efficiencies.

It has also been postulated that fluctuations of the In content in GaInN cause carriers to be localized in potential minima, thus preventing carriers from reaching dislocations. The carriers localized in potential minima will eventually recombine radiatively. Although the physical mechanisms of the high radiative efficiency of GaInN are not yet understood in detail, it is clear that the optical properties of III–V nitrides are much less affected by dislocations than those of III–V arsenides and III–V phosphides.

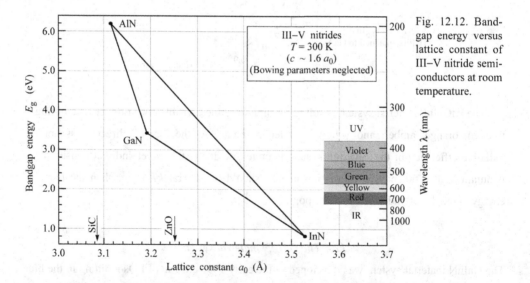

Fig. 12.12. Bandgap energy versus lattice constant of III–V nitride semiconductors at room temperature.

The bandgap energy versus the lattice constant in the nitride material family is shown in Fig. 12.12. Inspection of the figure indicates that GaInN is, in theory, suitable for covering the entire visible spectrum. However, the growth of high-quality GaInN becomes increasingly more difficult as the In composition is increased, in part due to re-evaporation of In from the growth surface. As a result, the GaInN material system is exclusively used for ultraviolet (UV), blue, and green LEDs at the present time and rarely for longer wavelengths.

Prior to the year 2002, the generally accepted value for the InN bandgap energy was 1.9 eV. However, Wu et al. (2002a, 2002b) showed by luminescence measurements that the bandgap of InN is lower, namely between 0.7 and 0.8 eV. Luminescence measurements also indicated that the InN bandgap exhibits an unusual blue shift with increasing temperature.

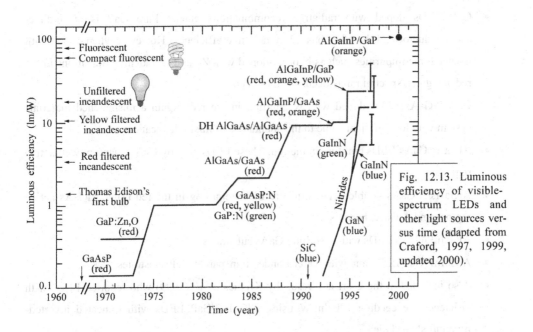

Fig. 12.13. Luminous efficiency of visible-spectrum LEDs and other light sources versus time (adapted from Craford, 1997, 1999, updated 2000).

12.5 General characteristics of high-brightness LEDs

The improvement in luminous efficiency of visible-spectrum LEDs has been truly breathtaking. The advancement of LED efficiency can be compared to the advancement made in Si integrated circuits where the performance increase versus time has been characterized by "Moore's law". This "law" states that the performance of Si integrated circuits doubles approximately every 18 months.

The historical development of the luminous efficiency of visible-spectrum LEDs is shown in Fig. 12.13 (Craford, 1997, 1999). The chart illustrates the modest beginnings of visible-spectrum LED technology which started in the 1960s. If the progress from 1960 to 2000 is assumed to be continuous, then the LED luminous efficiency has doubled every 4 years. The following types of LEDs are shown in the figure.

- GaAsP LEDs grown on GaAs substrates. The GaAsP/GaAs material system is lattice mismatched so that an abundance of misfit dislocations occurs in GaAsP epitaxial films. As a result, these LEDs have a low luminous efficiency (of the order of only 0.1 lm/W). Red GaAsP LEDs are still being manufactured due to the simple epitaxial growth and low fabrication cost.
- GaP LEDs doped with radiative recombination centers. Pure GaP is an indirect semiconductor and therefore has a low radiative efficiency. However, when doped with isoelectronic impurities such as N or co-doped with Zn and O, radiative transitions in the red and green spectral range occur via these centers.
- GaAsP/GaAs LEDs doped with N emitting in the red. Again a mismatched materials system with low efficiency due to the abundance of misfit dislocations.
- AlGaAs/GaAs LEDs emitting in the red. These LEDs employ GaAs quantum well active regions.
- AlGaAs/AlGaAs double heterostructure LEDs emitting in the red using AlGaAs active regions and AlGaAs barriers.
- AlGaInP/GaAs LEDs with absorbing GaAs substrates
- AlGaInP/GaP LEDs and with wafer-bonded transparent GaP substrates.
- Also included in the chart is a result by Krames *et al.* (1999) who reported LEDs with efficiencies exceeding 100 lm/W using AlGaInP/GaP LEDs with truncated inverted-pyramid-shaped dies.
- GaInN LEDs emitting in the blue and green wavelength range.

Figure 12.13 also shows the luminous efficiency of conventional light sources including Edison's first light bulb (1.4 lm/W) and red and yellow filtered incandescent lamps. Inspection of the figure reveals that LEDs outperform filtered red and yellow incandescent lights by a large margin.

The luminous efficiency of high-brightness LEDs and of some low-cost LEDs is shown versus wavelength in Fig. 12.14 (United Epitaxy Corp., 1999). The figure indicates that yellow (590 nm) and orange (605 nm) AlGaInP and green (525 nm) GaInN LEDs are excellent choices for high luminous efficiency devices.

12.5 General characteristics of high-brightness LEDs

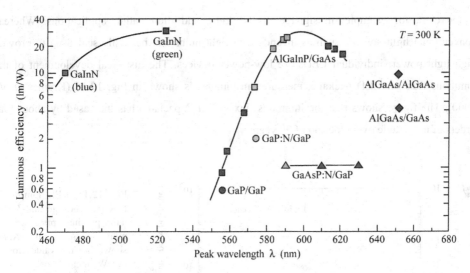

Fig. 12.14. Overview of luminous efficiency of visible LEDs made from the phosphide, arsenide, and nitride material system (adapted from United Epitaxy Corporation, 1999; updated 2000).

The fact that amber (dark yellow or orange–yellow) and orange AlGaInP LEDs provide an excellent luminous efficiency is also due to the high eye sensitivity in this wavelength range. Note that the maximum eye sensitivity occurs in the green at 555 nm, so that a green LED emitting at that wavelength appears brighter than an LED with the same optical power emitting at any other wavelength.

Owing to the high luminous efficiency and the lower manufacturing cost compared with green GaInN LEDs, amber AlGaInP LEDs are used in applications where high brightness and low power consumption are desirable, e.g. in highway signage applications. In the 1980s, such highway signs were made with power-hungry incandescent lamps powered by an electrical generator which, in turn, was powered by a gasoline engine. Today such signs use energy-saving amber LEDs powered by solar cells (during daytime) and batteries (at night).

Figures 12.13 and 12.14 also show low-power and low-cost LEDs such as GaAsP and GaP:N LEDs with much lower luminous efficiency. These LEDs are not suitable for high-brightness applications due to their inherently lower quantum efficiency. The GaAsP LEDs are mismatched to the GaAs substrate and therefore have a low internal efficiency. The GaP:N LEDs also have a low efficiency due to the nitrogen-impurity-assisted nature of the radiative transition.

Not only the *luminous efficiency* but also the *total power* emitted by an LED is of importance for many applications, in particular for applications where a high luminous flux is required. This is the case, for example, in signage, traffic light, and illumination applications. Whereas conventional light sources such as incandescent light bulbs can be easily scaled up to provide high light power, individual LEDs are low-power devices. The historical development of the luminous flux per LED package, measured in lumens, is shown in Fig. 12.15 (Krames et al., 2000). The figure shows that the luminous flux per LED package has increased by about four orders of magnitude over a period of 30 years.

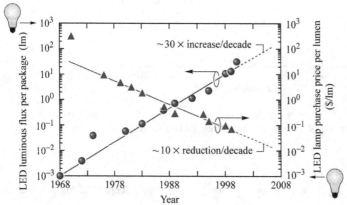

Fig. 12.15. LED luminous flux per package and LED lamp purchase price per lumen versus year. Also shown are the values for a 60 W incandescent tungsten-filament light bulb with a luminous efficiency of ~17 lm/W and a luminous flux of 1000 lm with an approximate price of 1.00 US$ (after Krames et al., 2000).

For comparison, Fig. 12.15 shows the luminous flux and an approximate purchase price of a 60 W incandescent light bulb with a luminous flux of 1000 lm. The figure illustrates that continued progress in the performance and manufacturing cost of LEDs is required to enable LEDs to enter into the general illumination market. Note that the cost shown in the figure is just the purchase price of the lamp and does not include the cost for the electrical power consumed over the lifetime of the lamp. The cost of the electrical power required to run an incandescent light bulb is much higher than the purchase price of the lamp so that efficient light sources can have a cost advantage over incandescent lights even if the initial purchase price is much higher.

12.6 Optical characteristics of high-brightness LEDs

Optical emission spectra of red AlGaInP and green and blue GaInN LEDs are shown in Fig. 12.16 (Toyoda Gosei, 2000). Comparison of the emission spectra reveals that the green LED has a wider emission spectrum than either the blue or the red LED. This can be attributed to the well-known difficulties of growing GaInN with a high In content. It has been found that **In-rich clusters** or **quantum dots** form during the growth of GaInN, especially in GaInN with a high In content. It is known that the formation of such In clusters depends strongly on the growth conditions.

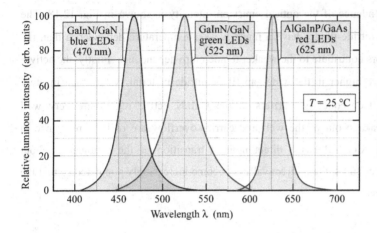

Fig. 12.16. Typical emission spectrum of GaInN/GaN blue, GaInN/GaN green, and AlGaInP/GaAs red LEDs at room temperature (after Toyoda Gosei Corp., 2000).

All LEDs shown in Fig. 12.16 have an active region composed of a ***semiconductor alloy***. Alloy broadening, i.e. the broadening of the emission band due to random fluctuations of the chemical composition of the active material, will lead to spectral broadening that goes beyond the 1.8 kT linewidth expected for thermally broadened emission bands.

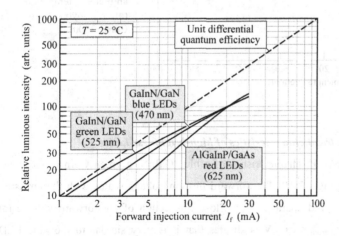

Fig. 12.17. Typical light output power vs. injection current of GaInN/GaN blue, GaInN/GaN green, and AlGaInP/GaAs red LEDs at room temperature (adopted from Toyoda Gosei Corp., 2000).

The light power versus injection current is shown in Fig. 12.17. A linear dependence with unit slope, i.e. unit differential quantum efficiency, is expected for light-versus-current curves in ideal LEDs. The unit-slope line is represented by the dashed line in Fig. 12.17. The mature

AlGaInP LED closely follows the unit-slope line. However, the green LED has a large deviation from the unit differential quantum efficiency slope due to the lower maturity of the GaInN material system, especially with high concentrations of In.

The temperature dependence of the optical emission intensity is shown in Fig. 12.18. The figure reveals that III–V nitride diodes have a much weaker temperature dependence than the AlGaInP LED. Two factors contribute to the weaker temperature dependence. Firstly, the active-to-confinement barriers are higher in the wide-gap III–V nitride material system than in other III–V material systems. Consequently, carriers in the GaInN active region are very well confined. Thus carrier leakage out of the well and carrier overflow are of little relevance in GaInN LEDs. Secondly, AlGaInP has a direct–indirect transition of the bandgap at about 555 nm. At elevated temperatures, the indirect valleys become increasingly populated so that the radiative efficiency decreases.

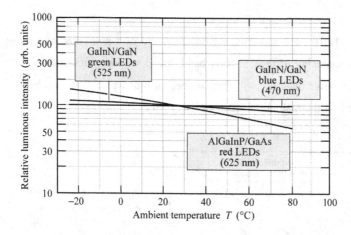

Fig. 12.18. Typical output intensity of GaInN/GaN blue, GaInN/GaN green, and AlGaInP/GaAs red LEDs versus ambient temperature (after Toyoda Gosei Corp., 2000).

12.7 Electrical characteristics of high-brightness LEDs

The forward current–voltage (I–V) characteristics of a blue GaInN, a green GaInN, and a red AlGaInP LED are shown in Fig. 12.19. The forward turn-on voltage scales with the emission energy, indicating a well-behaved characteristic. Closer inspection of the forward voltage (at 1 mA) of the green LED ($V_{f,\text{green}} = 2.65$ V) indicates that it is very similar to the blue LED ($V_{f,\text{blue}} = 2.75$ V) even though the emission energies of the blue and green LED are quite different ($\lambda_{\text{blue}} = 470$ nm, $h\nu_{\text{blue}} = 2.64$ eV; $\lambda_{\text{green}} = 525$ nm, $h\nu_{\text{green}} = 2.36$ eV). The small difference in forward voltage indicates that carriers probably lose energy by phonon emission when injected from the GaN barrier into the GaInN active region. They lose more energy when being injected from the GaN barrier into the In-richer active region of the green LED. The energy, dissipated by emission of phonons, is supplied by the external voltage applied to the LED.

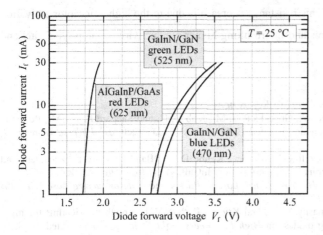

Fig. 12.19. Typical forward current–voltage (I–V) characteristic of GaInN/GaN blue, GaInN/GaN green, and AlGaInP/GaAs red LEDs at room temperature (after Toyoda Gosei Corporation, 2000).

The diode series resistance can be inferred from the slope of the I–V characteristics. The blue and green diodes have a higher series resistance than the red AlGaInP diode. The larger resistance in GaInN LEDs can be attributed to several factors including the "lateral" resistance in the n-type buffer layer for devices grown on sapphire substrates, strong polarization effects occurring in the nitride material family, lower p-type conductivity in the cladding layer, and higher p-type contact resistance. The lower p-type conductivity is due to the high acceptor activation energy (approximately 200 meV) in GaN and GaInN so that only a small fraction of acceptors is activated.

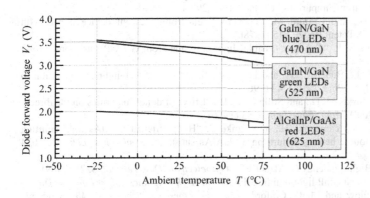

Fig. 12.20. Typical diode forward voltage at a current of 30 mA of GaInN/GaN blue, GaInN/GaN green, and AlGaInP/GaAs red LEDs versus temperature (after Toyoda Gosei Corp., 2000).

The temperature dependence of the forward voltage at a diode current of 30 mA is shown in Fig. 12.20. For all diodes shown, the forward voltage decreases as the temperature is increased. The decrease in forward voltage is due to the decrease of the bandgap energy. In the blue and green GaInN diodes, the lower forward voltage is also due to the decrease in series resistance occurring at high temperatures. This resistance decrease is due to the higher acceptor activation occurring at elevated temperatures and the resulting higher conductivity of the p-type GaN and GaInN layers.

References

Bradley R. R., Ash R. M., Forbes N. W., Griffiths R. J. M., Jebb D. P., and Shepard H. E. "Metalorganic chemical vapor deposition of junction isolated GaAlAs/GaAs LED structures" *J. Cryst. Growth* **77**, 629 (1986)

Campbell J. C., Holonyak Jr. N., Craford M. G., and Keune D. L. "Band structure enhancement and optimization of radiative recombination in GaAsP (and InGaP:N)" *J. Appl. Phys.* **45**, 4543 (1974)

Casey Jr. H. C. and Panish M. B. *Heterostructure Lasers, Part A* and *Heterostructure Lasers, Part B* (Academic Press, San Diego, 1978)

Chen C. H., Stockman S. A., Peanasky M. J., and Kuo C. P. "OMVPE growth of AlGaInP for high-efficiency visible light-emitting diodes" in *High Brightness Light Emitting Diodes* edited by G. B. Stringfellow and M. G. Craford, *Semiconductors and Semimetals* **48**, (Academic Press, San Diego, 1997)

Craford M. G., Shaw R. W., Herzog A. H., and Groves W. O. "Radiative recombination mechanisms in GaAsP diodes with and without nitrogen doping" *J. Appl. Phys.* **43**, 4075 (1972)

Craford M. G. "Overview of device issues in high-brightness light-emitting diodes" in *High Brightness Light Emitting Diodes* edited by G. B. Stringfellow and M. G. Craford, *Semiconductors and Semimetals* **48** (Academic Press, San Diego, 1997)

Craford M. G. "The bright future of light-emitting diodes" Plenary talk on light emitting diodes at the MRS Fall Meeting, Boston Massachusetts December (1999)

Dallesasse J. M., El-Zein N., Holonyak Jr. N., Hsieh K. C., Burnham R. D., and Dupuis R. D. "Environmental degradation of AlGaAs–GaAs quantum-well heterostructures" *J. Appl. Phys.* **68**, 2235 (1990)

Grimmeiss H. G. and Scholz H. "Efficiency of recombination radiation in GaP" *Phys. Lett.* **8**, 233 (1964)

Groves W. O. and Epstein A. S. "Epitaxial deposition of III–V compounds containing isoelectronic impurities" US Patent 4,001,056 (1977)

Groves W. O., Herzog A. H., and Craford M. G. "Process for the preparation of electroluminescent III–V materials containing isoelectronic impurities" US Patent Re. 29,648 (1978a)

Groves W. O., Herzog A. H., and Craford M. G. "GaAsP electroluminescent device doped with isoelectronic impurities" US Patent Re. 29,845 (1978b)

Holonyak Jr. N. and Bevacqua S. F. "Coherent (visible) light emission from Ga(AsP) junctions" *Appl. Phys. Lett.* **1**, 82 (1962)

Holonyak Jr. N., Bevacqua S. F., Bielan C. V., and Lubowski S. J. "The "direct–indirect" transition in Ga(AsP) p-n junctions," *Appl. Phys. Lett.* **3**, 47 (1963)

Holonyak Jr. N., Nuese C. J., Sirkis M. D., and Stillman G. E., "Effect of donor impurities on the direct–indirect transition in Ga(AsP)" *Appl. Phys. Lett.* **8**, 83 (1966)

Ishiguro H., Sawa K., Nagao S., Yamanaka H., and Koike S. "High efficient GaAlAs light emitting diodes of 660 nm with double heterostructure on a GaAlAs substrate" *Appl. Phys. Lett.* **43**, 1034 (1983)

Ishimatsu S. and Okuno Y. "High efficiency GaAlAs LED" *Optoelectron. Dev. Technol.* **4**, 21 (1989)

Kish F. A. and Fletcher R. M. "AlGaInP light-emitting diodes" in *High Brightness Light Emitting Diodes* edited by G. B. Stringfellow and M. G. Craford, *Semiconductors and Semimetals* **48** (Academic

Press, San Diego, 1997)
Krames M. R. et al. "High-power truncated-inverted-pyramid $(Al_xGa_{1-x})_{0.5}In_{0.5}P/GaP$ light emitting diodes exhibiting > 50% external quantum efficiency" *Appl. Phys. Lett.* **75**, 2365 (1999)
Krames M. R. et al. "High-brightness AlGaInN light emitting diodes" *Proc. SPIE* **3938**, 2 (2000)
Krames M. R., Amano H., Brown J. J., and Heremans P. L. "High-efficiency light-emitting diodes" Special Issue of *IEEE J. Sel. Top. Quantum Electron.* **8**, 185 (2002)
Logan R. A., White H. G., and Trumbore F. A. "P-n junctions in compensated solution grown GaP" *J. Appl. Phys.* **38**, 2500 (1967a)
Logan R. A., White H. G., and Trumbore F. A. "P-n junctions in GaP with external electrolumi-nescence efficiencies ~ 2% at 25 °C" *Appl. Phys. Lett.* **10**, 206 (1967b)
Logan R. A., White H. G., and Wiegmann W. "Efficient green electroluminescent junctions in GaP" *Solid State Electron.* **14**, 55 (1971)
Mueller G. (editor) *Electroluminescence I, Semiconductors and Semimetals* **64** (Academic Press, San Diego, 1999)
Mueller G. (editor) *Electroluminescence II, Semiconductors and Semimetals* **65** (Academic Press, San Diego, 2000)
Nakamura S. and Fasol G. *The Blue Laser Diode* (Springer, Berlin, 1997)
Nishizawa J., Koike M., and Jin C. C. "Efficiency of GaAlAs heterostructure red light emitting diodes" *J. Appl. Phys.* **54**, 2807 (1983)
Nuese C. J., Stillman G. E., Sirkis M. D., and Holonyak Jr. N., "Gallium arsenide-phosphide: crystal, diffusion, and laser properties" *Solid State Electron.* **9**, 735 (1966)
Nuese C. J., Tietjen J. J., Gannon J. J., and Gossenberger H. F. "Optimization of electroluminescent efficiencies for vapor-grown GaAsP diodes" *J. Electrochem. Soc.: Solid State Sci.* **116**, 248 (1969)
Pilkuhn M. and Rupprecht H. "Electroluminescence and lasing action in GaAsP" *J. Appl. Phys.* **36**, 684 (1965)
Prins A. D., Sly J. L., Meney A. T., Dunstan D. J., O'Reilly E. P., Adams A. R., and Valster A. J. *Phys. Chem. Solids* **56**, 349 (1995)
Steranka F. M., DeFevre D. C., Camras M. D., Tu C.-W., McElfresh D. K., Rudaz S. L., Cook L. W., and Snyder W. L. "Red AlGaAs light emitting diodes" *Hewlett-Packard Journal* p. 84 August (1988)
Steranka F. M. "AlGaAs red light-emitting diodes" in *High Brightness Light Emitting Diodes* edited by G. B. Stringfellow and M. G. Craford, *Semiconductors and Semimetals* **48** (Academic Press, San Diego, 1997)
Stringfellow G. B. and Craford M. G. (Editors) *High Brightness Light Emitting Diodes, Semiconductors and Semimetals* **48** (Academic Press, San Diego, 1997)
Strite S. and Morkoc H., "GaN, AlN, and InN: A review" *J. Vac. Sci. Technol.* **B 10**, 1237 (1992)
Tien P. K. Original version of the graph is courtesy of P. K. Tien of AT&T Bell Laboratories (1988)
Toyoda Gosei Corporation, Japan, LED product catalog (2000)
United Epitaxy Corporation, Taiwan, General LED and wafer product catalog (1999)
Wolfe C. M., Nuese C. J. and Holonyak Jr. N. "Growth and dislocation structure of single-crystal Ga(AsP)" *J. Appl. Phys.* **36**, 3790 (1965)
Wu J., Walukiewicz W., Yu K. M., Ager III J. W., Haller E. E., Lu H., Schaff W. J., Saito Y., and Nanishi Y. "Unusual properties of the fundamental bandgap of InN" *Appl. Phys. Lett.* **80**, 3967 (2002a)
Wu J., Walukiewicz W., Yu K. M., Ager III J. W., Haller E. E., Lu H., and Schaff W. J. "Small bandgap bowing in $In_{1-x}Ga_xN$ alloys" *Appl. Phys. Lett.* **80**, 4741 (2002b)

13

The AlGaInN material system and ultraviolet emitters

13.1 The UV spectral range

The ultraviolet–visible boundary is at about 390 nm, where the 1978 CIE eye sensitivity curve has a value of 0.1% of its maximum value. This chapter concentrates on materials issues of III–V nitrides, on devices emitting in the ultraviolet (UV, $\lambda < 390$ nm), and on devices emitting in the violet near the UV–visible boundary (390–410 nm). Although the latter devices emit in the visible spectrum, they are frequently classified as UV devices. For UV devices, we will differentiate between devices having a GaInN active region ($\lambda > 360$ nm) and devices having an AlGaN active region ($\lambda < 360$ nm).

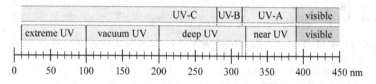

Fig. 13.1. Nomenclature of UV radiation versus wavelength (after International Congress on Light, 1932).

There are two classifications of the UV spectrum, both of which are shown in Fig. 13.1. The UV-A (315–390 nm), UV-B (280–315 nm), and UV-C (< 280 nm) classification is based on a convention established during the Second International Congress on Light in 1932 (International, 1932). UV-A radiation from the sun penetrates the earth's atmosphere (including clouds) and, due to the high energy of UV photons, creates damage to the skin, particularly to the deeper layers of the skin. UV-A radiation also causes cataracts (the clouding of the eye's lens), which can lead to total blindness. Both UV-B (partially absorbed by the earth's ozone layer) and UV-C radiation (mostly absorbed by ozone layer) create serious damage to skin and eyes. Because no natural light with $\lambda \leq 280$ nm exists at the earth's surface (i.e. in the UV-C range), this wavelength range is also referred to as the *solar-blind* range.

An alternative classification of UV radiation has evolved over time. This alternative classification has the following categories: Extreme UV (10–100 nm); vacuum UV (100–200 nm); deep UV (200–320 nm); and near UV (320–390 nm).

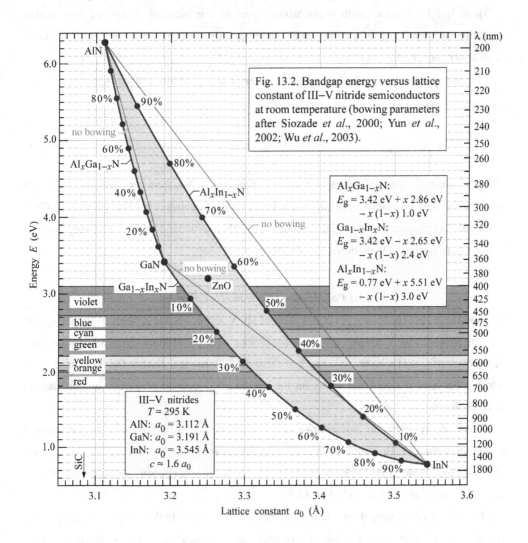

Fig. 13.2. Bandgap energy versus lattice constant of III–V nitride semiconductors at room temperature (bowing parameters after Siozade *et al.*, 2000; Yun *et al.*, 2002; Wu *et al.*, 2003).

13.2 The AlGaInN bandgap

The bandgap energy versus lattice constant of the AlGaInN material system is shown in Fig. 13.2. The AlGaInN material system spans a very wide range of wavelengths covering the deep UV, near UV, visible, and even the near infrared spectral range. Of the three binary semiconductors InN, GaN, and AlN, epitaxially grown GaN has been shown to be synthesizable with the highest quality. It has generally been difficult to synthesize Al-rich AlGaN alloys and In-rich GaInN alloys with high internal quantum efficiencies.

There has been some controversy with respect to the bandgap energy of InN, which originally had been found to be about 1.9 eV. More recently, the bandgap energy was found to be much lower, namely 0.77 eV.

The energy-gap bowing can be expressed in terms of a constant, a linear term ($\propto x$) and a non-linear term [$\propto x(1-x)$] according to

$$E_g^{AB} = E_g^A + (E_g^B - E_g^A)x + x(1-x)E_b \tag{13.1}$$

with E_b called the **bowing energy** or **bowing parameter**. The bowing parameters used in Fig. 13.2 for AlGaN, GaInN, and AlInN are based on data published by Siozade *et al.* (2000), Yun *et al.* (2002), and Wu *et al.* (2002; 2003). Recently additional data on the bowing energies have become available (Walukiewicz *et al.*, 2004).

13.3 Polarization effects in III–V nitrides

The most common epitaxial growth direction of III–V nitrides is the *c*-plane of the hexagonal wurtzite structure. III–V nitrides grown on the *c*-plane have polarization charges located at each of the two surfaces of a layer. As a result of these charges, internal electric fields occur in III–V nitrides that have a significant effect on the optical and electrical properties of this class of semiconductors.

There are **spontaneous polarization** charges as well as strain-induced or **piezoelectric polarization charges** (Bernardini *et al.*, 1997; Ambacher *et al.*, 1999; 2000; 2002). The direction of the internal electric field depends on the strain and the growth orientation (Ga face or N face) and is shown for different cases in Fig. 13.3.

The strain in the epitaxial layer can be compressive or tensile. In the compressive-strain case, the epitaxial layer of interest is laterally compressed ("laterally" meaning "in the plane of the wafer"). For example, GaInN is compressively strained when grown on a thick relaxed GaN buffer layer. In the tensile-strain case, the epitaxial layer of interest is expanded along the lateral direction. For example, AlGaN is under tensile strain when grown on a thick relaxed GaN buffer layer.

The calculated magnitude of the electric field for common III–V nitride alloys grown on relaxed GaN is shown in Fig. 13.4 (Gessmann et al., 2002).

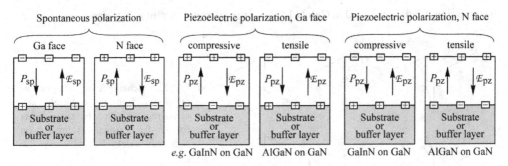

Fig. 13.3. Surface charges and direction of electric field and polarization field for spontaneous and piezoelectric polarization in III–V nitrides for Ga and N face orientation.

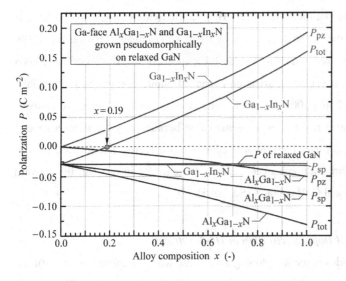

Fig. 13.4. Magnitude and direction of spontaneous and piezoelectric polarization in GaInN and AlGaN grown pseudomorphically on relaxed GaN. Relaxed GaN has a spontaeous polarization, but no piezoelectric polarization (after Gessmann et al., 2002).

A consequence of the polarization fields for quantum well structures is shown in Fig. 13.5. The quantum well layers have an internal electric field that spatially separates electrons and holes thereby preventing efficient radiative recombination. This is particularly true for thick quantum wells, e.g. > 100 Å. To avoid this deleterious effect, it is imperative that the quantum well layers are kept very thin. Quantum well thicknesses of 20–30 Å are typically used to minimize such electron–hole separation effects.

The large electric fields caused by the polarization effects can be screened by a high free-carrier concentration, which can be attained through either (*i*) high doping of the active region or (*ii*) a high injection current. Screening of the internal electric field also results in a blue-shift of the emission, frequently found in GaInN LEDs as the injection current is increased.

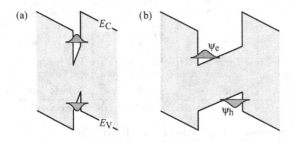

Fig. 13.5. Schematic band diagram of (a) thin and (b) thick AlGaN/GaN active regions with polarization fields for Ga-face growth (substrate on right-hand side).

Polarization effects can be used advantageously to reduce ohmic contact resistances in GaN devices. Polarization-enhanced contacts to p-type GaN employ a thin, compressively strained GaInN cap layer that is deposited on the p-type GaN. The electric field in the GaInN cap layer is polarized in such a way that the tunneling probability of holes is enhanced (Li *et al.*, 2000; Gessmann *et al.*, 2002). Contact resistances as low as 1.1×10^{-6} and 2×10^{-7} $\Omega\,\text{cm}^2$ have been reported for non-annealed and annealed polarization-enhanced ohmic contacts to p-type GaN, respectively (Kumakura *et al.*, 2001; 2003). Both the concept of polarization-enhanced contacts and the very low specific contact resistances are remarkable because they represent a several orders-of-magnitude improvement over conventional contact technologies. The forward voltage of GaN-based LEDs with polarization-enhanced contacts is generally lower than the forward voltage of LEDs with conventional GaN contacts (Su, 2005).

13.4 Doping activation in III–V nitrides

Another problem in III–V nitrides is the low doping activation, which is caused by two effects: (*i*) Chemical deactivation of acceptors by hydrogen atoms bonding to the acceptors. The missing electron that acceptors strive to capture is provided by a hydrogen atom. Hydrogen is available in abundance during epitaxial growth and possible sources for hydrogen passivation include the methyl ($-CH_3$) and ethyl groups ($-C_2H_5$) of the organo-metallics, ammonia (NH_3), and hydrogen from the H_2 carrier gas. (*ii*) Acceptors in III–V nitrides have a high thermal activation energy which is $\gg kT$ at 300 K. As a result, only a small percentage of acceptors are ionized at room temperature.

Exercise: *Activation of Mg acceptors in GaN*. Mg acceptors in GaN have an activation energy of $E_a = 200$ meV. (a) Calculate the fraction of acceptors that are ionized at 300 K for an acceptor concentration of $N_{Mg} = 10^{18}$ cm^{-3} using the formula $p = (g^{-1} N_{Mg} N_v)^{1/2} \exp(-E_a/2kT)$ where g is the acceptor ground state degeneracy ($g = 4$) and N_v is the effective density of states at the valence band edge of GaN. (b) What would be the activation of acceptors if a hydrogen atom were bonded to each acceptor?

Solution: (a) Using the formula given above, one obtains that only about 6% of the acceptors are ionized. (b) If acceptors are passivated, p-type conductivity cannot be established.

Amano et al. (1989) discovered that acceptor dopants can be activated by low-energy electron-beam irradiation (LEEBI). Nakamura et al. (1991; 1992) showed that acceptors can be activated by LEEBI as well as by thermal annealing, with thermal annealing being the preferred method. Typical annealing conditions for MOCVD-grown p-type GaN are 675–725 °C for 5 minutes in an N$_2$ atmosphere; p-type Al$_{0.30}$Ga$_{0.70}$N is annealed at higher temperatures, typically at 850 °C for 1–2 minutes. It is believed that during thermal annealing the relatively weak acceptor–hydrogen bond is broken and that the hydrogen atoms are driven out of the epitaxial film. Hydrogen atoms are small and generally diffuse easily through the interstitial sites of a crystalline material. The heating caused by LEEBI is believed to have a similar effect.

The doping of AlGaN/GaN and AlGaN/AlGaN superlattices (rather than GaN or AlGaN bulk material) was postulated to strongly increase the electrical activation of deep acceptors such as Mg (Schubert et al., 1996). Experimental increases in conductivity by a factor of 10 and more have indeed been reported by several research groups (Goepfert et al., 1999, 2000; Kozodoy et al. 1999a, b; Kipshidze et al., 2002, 2003). Figure 13.6 summarizes the problems and solutions to the doping activation in GaN and AlGaN.

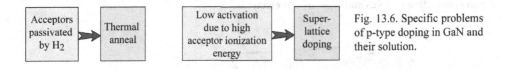

Fig. 13.6. Specific problems of p-type doping in GaN and their solution.

13.5 Dislocations in III–V nitrides

The most common substrate for GaN epitaxial growth, sapphire, is a very stable substrate in terms of its thermal, chemical, and mechanical properties. However, sapphire has the complex **corundum structure** whereas III–V nitrides crystallize in the **wurtzite structure**. Furthermore, the lattice constants of sapphire and GaN are different. As a result, GaN epitaxial films have misfit dislocations (threading and edge dislocations) that are typically on the order of 10^8–

10^9 cm^{-2}.

The initial stages of GaN growth on sapphire are shown schematically in Fig. 13.7 (Nakamura and Fasol, 1997). The initial layer (called the faulted zone), grown at low temperatures (~ 500 °C) and subsequently annealed, is highly dislocated. However, dislocations undergo self-annihilation during anneal, so that subsequent layers (called the semi-sound zone and the sound zone) have much lower dislocation densities. A thorough review of the initial stages of GaN epitaxial growth on sapphire and an analysis of these initial stages using atomic-force microscopy and optical reflectometry was published by Koleske *et al.* (2004).

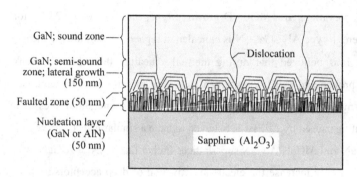

Fig. 13.7. Dislocation structure of a GaN epitaxial layer grown on sapphire by means of the GaN or AlN nucleation layer developed by Amano *et al.* (1986). The nucleation layer is normally grown at 500 °C, much lower than the following GaN epitaxial layers (after Nakamura and Fasol, 1997).

Generally, dislocation lines are electrically charged so that the region surrounding a dislocation line is either coulombically attractive or repulsive to a free carrier. The nature of the coulombic interaction (attractive or repulsive) depends on the polarity of the dislocation line and the polarity of the carrier. As an example, Fig. 13.8 shows a negatively charged dislocation line which is attractive to holes and repulsive to electrons.

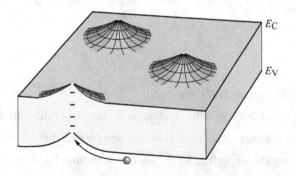

Fig. 13.8. Band diagram of semiconductor having negatively charged dislocations. Holes are attracted to dislocation lines where they must ultimately recombine with electrons.

Figure 13.9 shows the temporal development of the carrier dynamics of a positively charged dislocation line. Initially, electrons are attracted but holes are repelled due to the potential created by the dislocation. However, the continued collection of electrons will screen the dislocation potential thereby reducing the repulsive barrier for holes. As a result, electrons and holes will recombine non-radiatively via electron states of the dislocation line. In Fig. 13.9, the electronic states of the dislocation are located within the bandgap.

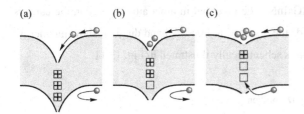

Fig. 13.9. Recombination in a positively charged dislocation. (a)–(c) Sequence shows electrons accumulating in the potential minimum thereby screening the dislocation potential and allowing holes to recombine.

A puzzling question is as to why the radiative recombination efficiency in III–V nitrides is so high despite the high density of dislocations? Several possible explanations are discussed below. However, none of these explanations has gained general acceptance.

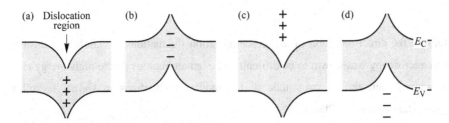

Fig. 13.10. Band diagram of dislocation: (a) donor states in gap, (b) acceptor states in gap, (c) donor states in conduction band, and (d) acceptor states in valence band. Whereas (a) and (b) lead to non-radiative recombination, (c) and (d) no not.

One possible explanation for high radiative rates in III–V nitrides is that the electronic states of the dislocation lie *outside* the forbidden gap, i.e. within the allowed bands of the semiconductor, as shown in Fig. 13.10 (c) and (d). This explanation is not in conflict with dark regions surrounding dislocations, observed in cathodoluminescence experiments (Rosner et al., 1997; Albrecht et al., 2002). Such dark regions unequivocally show the *absence* of *radiative* recombination but do not necessarily prove the *presence* of *non-radiative* recombination. The dark regions observed in cathodoluminescence could be explained by the incomplete screening of the dislocation potential, which would result in the repulsion of either electrons or holes, and thus result in the absence of radiative recombination.

Other possible explanations for high radiative rates in III–V nitrides are the compositional alloy fluctuations, alloy clustering effects, and phase separation effects that necessarily result in a variation of the bandgap energy and lead to local potential minima, which carriers are attracted to and could be confined to (Nakamura and Fasol, 1997; Chichibu et al., 1996; Narukawa et al., 1997a, b). This explanation would be particularly suitable for ternary and quaternary alloy semiconductors such as GaInN and AlGaInN. The potential minima attract and confine carriers and prevent them from diffusing towards the dislocation lines. A band diagram showing energy-gap fluctuations and *carrier localization* is schematically illustrated in Fig. 13.11.

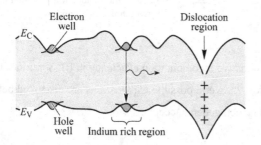

Fig. 13.11. Band diagram of GaInN having clusters of In rich regions which spatially localize carriers and prevent them from diffusing to dislocations.

Because the direct detection of indium-composition fluctuations in GaInN by transmission-electron microscopy was shown to be difficult and even ambiguous due to high-energy electron-beam induced damage, the magnitude of the indium fluctuations in GaInN is still under discussion (Smeeton et al., 2003).

Recently, Hangleiter et al. (2005) pointed out that a reduction of GaInN/GaN quantum-well thickness occurs in the vicinity of V-shaped defects of III–V nitride epitaxial layers. The authors proposed that the higher bandgap energy associated with the thinner GaInN quantum wells shields the dislocation line-defect from mobile carriers located in the thicker (planar) GaInN quantum wells. As a result, a high radiative efficiency would be maintained in quantum-well structures despite the presence of dislocation line-defects. However, this model cannot explain the high radiative efficiency of GaN thin films.

Although a generally accepted explanation has not yet been established, it remains a fact that the radiative efficiency in III–V nitrides, in particular GaInN/GaN blue emitters, exhibits low sensitivity to the presence of dislocations. That is, high radiative efficiencies are obtained in GaInN/GaN blue emitters despite high dislocation densities. This is illustrated in Fig. 13.12 which compares the normalized efficiency of different III–V semiconductors as a function of the dislocation density in GaAs, AlGaAs, GaP, and GaAsP. Lester *et al.* (1995) estimated a 4% radiative efficiency for GaN with a dislocation density of 10^{10} cm^{-2}. The shaded region is the estimation of the author of this book based on blue GaInN/GaN emitters. The data shown in the figure elucidate that III–V nitrides have a much higher tolerance towards dislocations as compared to III–V arsenides and phosphides.

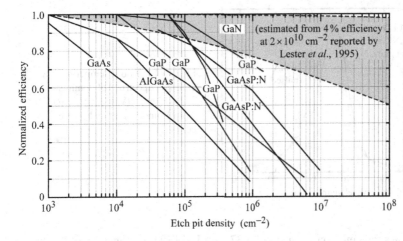

Fig. 13.12. Dependence of radiative efficiency on etch pit density (III–V arsenide and phosphide data adopted from Lester *et al.*, 1995; III–V nitride data estimated by the author of this book based on data published by Lester *et al.*, 1995).

13.6 UV devices emitting at wavelengths longer than 360 nm

UV devices emitting at wavelengths longer than 360 nm generally have GaN or GaInN active regions. GaInN LEDs with peak wavelengths ranging from 400 nm to 410 nm were reported as early as 1993 (Nakamura *et al.*, 1993a, b, 1994). The design of an early GaInN UV LED emitting at 370 nm is shown in Fig. 13.13. The device has a GaInN/AlGaN double heterostructure active region and AlGaN electron- and hole-blocking layers (Mukai *et al.*, 1998). The active region and the carrier-blocking layers are clad by p-type and n-type GaN.

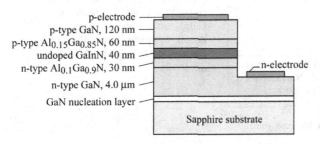

Fig. 13.13. Layer structure of GaInN UV LED grown on sapphire substrate emitting at 370 nm (after Mukai *et al.*, 1998).

Figure 13.14 shows the emission spectrum of a 375 nm UV LED under pulsed and continuous wave (cw) injection conditions. Inspection of the figure reveals a small red-shift of the peak wavelength when going from pulsed to cw current injection. This shift is likely caused by junction heating, which generally leads to a lower bandgap energy and a red-shift of the emission wavelength.

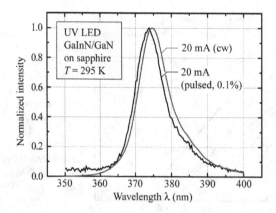

Fig. 13.14. Room temperature emission spectrum of commercial 375 nm UV LED (Nichia Corp.) under cw and pulsed conditions.

Figure 13.15 shows the emission intensity of a 375 nm UV LED as a function of the active layer thickness. The figure reveals that optimum output power is attained at 30–50 nm active layer thickness. It can be assumed that the active region is heavily doped in order to screen the polarization fields. Quantum well active regions with quantum well thicknesses < 5 nm reduce the effect of spatial electron–hole separation and for this reason, quantum well active regions have superseded the double heterostructure designs.

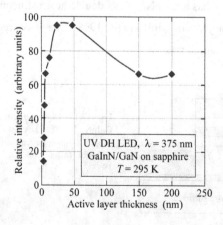

Fig. 13.15. Room temperature emission intensity as a function of GaInN active layer thickness for double heterostructure UV LEDs emitting at 375 nm (after Mukai et al., 1998).

13.7 UV devices emitting at wavelengths shorter than 360 nm

Figure 13.16 shows the dependence of the output power as a function of wavelength. As the emission wavelength decreases, a pronounced drop in the emission intensity is found. This has been attributed to the very positive influence of indium (In) incorporation on the internal quantum efficiency. The positive influence is reduced as less indium is incorporated in the active region with pure GaN active regions ($\lambda \approx 360$ nm) not benefiting at all.

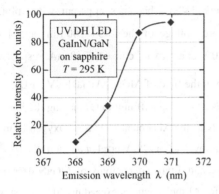

Fig. 13.16. Room temperature intensity as a function of emission wavelength for GaInN double heterostructure UV LEDs (after Mukai et al., 1998).

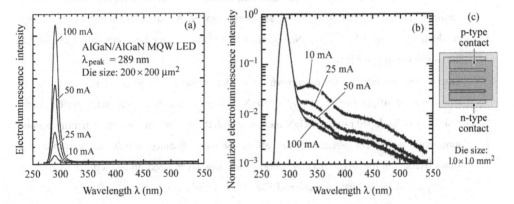

Fig. 13.17. Emission spectrum of deep-UV AlGaN/AlGaN multiple-quantum well LED for different injection currents on (a) linear and (b) logarithmic scales. An interdigitated contact geometry, as shown in (c), was used for large-area dies (after Fischer et al., 2004).

13.7 UV devices emitting at wavelengths shorter than 360 nm

Diodes emitting at wavelengths less than 360 nm have AlGaN active regions or $Al_xGa_{1-x}N/Al_yGa_{1-y}N$ multiple-quantum well (MQW) active regions. The power efficiency of these devices is generally low, i.e. less than 1%, although substantial progress has been made in recent years

(Zhang et al., 2002a, 2003; Yasan et al., 2002; Kipshidze et al., 2003; Fischer et al., 2004; Kim et al., 2004; Oder et al., 2004; Razeghi and Henini, 2004; Shakya et al., 2004;).

The emission spectrum of an AlGaN/AlGaN deep-UV LED with interdigitated contact geometry is shown in Fig. 13.17 for different injection currents (Fischer et al., 2004). The active region of the device is composed of three $Al_{0.36}Ga_{0.64}N$ quantum wells with $Al_{0.48}Ga_{0.52}N$ barriers for emission at 290 nm. The spectrum displays one clean emission line with a peak wavelength of 289 nm. Some sub-bandgap emission near 330 nm becomes apparent when plotting the spectrum on a logarithmic scale. The forward voltage of the 200×200 μm^2 and 1×1 mm^2 devices was reported to be about 7.0 V and 6.0 V at a forward current of 20 mA, respectively.

The following issues deserve special attention in the field of AlGaN/AlGaN UV LEDs:

- *Affinity of aluminum to oxygen*: Al has a very high affinity to O_2 making the incorporation of oxygen into AlGaN increasingly likely as the Al content increases. Oxygen forms a deep, DX-like level in Al-rich AlGaN (McCluskey et al., 1998; Wetzel et al., 2001).

- *Conductivity of AlGaN*: Both the p-type and n-type conductivity of AlGaN decrease as the Al mol fraction increases, particularly for Al mol fractions exceeding 30% (Katsuragawa et al., 1998; Goepfert et al., 2000; Jiang and Lin, 2002). This leads to a higher resistivity in the confinement layers and higher device series resistances. A particular problem is the p-type conductivity in AlGaN. $Al_xGa_{1-x}N/Al_yGa_{1-y}N$ superlattices have been employed to alleviate the p-type doping problem.

- *Lateral conductivity*: In devices grown on insulating substrates, with the standard side-by-side contact configuration, the n-type AlGaN layer provides the lateral conductivity. As the resistivity of Si-doped n-type AlGaN increases with the Al content, devices generally become more resistive. To compensate for this effect, the mean distance, which the electron current flows laterally in the n-type AlGaN lateral-conduction layer, needs to be reduced. This can be accomplished by an array of micro-LEDs (Kim et al., 2003; Khan, 2004) or by closely spaced fingers in interdigitated-contact geometries.

- *Contact resistance*: Due to the high bandgap of AlGaN, contact-barrier heights are generally higher, which makes the attainment of low-resistance contacts increasingly difficult as the Al content increases.

- *Diffusion of acceptors*: During the epitaxial growth of the top cladding layer, Mg acceptors may diffuse back into the active region, thereby lowering its radiative efficiency. Acceptor diffusion and the associated decrease in radiative efficiency may impose a limit on the maximum thickness of the p-type cladding layer.

- *Heterojunction barriers*: Due to the larger bandgap energies, the conduction- and valence-band discontinuities of heterojunctions are generally larger than for smaller-bandgap semiconductors. Compositional grading at the heterojunction interfaces reduces the resistance of heterojunction barriers.

- **Light extraction**: To reduce reabsorption effects, all device layers should have an Al content sufficiently high to be transparent to the emitted light.
- **Cracking**: AlGaN films grown on relaxed GaN are, due to the smaller lattice constant of AlGaN, under tensile strain. If the films are sufficiently thick, they crack. However, cracking can be strongly reduced or even eliminated by using Al-rich strain-compensating superlattices (Hearne et al., 2000; Han et al., 2001; Zhang et al., 2002b). Such strain-compensating superlattices reduce the lattice constant so that subsequent epitaxial layers have a much reduced tensile strain or are even under compressive strain. Hearne et al. (2000) provided a quantitative analysis of the maximum attainable thickness of a crack-free layer under tensile strain.

Optical micrographs of a 0.9 μm thick $Al_{0.15}Ga_{0.85}N$ film are shown in Fig. 13.18 (a) and (b) for growth without and with a strain-compensating superlattice, respectively. Figure 13.18 (b) shows a virtually crack-free AlGaN layer that was grown on a strain-compensating $AlN/Al_{0.45}Ga_{0.55}N$ superlattice with 10 periods and equal well- and barrier-layer thicknesses of 100 Å.

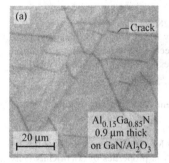
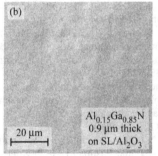

Fig. 13.18. Optical micrographs of $Al_{0.15}Ga_{0.85}N$ layer grown (a) without and (b) with a strain-compensating $AlN/Al_{0.45}Ga_{0.55}N$ superlattice (SL). The SL has 10 periods and equal well and barrier thicknesses of 10 nm. Angles between crack lines frequently are 60° or 120°.

Exercise: Cracking. Why does cracking occur in epitaxial layers that are under biaxial tensile strain but not in epilayers that are under biaxial compressive strain?
Solution: Wafer bowing and ultimately cracking of an epitaxial film that is under biaxial tensile strain releases the strain energy stored in the film. For epitaxial layers that are under compressive strain, the strain energy can be released by wafer bowing, film buckling, and film delamination. Due to the compressive strain, there is "no room" for fissures or cracks, so that cracks generally do not form in compressively strained films.
The strain energy stored in a homo-epitaxial film, that is lattice mismatched to the substrate, is proportional to the thickness of the film. As the thickness of a strained film increases, it will at some point become energetically more favorable to reduce the strain energy by creating misfit dislocations and cracks. Thus, at a certain thickness, the film will form misfit dislocations to release the strain energy. The

critical thickness, at which a homo-epitaxial film starts forming misfit dislocations, is given by the *Matthews–Blakeslee law* (Matthews and Blakeslee, 1976). As the film thickness increases further, misfit dislocations do not suffice to release the strain energy, so that at some point the film will start to crack. A formula for the critical thickness at which a film under biaxial tensile strain starts to crack was given by Hearne *et al.* (2000).

AlGaN UV LEDs frequently have forward voltages $V_f \gg h\nu/e$. Depending on the device structure, the excess forward voltage may originate from the p-type contact, the p-type AlGaN confinement layer, the n-type AlGaN layer providing lateral conduction, or from unipolar heterojunctions.

For devices with low or moderate efficiency as well as for high-power devices, device packages with low thermal resistance are desirable. The heat resulting from an excess forward voltage and low quantum efficiency must be removed to avoid excessively high junction temperatures. Morita *et al.* (2004) reported a particularly well heat-sunk device, namely a structure in which the sapphire substrate was removed by laser lift-off, and, using an AuSn solder, the epilayer was directly bonded to a CuW heat sink.

References

Albrecht M., Strunk H. P., Weyher J. L., Grzegory I., Porowski S., and Wosinski T. "Carrier recombination at single dislocations in GaN measured by cathodoluminescence in a transmission electron microscope" *J. Appl. Phys.* **92**, 2000 (2002)

Amano H., Kito M., Hiramatsu K., and Akasaki I. "P-type conduction in Mg-doped GaN treated with low-energy electron beam irradiation (LEEBI)" *Jpn. J. Appl. Phys.* **28**, L2112 (1989)

Ambacher O., Smart J., Shealy J. R., Weimann N. G., Chu K., Murphy M., Schaff W. J., Eastman L. F. Dimitrov R., Wittmer L., Stutzmann M., Rieger W. and Hilsenbeck J. "Two-dimensional electron gases induced by spontaneous and piezoelectric polarization charges in N- and Ga-face AlGaN/GaN heterostructures" *J. Appl. Phys.* **85**, 3222 (1999)

Ambacher O., Foutz B., Smart J., Shealy J. R., Weimann N. G., Chu K., Murphy M., Sierakowski A. J., Schaff W. J., Eastman L. F., Dimitrov R., Mitchell A., and Stutzmann M. "Two dimensional electron gases induced by spontaneous and piezoelectric polarization in undoped and doped AlGaN/GaN heterostructures" *J. Appl. Phys.* **87**, 334 (2000)

Ambacher O., Majewski J., Miskys C., Link A., Hermann M., Eickhoff M., Stutzmann M., Bernardini F., Fiorentini V., Tilak V., Schaff W. J, and Eastman L. F. "Pyroelectric properties of Al(In)GaN/GaN hetero- and quantum well structures" *J. Phys.: Condens. Matter* **14**, 3399 (2002)

Bernardini F., Fiorentini V., and Vanderbilt D. "Spontaneous polarization and piezoelectric constants of III–V nitrides" *Phys. Rev. B* **56**, R10 024 (1997)

Chichibu S., Azuhata T., Sota T., and Nakamura S. "Spontaneous emission of localized excitons in InGaN single and multiquantum well structures" *Appl. Phys. Lett.* **69**, 4188 (1996)

Fischer A. J., Allerman A. A., Crawford M. H., Bogart K. H. A., Lee S. R., Kaplar R. J., Chow W. W., Kurtz S. R., Fullmer K. W., and Figiel J. J. "Room-temperature direct current operation of 290 nm light-emitting diodes with milliwatt power levels" *Appl. Phys. Lett.* **84**, 3394 (2004)

Gessmann T., Li Y.-L., Waldron E. L., Graff J. W., Schubert E. F., and Sheu J. K. "Ohmic contacts to p-type GaN mediated by polarization fields in thin $In_xGa_{1-x}N$ capping layers" *Appl. Phys. Lett.* **80**, 986 (2002)

Goepfert I. D., Schubert E. F., Osinsky A., and Norris P. E. "Demonstration of efficient p-type doping in AlGaN/GaN superlattice structures" *Electron. Lett.* **35**, 1109 (1999)

Goepfert I. D., Schubert E. F., Osinsky A., Norris P. E., and Faleev N. N. "Experimental and theoretical study of acceptor activation and transport properties in p-type $Al_xGa_{1-x}N/GaN$ superlattices" *J. Appl. Phys.* **88**, 2030 (2000)

Hangleiter A., Hitzel F., Netzel C., Fuhrmann D., Rossow U., Ade G., and Hinze P. "Suppression of nonradiative recombination by V-shaped pits in GaInN/GaN quantum wells produces a large increase in the light emission efficiency" *Phys. Rev. Lett.* **95**, 127402 (2005)

Han J., Waldrip K. E., Lee S. R., Figiel J. J., Hearne S. J., Petersen G. A., and Myers S. M. "Control and elimination of cracking of AlGaN using low-temperature AlGaN interlayers" *Appl. Phys. Lett.* **78**, 67 (2001)

Hearne S. J., Han J., Lee S. R., Floro J. A., Follstaedt D., Chason M. E., and Tsong I. S. T. "Brittle-ductile relaxation kinetics of strained AlGaN/GaN heterostructures" *Appl. Phys. Lett.* **76**, 1534 (2000)

International Congress on Light, Copenhagen, Denmark (1932)

Jiang H. X. and Lin J. Y. "AlGaN and InAlGaN alloys – epitaxial growth, optical and electrical properties, and applications" *Opto-Electron. Rev.* **10**, 271 (2002)

Katsuragawa M., Sota S., Komori M., Anbe C., Takeuchi T., Sakai H., Amano H., and Akasaki I. "Thermal ionization energy of Si and Mg in AlGaN" *J. Cryst. Growth* **189/190**, 528 (1998)

Khan M. A. "Deep-ultraviolet LEDs fabricated in AlInGaN using MEMOCVD" *Proc. SPIE* **5530**, 224 (2004)

Kim K. H., Li J., Jin S. X., Lin J. Y., and Jiang H. X. "III-nitride ultraviolet light-emitting diodes with delta doping" *Appl. Phys. Lett.* **83**, 566 (2003)

Kim K. H., Fan Z. Y., Khizar M., Nakarmi M. L., Lin J. Y., and Jiang H. X. "AlGaN-based ultraviolet light-emitting diodes grown on AlN epilayers" *Appl. Phys. Lett.* **85**, 4777 (2004)

Kipshidze G., Kuryatkov V., Borisov B., Holtz M., Nikishin S., and Temkin H. "AlGaInN-based ultraviolet light-emitting diodes grown on Si $\langle 111 \rangle$" *Appl. Phys. Lett.* **80**, 3682 (2002)

Kipshidze G., Kuryatkov V., Zhu K., Borisov B., Holtz M., Nikishin S., and Temkin H. "AlN/AlGaInN superlattice light-emitting diodes at 280 nm" *J. Appl. Phys.* **93**, 1363 (2003)

Koleske D. D., Coltrin M. E., Cross K. C., Mitchell C. C., and Allerman A. A. "Understanding GaN nucleation layer evolution on sapphire" *J. Cryst. Growth* **273**, 86 (2004)

Kozodoy P., Hansen M., DenBaars S. P., and Mishra U. K. "Enhanced Mg doping efficiency in $Al_{0.2}Ga_{0.8}N/GaN$ superlattices" *Appl. Phys. Lett.* **74**, 3681 (1999a)

Kozodoy P., Smorchkova Y. P., Hansen M., Xing H., DenBaars S. P., Mishra U. K., Saxler A. W., Perrin R., and Mitchel W. C. "Polarization-enhanced Mg doping of AlGaN/GaN superlattices" *Appl. Phys. Lett.* **75**, 2444 (1999b)

Kumakura K., Makimoto T., and Kobayashi N. "Low resistance non-alloy ohmic contact to p-type GaN using Mg-doped InGaN contact layer" *Phys. Stat. Sol. (a)* **188**, 363 (2001)

Kumakura K., Makimoto T., and Kobayashi N. "Ohmic contact to p-GaN using a strained InGaN contact layer and its thermal stability" *Jpn. J. Appl. Phys.* **42**, 2254 (2003)

Lester S. D., Ponce F. A., Craford M. G., and Steigerwald D. A. "High dislocation densities in high efficiency GaN-based light-emitting diodes" *Appl. Phys. Lett.* **66**, 1249 (1995)

Li Y.-L., Schubert E. F., Graff J. W., Osinsky A., and Schaff W. F. "Low-resistance ohmic contacts to p-type GaN" *Appl. Phys. Lett.* **76**, 2728 (2000)

Matthews J. W. and Blakeslee A. E. "Defects in epitaxial multilayers – III. Preparation of almost perfect multilayers" *J. Cryst. Growth* **32**, 265 (1976)

McCluskey M. D., Johnson N. M., Van de Walle C. G., Bour D. P., Kneissl M., and Walukiewicz W. "Metastability of oxygen donors in AlGaN" *Phys. Rev. Lett.* **80**, 4008 (1998)

Morita D., Yamamoto M., Akaishi K., Matoba K., Yasutomo K., Kasai Y., Sano M., Nagahama S.-I., and Mukai T. "Watt-class high-output-power 365 nm ultraviolet light-emitting diodes" *Jpn. J. Appl. Phys.* **43**, 5945 (2004)

Mukai T., Morita D., and Nakamura S. "High-power UV GaInN/AlGaN double-heterostructure LEDs" *J. Cryst. Growth* **189**, 778 (1998)

Nakamura S., Senoh M., and Mukai T. "Highly p-type Mg doped GaN films grown with GaN buffer layers" *Jpn. J. Appl. Phys.* **30**, L1708 (1991)

Nakamura S., Mukai T., Senoh M., and Iwasa N. "Thermal annealing effects on p-type Mg-doped GaN films" *Jpn. J. Appl. Phys.* **31**, L139 (1992)

Nakamura S., Senoh M., and Mukai T. "P-GaN/N-InGaN/N-InGaN double heterostructure blue-light-emitting diodes" *Jpn. J. Appl. Phys.* **32**, L8 (1993a)

Nakamura S., Senoh M., and Mukai T. "High-power InGaN/GaN double-heterostructure violet light emitting diodes" *Appl. Phys. Lett.* **62**, 2390 (1993b)

Nakamura S. "Growth of InGaN compound semiconductors and high-power InGaN/AlGaN double heterostructure violet-light-emitting diode" *Microelectron. J.* **25**, 651 (1994)

Nakamura S. and Fasol G. *The Blue Laser Diode* (Springer, Berlin, 1997). The authors discuss carrier localization effects on page 305 of the book. The authors discuss the initial stages of epitaxial growth on page 13 of the book.

Narukawa Y., Kawakami Y., Fujita S., Fujita S., and Nakamura S. "Recombination dynamics of localized excitons in $In_{0.20}Ga_{0.80}N$ - $In_{0.05}Ga_{0.95}N$ multiple quantum wells" *Phys. Rev.* **B 55**, R 1938 (1997a)

Narukawa Y., Kawakami Y., Funato M., Fujita S., Fujita S., and Nakamura S. "Role of self-formed InGaN quantum dots for exciton localization in the purple laser diode emitting at 420 nm" *Appl. Phys. Lett.* **70**, 981 (1997b)

Oder T. N., Kim K. H., Lin J. Y., and Jiang H. X. "III-nitride blue and ultraviolet photonic crystal light emitting diodes" *Appl. Phys. Lett.* **84**, 466 (2004)

Razeghi M. and Henini M. "Optoelectronic devices: III–nitrides" (Elsevier, Amsterdam, 2004)

Rosner S. J., Carr E. C., Ludowise M. J., Girolami G., and Erikson H. I. "Correlation of cathodoluminescence inhomogeneity with microstructural defects in epitaxial GaN grown by metal-organic chemical-vapor deposition" *Appl. Phys. Lett.* **70**, 420 (1997)

Schubert E. F., Grieshaber W., and Goepfert I. D. "Enhancement of deep acceptor activation in semiconductors by superlattice doping" *Appl. Phys. Lett.* **69**, 3737 (1996)

Shakya J., Kim K. H., Lin J. Y., and Jiang H. X. "Enhanced light extraction in III-nitride ultraviolet photonic crystal light-emitting diodes" *Appl. Phys. Lett.* **85**, 142 (2004)

Siozade L., Leymarie J., Disseix P., Vasson A., Mihailovic M., Grandjean N., Leroux M., and Massies J. "Modelling of thermally detected optical absorption and luminescence of (In,Ga)N/GaN heterostructures" *Solid State Commun.* **115**, 575 (2000)

Smeeton T. M., Kappers M. J., Barnard J. S., Vickers M. E., and Humphreys C. J. "Electron-beam-induced strain within InGaN quantum wells: False indium "cluster" detection in the transmission electron microscope" *Appl. Phys. Lett.* **83**, 5419 (2003)

Su Y. K., personal communication at the *Second Asia-Pacific Workshop on Widegap Semiconductors* (APWS), Hsinchu, Taiwan, March 7–9 (2005)

Walukiewicz W., Li S. X., Wu J., Yu K. M., Ager III J. W., Haller E. E., Lu K., and Schaff W. J. "Optical properties and electronic structure of InN and In-rich group III-nitride alloys" *J. Cryst. Growth* **269**, 119 (2004)

Wetzel C., Amano H., Akasaki I., Ager III J. W., Grzegory I., and Meyer B. K. "DX-like behavior of oxygen in GaN" *Physica B* **302–303**, 23 (2001)

Wu J., Walukiewicz W., Yu K. M., Ager J. W., Haller E. E., Lu H., and Schaff W. J. "Small band gap bowing in $In_{1-x}Ga_xN$ alloys" *Appl. Phys. Lett.* **80**, 4741 (2002)

Wu J, Walukievicz W., Yu K. M., Ager III J. W., Li S. X., Haller E. E., Lu H., and Schaff W. J. "Universal bandgap bowing in group-III nitride alloys" *Solid State Commun.* **127**, 411 (2003)

Yasan A., McClintock R., Mayes K., Darvish S. R., Kung P., and Razeghi M. "Top-emission ultraviolet light-emitting diodes with peak emission at 280 nm" *Appl. Phys. Lett.* **81**, 801 (2002)

Yun F., Reshchikov M. A., He L., King T., Morkoç H., Novak S. W., and Wei L. "Energy band bowing parameter in $Al_xGa_{1-x}N$ alloys" *J. Appl. Phys.* **92**, 4837 (2002)

Zhang J. P., Chitnis A., Adivarahan V., Wu S., Mandavilli V., Pachipulusu R., Shatalov M., Simin G., Yang J. W., and Khan M. A. "Milliwatt power deep ultraviolet light-emitting diodes over sapphire with emission at 278 nm" *Appl. Phys. Lett.* **81**, 4910 (2002a)

Zhang J. P., Wang H. M., Gaevski M. E., Chen C. Q., Fareed Q., Yang J. W., Simin G., and Khan M. A. "Crack-free thick AlGaN grown on sapphire using AlN/AlGaN superlattices for strain management" *Appl. Phys. Lett.* **80**, 3542 (2002b)

Zhang J. P., Wu S., Rai S., Mandavilli V., Adivarahan V., Chitnis A., Shatalov M., and Khan M. A. "AlGaN multiple-quantum-well-based deep ultraviolet light-emitting diodes with significantly reduced long-wave emission" *Appl. Phys. Lett.* **83**, 3456 (2003)

14

Spontaneous emission from resonant cavities

14.1 Modification of spontaneous emission

Radiative transitions, i.e. transitions of electrons from an initial quantum state to a final state and the simultaneous emission of a light quantum, are one of the most fundamental processes in optoelectronic devices. There are two distinct ways by which the emission of a photon can occur, namely by *spontaneous* and *stimulated* emission. These two processes were first postulated by Einstein (1917).

Stimulated emission is employed in semiconductor lasers and superluminescent LEDs. It was realized in the 1960s that the stimulated emission mode can be used in semiconductors to drastically change the radiative emission characteristics. The efforts to harness stimulated emission resulted in the first room-temperature operation of semiconductor lasers (Hayashi *et al.*, 1970) and the first demonstration of a superluminescent LED (Hall *et al.*, 1962).

Spontaneous emission implies the notion that the recombination process occurs *spontaneously*, that is without a means to influence this process. In fact, spontaneous emission has long been believed to be uncontrollable. However, research in microscopic optical resonators, where spatial dimensions are of the order of the wavelength of light, showed the possibility of controlling the spontaneous emission properties of a light-emitting medium. The changes of the emission properties include the spontaneous emission rate, spectral purity, and emission pattern. These changes can be employed to make more efficient, faster, and brighter semiconductor devices. The changes in spontaneous emission characteristics in resonant cavity (RC) and photonic crystal (PC) structures were reviewed by Joannopoulos (1995).

Microcavity structures have been demonstrated with different active media and different microcavity structures. The first microcavity structure was proposed by Purcell (1946) for emission frequencies in the radio frequency (rf) regime. Small metallic spheres were proposed as the resonator medium. However, no experimental reports followed Purcell's theoretical publication. In the 1980s and 1990s, several microcavity structures were realized with different types of optically active media. The emission media included organic dyes (De Martini et al., 1987; Suzuki et al., 1991), semiconductors (Yablonovitch et al., 1988; Yokoyama et al., 1990), rare-earth-doped silica (Schubert et al., 1992b; Hunt et al., 1995b), and organic polymers (Nakayama et al., 1993; Dodabalapur et al., 1994). In these publications, clear changes in spontaneous emission were demonstrated including changes in spectral, spatial, and temporal emission characteristics.

At the beginning of the 1990s, current-injection **resonant-cavity light-emitting diodes** (RCLEDs) were first demonstrated in the GaAs material system (Schubert et al., 1992a) and subsequently in organic light-emitting materials (Nakayama et al., 1993). Both publications reported an emission line narrowing due to the resonant cavities. RCLEDs have many advantageous properties when compared with conventional LEDs, including higher brightness, increased spectral purity, and higher efficiency. For example, the spectral power density in RCLEDs was shown to be enhanced by more than one order of magnitude (Hunt et al., 1992, 1995a).

The changes in optical gain in VCSELs due to the enhancement in spontaneous emission was analyzed by Deppe and Lei (1992). The comparison of a *macro*cavity, in which the cavity is much *longer* than the emission wavelength ($\lambda \ll L_{cav}$), with a *micro*cavity ($\lambda \approx L_{cav}$) revealed that the gain can be enhanced by factors of 2–4 for typical GaAs emission linewidths at room temperature (50 nm). Thus laser threshold currents can be lower in microcavity structures due to the higher gain.

It is important to distinguish between emission *inside* the cavity and emission *out of* the cavity. The enhancement of the spontaneous emission *inside* the cavity and emission through one of the mirrors *out of* the cavity can be very different. At moderate values of the cavity finesse, the spontaneous emission inside and out of the cavity is enhanced. However, for very high finesse cavities (see, for example, Jewell et al., 1988), the overall emission out of the cavity *decreases* (Schubert et al., 1996). In the limit of very high reflectivity reflectors ($R_1 = R_2 \rightarrow 100\%$), the emission out of the cavity becomes zero. This effect will be discussed in detail below.

A device in which all the spontaneous emission occurs into a single optical mode has been proposed by Kobayashi et al. (1982, 1985). This device has been termed a **zero-threshold laser** (Yokoyama, 1992) and a **single-mode LED** (Yablonovitch, 1994). In a conventional laser, only a small portion of the spontaneous emission couples into a single state of the electromagnetic field controlled by the laser cavity. The rest is lost to free-space modes that radiate out of the side of the laser. The idea of a thresholdless laser is simple. It assumes a wavelength-size cavity in which only one optical mode exists. Thus spontaneous as well as stimulated emission couples to this optical mode. The thresholdless laser should lack a threshold, i.e. the clear distinction between the spontaneous and the lasing regime which is observed in the light-output versus current characteristic of conventional lasers. Clearly, the prospects of such a device are intriguing. Even though several attempts to demonstrate a thresholdless laser have been reported (Yokoyama et al., 1990; Yokoyama 1992; Numai et al., 1993), a thresholdless laser has not yet been demonstrated.

14.2 Fabry–Perot resonators

The simplest form of optical cavity consists of two coplanar mirrors separated by a distance L_{cav}. About one century ago, Fabry and Perot (1899) were the first to build and analyze optical cavities with coplanar reflectors. These cavities had a large separation between the two reflectors, i.e. $L_{cav} \gg \lambda$. However, if the distance between the two reflectors is of the order of the wavelength of light, new physical phenomena occur, including the enhancement of the optical emission from an active material inside the cavity. Very small cavities, with typical dimensions of $L_{cav} \approx \lambda$, will be denoted as *microcavities*.

Coplanar microcavities are the simplest form of optical microcavities and their properties are summarized below. For a detailed discussion of the optical properties of Fabry–Perot cavities, the reader is referred to the literature (Coldren and Corzine, 1995; Saleh and Teich, 1991). Fabry–Perot cavities with two reflectors of reflectivity R_1 and R_2 are shown in Figs. 14.1 (a) and (b). Plane waves propagating inside the cavity can interfere constructively and destructively resulting in stable (allowed) optical modes and attenuated (disallowed) optical modes, respectively. (Note that the photon length is much longer than the microcavity length.) For lossless (non-absorbing) reflectors, the transmittance through the two reflectors is given by $T_1 = 1 - R_1$ and $T_2 = 1 - R_2$. Taking into account multiple reflections inside the cavity, the transmittance through a Fabry–Perot cavity can be expressed in terms of a geometric series. The transmitted light intensity (transmittance) is then given by

$$T = \frac{T_1 T_2}{1 + R_1 R_2 - 2\sqrt{R_1 R_2} \cos 2\phi} \tag{14.1}$$

where ϕ is the phase change of the optical wave for a single pass between the two reflectors. Phase changes at the reflectors are neglected. The maxima of the transmittance occur if the

14 Spontaneous emission from resonant cavities

condition of constructive interference is fulfilled, i.e. if $2\phi = 0, 2\pi, ...$. Insertion of these values into Eq. (14.1) yields the transmittance maxima as

$$T_{max} = \frac{T_1 T_2}{\left(1 - \sqrt{R_1 R_2}\right)^2}. \tag{14.2}$$

For asymmetric cavities ($R_1 \neq R_2$), $T_{max} < 1$. For symmetric cavities ($R_1 = R_2$), the transmittance maxima are unity, $T_{max} = 1$.

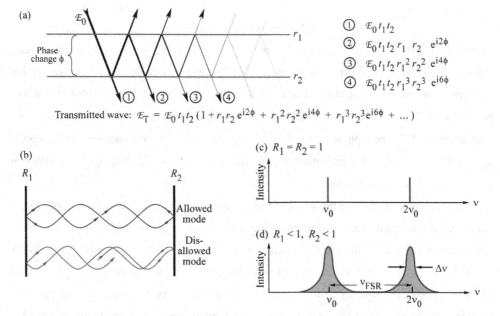

Fig. 14.1. (a) Transmission of a light wave with electric field ampltitude E_0 through a Fabry–Perot resonator. (b) Schematic illustration of allowed and disallowed optical modes in a Fabry–Perot cavity consisting of two coplanar reflectors. Optical mode density for a resonator with (c) no mirror losses ($R_1 = R_2 = 100\%$) and (d) mirror losses.

Exercise: *Transmission through a Fabry–Perot cavity*. Derive Eq. (14.1) by calculating the transmitted wave intensity in terms of a geometric series as illustrated in Fig. 14.1 (a).

Near $\phi = 0, 2\pi, 4\pi ...$, the cosine term in Eq. (14.1) can be expanded into a power series ($\cos 2\phi \approx 1 - 2\phi^2$). One obtains

$$T = \frac{T_1 T_2}{\left(1 - \sqrt{R_1 R_2}\right)^2 + \sqrt{R_1 R_2}\, 4\phi^2}. \tag{14.3}$$

Equation (14.3) indicates that near the maxima, the transmittance can be approximated by a lorentzian function. The transmittance T in Eq. (14.3) has a maximum at $\phi = 0$. The transmittance decreases to half of the maximum value at $\phi_{1/2} = [1 - (R_1 R_2)^{1/2}] / [4 (R_1 R_2)^{1/2}]^{1/2}$. For high values of R_1 and R_2, i.e. $R_1 \approx 1$ and $R_2 \approx 1$, it is $\phi_{1/2} = (1/2)[1 - (R_1 R_2)^{1/2}]$.

The cavity finesse, F, is defined as the ratio of the transmittance peak separation to the transmittance full-width at half-maximum, i.e.

$$\boxed{F = \frac{\text{peak separation}}{\text{peak width}} = \frac{\pi}{2\phi_{1/2}} = \frac{\pi \sqrt[4]{R_1 R_2}}{1 - \sqrt{R_1 R_2}} \approx \frac{\pi}{1 - \sqrt{R_1 R_2}}}. \tag{14.4}$$

Inspection of Eq. (14.4) shows that the finesse becomes very large for high values of R_1 and R_2.

The wavelength and frequency of light are practically more accessible than the phase. Equations (14.1)–(14.4) can be converted to wavelength and frequency using

$$\phi = 2\pi \frac{\bar{n} L_{\text{cav}}}{\lambda} = 2\pi \frac{\bar{n} L_{\text{cav}} \nu}{c} \tag{14.5}$$

where L_{cav} is the length of the cavity, λ is the wavelength of light in vacuum, ν is the frequency of light, and $\bar{n}$ is the refractive index inside the cavity. Figures 14.1 (c) and (d) show the transmittance through a cavity with $R = 1$ and $R < 1$, respectively. In the frequency domain, the transmittance peak separation is called the *free spectral* range ν_{FSR}, as shown in Fig. 14.1 (d). The finesse of the cavity in the frequency domain is then given by $F = \nu_{\text{FSR}} / \Delta\nu$.

Frequently the *cavity quality factor* Q rather than the finesse is used. The cavity Q is defined as the ratio of the transmittance peak frequency to the peak width. Using this definition and Eq. (14.4), one obtains

$$\boxed{Q = \frac{\text{peak frequency}}{\text{peak width}} = \frac{2\bar{n} L_{\text{cav}}}{\lambda} \frac{\pi \sqrt[4]{R_1 R_2}}{1 - \sqrt{R_1 R_2}} \approx \frac{2\bar{n} L_{\text{cav}}}{\lambda} \frac{\pi}{1 - \sqrt{R_1 R_2}}} \tag{14.6}$$

where the peak width is measured in units of frequency.

Figure 14.2 shows an example of a reflectance spectrum of a microcavity consisting of a four-pair Si/SiO$_2$ distributed Bragg reflector (DBR) deposited on a Si substrate, a SiO$_2$ center region, a 2.5-pair Si/SiO$_2$ DBR top reflector. The resonance wavelength of the cavity is approximately 1.0 µm. The reflectance of the cavity does not approach zero at the resonance wavelength due to the unequalness of the reflectivities of the two reflectors.

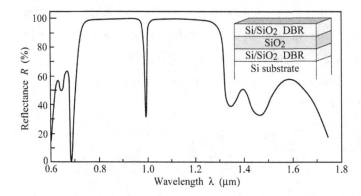

Fig. 14.2. Reflectance of a Fabry–Perot cavity consisting of two Si/SiO$_2$ reflectors and a SiO$_2$ center region. At the resonance wavelength ($\lambda \approx$ 1000 nm), the reflectivity has a sharp dip.

14.3 Optical mode density in a one-dimensional resonator

In this section, the enhancement of spontaneous emission will be calculated based on the changes of the *optical mode density* in a one-dimensional (1D) resonator, i.e. a coplanar Fabry–Perot microcavity. We first discuss the basic physics causing the changes of the spontaneous emission from an optically active medium located inside a microcavity and then derive analytical formulas for the spectral and integrated emission enhancement. The spontaneous radiative transition rate in an optically active, homogeneous medium is given by (see, for example, Yariv, 1982)

$$W_{\text{spont}} = \tau_{\text{spont}}^{-1} = \int_0^\infty W_{\text{spont}}^{(\ell)} \rho(\nu_\ell) \, d\nu_\ell \tag{14.7}$$

where $W_{\text{spont}}^{(\ell)}$ is the spontaneous transition rate into the optical mode ℓ and $\rho(\nu_\ell)$ is the optical mode density. Assuming that the optical medium is homogeneous, the spontaneous emission lifetime, τ_{spont}, is the inverse of the spontaneous emission rate. However, if the optical mode density in the device depends on the spatial direction, as in the case of a cavity structure, then the emission rate given in Eq. (14.7) depends on the direction. Equation (14.7) can be applied to some small range of solid angle along a certain direction, for example the direction perpendicular to the reflectors of a Fabry–Perot cavity. Thus, Eq. (14.7) can be used to calculate the emission rate along a specific direction, in particular the optical axis of a cavity.

14.3 Optical mode density in a one-dimensional resonator

The spontaneous emission rate into the optical mode ℓ, $W_{\text{spont}}^{(\ell)}$, contains the dipole matrix element of the two electronic states involved in the transition (Yariv, 1982). Thus $W_{\text{spont}}^{(\ell)}$ will *not* be changed by placing the optically active medium inside an optical cavity. However, the optical mode density, $\rho(v_\ell)$, is strongly modified by the cavity. Next, the changes in optical mode density will be used to calculate the changes in spontaneous emission rate.

We first compare the optical mode density in free space with the optical mode density in a microcavity. For simplicity, we restrict our considerations to the one-dimensional case, i.e. to the case of a coplanar Fabry–Perot microcavity. Furthermore, we restrict our considerations to the emission along the optical axis of the cavity.

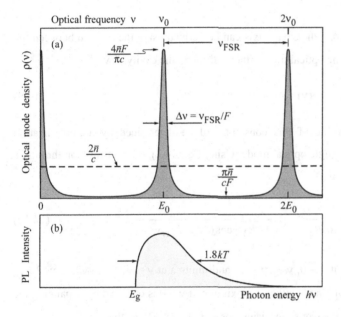

Fig. 14.3. (a) Optical mode density of a one-dimensional planar microcavity (solid line) and of homogeneous one-dimensional space. (b) Theoretical shape of the luminescence spectrum of bulk semiconductors.

In a one-dimensional homogeneous medium, the density of optical modes per unit length per unit frequency is given by

$$\rho^{1D}(v) = \frac{2\bar{n}}{c} \tag{14.8}$$

where $\bar{n}$ is the refractive index of the medium. Equation (14.8) can be derived using a similar formalism commonly used for the derivation of the mode density in free space. The constant optical mode density given by Eq. (14.8) is shown in Fig. 14.3 (a).

245

14 Spontaneous emission from resonant cavities

Exercise: *Optical mode density*. Derive Eq. (14.8), i.e. optical mode density in a 1D space.

In planar microcavities, the optical modes are discrete and the frequencies of these modes are integer multiples of the fundamental mode frequency. The optical mode density of a planar microcavity is shown schematically in Fig. 14.3 (a). The fundamental and first excited mode occur at frequencies of v_0 and $2v_0$, respectively. For a cavity with two metallic reflectors (no distributed Bragg reflectors) and a π phase shift of the optical wave upon reflection, the fundamental frequency is given by $v_0 = c/(2\bar{n} L_{cav})$, where c is the velocity of light in vacuum and L_{cav} is the length of the cavity. In a *resonant microcavity*, the emission frequency of an optically active medium located inside the cavity equals the frequency of one of the cavity modes.

The optical mode density along the cavity axis can be derived using the relation between the mode density in the cavity and the optical transmittance through the cavity, $T(v)$,

$$\rho(v) = K T(v) \tag{14.9}$$

where K is a constant. The value of this constant will be determined by a normalization condition, i.e. by considering a single optical mode. Using Eq. (14.1), the transmission through a Fabry–Perot cavity can be written as

$$T(v) = \frac{T_1 T_2}{1 + R_1 R_2 - 2\sqrt{R_1 R_2} \cos(4\pi \bar{n} L_{cav} v/c)}. \tag{14.10}$$

The transmittance has maxima at $v = 0, v_0, 2v_0 \ldots$, and minima at $v = v_0/2, 3v_0/2, 5v_0/2 \ldots$. The lorentzian approximation of a transmittance maximum at $v = 0$ is obtained by expanding the cosine term in Eq. (14.10) into a power series using $\cos x \approx 1 - (x^2/2)$, so that

$$T(v) = \frac{T_1 T_2 \left(\sqrt{R_1 R_2}\right)^{-1} (4\pi \bar{n} L_{cav}/c)^{-2}}{\frac{\left(1 - \sqrt{R_1 R_2}\right)^2}{\left(\sqrt{R_1 R_2}\right)^{-1} (4\pi \bar{n} L_{cav}/c)^2} + v^2}. \tag{14.11}$$

Integrating $\rho(v)$ over all frequencies and the cavity length yields a single optical mode, i.e.

$$K \int_0^{L_c} \int_{-\infty}^{\infty} \rho(v) \, dv \, dL = 1. \tag{14.12}$$

The lower and upper limit of the frequency integration can be chosen to be $\pm\infty$ since the lorentzian approximation of Eq. (14.11) has only one maximum at $v = 0$. Using Eqs. (14.9), (14.11), and the integration formula $\int_{-\infty}^{\infty} (a^2 + x^2)^{-1} dx = \pi/a$ yields

$$K = \frac{(R_1 R_2)^{3/4}}{T_1 T_2} \frac{4\bar{n}}{c} \left(1 - \sqrt{R_1 R_2}\right). \tag{14.13}$$

Using Eq. (14.9), the optical mode density of a one-dimensional cavity for emission along the cavity axis is then given by

$$\boxed{\rho(v) = \frac{(R_1 R_2)^{3/4}}{T_1 T_2} \frac{4\bar{n}}{c} \left(1 - \sqrt{R_1 R_2}\right) T(v)}. \tag{14.14}$$

Equation (14.14) allows one to calculate the density of optical modes at the maxima and minima. At the *maxima*, the mode density is given by

$$\rho_{max} = \frac{(R_1 R_2)^{3/4}}{1 - \sqrt{R_1 R_2}} \frac{4\bar{n}}{c}. \tag{14.15}$$

Using $(R_1 R_2)^{3/4} \approx 1$ and the expression derived for the finesse F (see Eq. 14.4), one obtains an approximate expression for the mode density at the maxima

$$\rho_{max} \approx \frac{4}{\pi} \frac{\bar{n} F}{c}. \tag{14.16}$$

That is, the mode density at the maxima is proportional to the finesse of the cavity. At the *minima*, the mode density is given by

$$\rho_{min} = \frac{(R_1 R_2)^{3/4} \left(1 - \sqrt{R_1 R_2}\right)}{\left(1 + \sqrt{R_1 R_2}\right)^2} \frac{4\bar{n}}{c}. \tag{14.17}$$

Using $(R_1 R_2)^{3/4} \approx 1$ and the expression derived for the finesse F (see Eq. 14.4), one obtains an approximate expression for the mode density at the minima

$$\rho_{min} \approx \pi \frac{\bar{n}}{cF}. \tag{14.18}$$

14 Spontaneous emission from resonant cavities

That is, the mode density at the minima is *inversely* proportional to the finesse of the cavity.

The comparison of the optical mode densities of a one-dimensional (1D) free space and a 1D planar cavity is shown in Fig. 14.3. Note that the *mode density is conserved*, i.e. the areas below the 1D mode density and the 1D cavity mode density are the same.

14.4 Spectral emission enhancement

Because the emission rate at a given wavelength is directly proportional to the optical mode density (see Eq. 14.7), the emission rate *enhancement spectrum* is given by the ratio of the 1D cavity mode density to the 1D free space mode density. As calculated earlier, the cavity enhancement spectrum has a lorentzian lineshape. The enhancement factor at the resonance wavelength is thus given by the ratio of the optical mode densities with and without a cavity, i.e.

$$G_e = \frac{\rho_{max}}{\rho^{1D}} \approx \frac{2}{\pi} F \approx \frac{2}{\pi} \frac{\pi (R_1 R_2)^{1/4}}{1 - \sqrt{R_1 R_2}} . \tag{14.19}$$

The equation shows that a strong enhancement of the spontaneous emission rate along the cavity axis can be achieved with microcavities.

Equation (14.19) represents the *average* emission rate enhancement out of *both* reflectors of the cavity. To find the enhancement in a *single* direction, we multiply the enhancement given by Eq. (14.19) by the fraction of the light exiting the mirror with reflectivity R_1 (i.e. $1 - R_1$) divided by the average loss of the two mirrors for one round trip in the cavity (i.e. $(1/2) [(1 - R_1) + (1 - R_2)]$). For large R_1 and R_2, this gives for the enhancement of the emission exiting R_1

$$G_e \approx \frac{2(1-R_1)}{2-R_1-R_2} \frac{2F}{\pi} \approx \frac{1-R_1}{1-\sqrt{R_1 R_2}} \frac{2F}{\pi} \approx \frac{2}{\pi} \frac{\pi (R_1 R_2)^{1/4} (1-R_1)}{\left(1-\sqrt{R_1 R_2}\right)^2} \tag{14.20}$$

where we used the approximation $1 - (R_1 R_2)^{1/2} \approx (1/2)(1 - R_1 R_2) \approx (1/2)(2 - R_1 - R_2)$. Equation (14.20) represents the emission rate enhancement from a *single* reflector with reflectivity R_1.

Next we take into account the standing wave effect, that is, the distribution of the optically active material relative to the nodes and antinodes of the optical wave. The antinode enhancement factor ξ has a value of 2, if the active region is located exactly at an antinode of the standing wave inside the cavity. The value of ξ is unity if the active region is smeared out over many periods of the standing optical wave. Finally, $\xi = 0$ if the active material is located at a node.

The emission rate enhancement is then given by

$$G_e = \frac{\xi}{2} \frac{2}{\pi} \frac{\pi(R_1 R_2)^{1/4}(1-R_1)}{\left(1 - \sqrt{R_1 R_2}\right)^2} \frac{\tau_{cav}}{\tau} \qquad (14.21)$$

where R_1 is the reflectivity of the light-exit mirror and therefore $R_1 < R_2$. Equation (14.21) also takes into account changes in the spontaneous emission lifetime in terms of τ, the lifetime without cavity, and τ_{cav}, the lifetime with cavity. The factor of τ_{cav}/τ ensures that the enhancement decreases if the cavity lifetime is reduced as a result of the cavity. For planar microcavities, the ratio of the spontaneous lifetime with a cavity, τ_{cav}, and the lifetime without a cavity, τ, is $\tau_{cav}/\tau \geq 0.9$ (Vredenberg et al., 1993). Thus, the emission lifetime is changed by only a minor amount in a planar microcavity.

14.5 Integrated emission enhancement

The total enhancement *integrated over wavelength*, rather than the enhancement at the resonance wavelength, is relevant for many practical devices. *On resonance*, the emission is enhanced along the axis of the cavity. However, sufficiently far *off resonance*, the emission is suppressed. Because the natural emission spectrum of the active medium (without a cavity) can be much broader than the cavity resonance, it is, *a priori*, not clear whether the integrated emission is enhanced at all. To calculate the wavelength-integrated enhancement, the spectral width of the cavity resonance and the spectral width of the natural emission spectrum must be determined. The resonance spectral width can be calculated from the finesse of the cavity or the cavity quality factor.

The theoretical width of the emission spectrum of bulk semiconductors is $1.8kT$ (see, for example, Schubert, 1993), where k is Boltzmann's constant and T is the absolute temperature. At room temperature, $1.8kT$ corresponds to an emission linewidth of $\Delta\lambda_n = 31$ nm for an emission wavelength of 900 nm. For a cavity resonance width of 5–10 nm, one part of the spectrum is strongly enhanced, whereas the rest of the spectrum is suppressed. The integrated enhancement ratio (or suppression ratio) can be calculated analytically by assuming a gaussian natural emission spectrum. For semiconductors at 300 K, the linewidth of the natural emission is, in the case of high-finesse cavities, *larger* than the width of the cavity resonance. The gaussian

14 Spontaneous emission from resonant cavities

emission spectrum has a width of $\Delta\lambda_n = 2\sigma(2\ln 2)^{1/2}$ and a peak value of $(\sigma(2\pi)^{1/2})^{-1}$, where σ is the standard deviation of the gaussian function. The integrated enhancement ratio (or suppression ratio) is then given by (Hunt et al., 1993)

$$G_{int} = \frac{\pi}{2} G_e \Delta\lambda \frac{1}{\sigma\sqrt{2\pi}} = G_e \sqrt{\pi \ln 2} \frac{\Delta\lambda}{\Delta\lambda_n} \tag{14.22}$$

where the factor of $\pi/2$ is due to the lorentzian lineshape of the enhancement spectrum. Hence, the integrated emission enhancement depends on the natural emission linewidth of the active material. The value of G_{int} can be quite different for different types of optically active materials. Narrow atomic emission spectra can be enhanced by several orders of magnitude (Schubert et al., 1992b). On the other hand, materials having broad emission spectra such as dyes or polymers (de Martini et al., 1987; Suzuki et al., 1991) may not exhibit any integrated enhancement at all. Equation (14.22) also shows that the width of the resonance has a profound influence on the integrated enhancement. Narrow resonance spectral widths, i.e. high finesse values or long cavities (Hunt et al., 1992), reduce the integrated enhancement.

Example: *Spectral enhancement and integrated enhancement of a resonant-cavity structure.* As an example, we calculate the spectral and wavelength-integrated enhancement of a semiconductor resonant-cavity structure using Eqs. (14.21) and (14.22). With the reflectivities $R_1 = 90\%$ and $R_2 = 97\%$, an antinode enhancement factor of $\xi = 1.5$ and $\tau_{cav}/\tau \approx 1$, one obtains a finesse of $F = 46$, and a peak enhancement factor of $G_e = 68$ using Eq. (14.21). Insertion of this value into Eq. (14.22), using a cavity resonance bandwidth of $\Delta\lambda = 6.5$ nm (Schubert et al., 1994), and the theoretical 300 K natural emission linewidth of $\Delta\lambda = 31$ nm, one obtains a theoretical integrated enhancement factor of $G_{int} = 13$. Experimental enhancement factors of 5 have been demonstrated (Schubert et al., 1994) for the reflectivity values assumed above. The lower experimental enhancement is in part due to a broader natural emission linewidth, which exceeds the theoretical value of $1.8\,kT$.

The spontaneous emission spectrum of a bulk semiconductor is shown schematically in Fig. 14.3 (b). For maximum enhancement along the cavity axis, the cavity must be in resonance with the natural emission spectrum. Note that additional broadening mechanisms, such as alloy broadening, will broaden the natural emission spectrum over its theoretical value of $1.8\,kT$. Quantum well structures have inherently narrower spectra ($0.7\,kT$), due to the step-function-like density of states. Low temperatures and excitonic effects can further narrow the natural emission linewidth. Thus, higher enhancements are expected for low temperatures and quantum well active regions.

14.6 Experimental emission enhancement and angular dependence

Particularly high spontaneous emission enhancements can be attained with emitters that have naturally narrow emission lines. Atomic transitions, as in rare-earth elements, have such narrow emission lines. For this reason, rare-earth doped cavities are a great system to study the emission enhancement. The emission spectrum of a high-finesse Si/SiO_2 cavity with an erbium-doped SiO_2 active layer is shown in Fig. 14.4 (Schubert et al., 1992b). A distinct narrowing of the Er emission spectrum and a giant enhancement of the emission was found. The enhancement factor was greater than 50, when compared to a non-cavity structure.

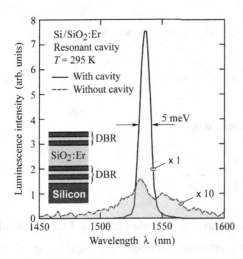

Fig. 14.4. Photoluminescence spectra of Er-doped SiO_2. One of the spectra is for Er-doped SiO_2 located in a cavity resonant at 1540 nm. The other spectrum is without a cavity. An emission enhancement factor greater than 50 is found (after Schubert et al., 1992b).

The cavity consisted of a 4-pair Si/SiO_2 bottom DBR with a calculated reflectivity of 99.8%, the Er-doped $\lambda/2$ thick SiO_2 active layer, and a 2.5-pair Si/SiO_2 top DBR with a calculated reflectivity of 98.5%. The SiO_2 active layer was doped with Er of dose 7.7×10^{15} cm^{-2}. The measured cavity finesse was $F = 310$, slightly lower than the calculated finesse of 370.

The peak emission wavelength depends on the emission angle. The angular dependence can be derived from the condition that the normal wave vector, $k_\perp$, must be constant (in resonance with the cavity), independent of the k propagation direction. This condition can be written as

$$k \cos\theta = k_\perp \quad (14.23)$$

or

$$\frac{2\pi}{\lambda_e(\theta)} \cos\theta = \frac{2\pi}{\lambda_{res}} \quad (14.24)$$

where λ_{res} and λ_e are the resonance and emission wavelengths, respectively, and θ is the angle inside the cavity with respect to the surface normal (polar angle), as illustrated in Fig. 14.5.

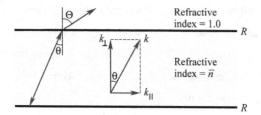

Fig. 14.5. Perpendicular and parallel component of wave vector k for light propagating in a resonant cavity.

Using Eq. (14.24) and Snell's law ($\bar{n} \sin \theta = \sin \Theta$), the emission wavelength is given by

$$\lambda_e = \lambda_{res} \cos\left[\arcsin\left(\frac{1}{\bar{n}} \sin \Theta\right)\right]. \tag{14.25}$$

For small angles, the equation can be approximated by

$$\lambda_e \approx \lambda_{res} \left(1 - \frac{\Theta^2}{2\bar{n}^2}\right). \tag{14.26}$$

The angular dependence of the peak emission wavelength is shown in Fig. 14.6. The solid and dashed lines are the exact (Eq. 14.25) and approximate solutions (Eq. 14.26), respectively. A refractive index of 1.5 is used in the calculation.

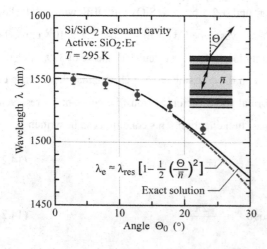

Fig. 14.6. Peak emission wavelength as a function of polar angle for a planar Si/SiO$_2$:Er resonant cavity (after Schubert et al., 1992b).

A similar angular dependence is found for semiconductors. The emission spectra of an AlGaInP cavity are shown in Fig. 14.7 for different emission angles (Streubel *et al.*, 2002). Note that the highest intensity is found at an emission angle of 30°, indicating that the resonance wavelength of the cavity is located at the long-wavelength end of the semiconductor natural emission spectrum. This results in the highest angle-integrated emission power (Streubel *et al.*, 2002).

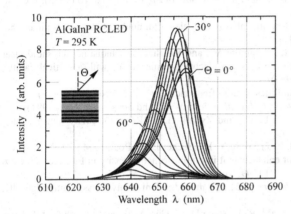

Fig. 14.7. Emission spectra of AlGaInP RCLED for different polar angles. The long-wavelength part of the QW emission is emitted in the forward direction (0°). The shorter wavelengths are emitted off-axis. When measured with an integrating sphere, an 18 nm wide spectrum (FWHM) is found (after Streubel *et al.*, 2002).

References

Coldren L. A. and Corzime S. W. *Diode Lasers and Photonic Integrated Circuits* (John Wiley and Sons, New York, 1995)

De Martini F., Innocenti G., Jacobovitz G. R., and Mataloni P. "Anomalous spontaneous emission time in a microscopic optical cavity" *Phys. Rev. Lett.* **59**, 2955 (1987)

Deppe D. G. and Lei C. "Spontaneous emission and optical gain in a Fabry–Perot microcavity" *Appl. Phys. Lett.* **60**, 527 (1992)

Dodabalapur A., Rothberg L. J., and Miller T. M. "Color variation with electroluminescent organic semiconductors in multimode resonant cavities" *Appl. Phys. Lett.* **65**, 2308 (1994)

Einstein A. "On the quantum theory of radiation" (translated from German) *Z. Phys.* **18**, 121 (1917)

Fabry G. and Perot A. "Theory and applications of a new interference method for spectroscopy" (translated from French) *Ann. Chim. Phys.* **16**, 115 (1899)

Hall R. N., Fenner G. E., Kingsley J. D., Soltys T. J., and Carlson R. O. "Coherent light emission from GaAs junctions" *Phys. Rev. Lett.* **9**, 366 (1962)

Hayashi I., Panish M. B., Foy P. W., and Sumski S. "Junction lasers which operate continuously at room temperature" *Appl. Phys. Lett.* **17**, 109 (1970)

Hunt N. E. J., Schubert E. F., Logan R. A., and Zydzik G. J. "Enhanced spectral power density and reduced linewidth at 1.3 µm in an InGaAsP quantum well resonant cavity light emitting diode" *Appl. Phys. Lett.* **61**, 2287 (1992)

Hunt N. E. J., Schubert E. F., Kopf R. F., Sivco D. L., Cho A. Y., and Zydzik G. J. "Increased fiber communications bandwidth from a resonant cavity light emitting diode emitting at λ = 940 nm" *Appl. Phys. Lett.* **63**, 2600 (1993)

Hunt N. E. J., Schubert E. F., Sivco D. L., Cho A. Y., Kopf R. F., Logan R. A., and Zydzik G. J. "High efficiency, narrow spectrum resonant cavity light-emitting diodes" in *Confined Electrons and Photons* edited by E. Burstein and C. Weisbuch (Plenum Press, New York, 1995a)

Hunt N. E. J., Vredenberg A. M., Schubert E. F., Becker P. C., Jacobson D. C., Poate J. M., and Zydzik G. J. "Spontaneous emission control of Er^{3+} in Si/SiO_2 microcavities" in *Confined Electrons and Photons* edited by E. Burstein and C. Weisbuch (Plenum Press, New York, 1995b)

Jewell J. L., Lee Y. H., McCall S. L., Harbison J. P., and Florez L. T. "High-finesse AlGaAs interference filters grown by molecular beam epitaxy" *Appl. Phys. Lett.* **53**, 640 (1988)

Joannopoulos J. D., Meade R. D., and Winn J. N. *Photonic Crystals* (Princeton University Press, Princeton, 1995)

Kobayashi T., Segawa T., Morimoto A., and Sueta T. *Meeting of the Jpn. Soc. of Appl. Phys.* Tokyo (1982); see also Yokoyama (1992)

Kobayashi T., Morimoto A., and Sueta T. *Meeting of the Jpn. Soc. of Appl. Phys.* Tokyo (1985); see also Yokoyama (1992)

Nakayama T., Itoh Y., and Kakuta A. "Organic photo- and electroluminescent devices with double mirrors" *Appl. Phys. Lett.* **63**, 594 (1993)

Numai T., Kosaka H., Ogura I., Kurihara K., Sugimoto M., and Kasahara K. "Indistinct threshold laser operation in a pnpn vertical to surface transmission electrophotonic device with a vertical cavity" *IEEE J. Quantum Electron.* **29**, 403 (1993)

Purcell E. M. "Spontaneous emission probabilities at radio frequencies" *Phys. Rev.* **69**, 681 (1946)

Saleh B. E. A. and Teich M. C. *Fundamentals of Photonics* (John Wiley and Sons, New York, 1991)

Schubert E. F., Wang Y. H., Cho A. Y., Tu L. W., and Zydzik G. J. "Resonant cavity light emitting diode" *Appl. Phys. Lett.* **60**, 921 (1992a)

Schubert E. F., Vredenberg A. M., Hunt N. E. J., Wong Y. H., Becker P. C., Poate J. M., Jacobson D. C., Feldman L. C., and Zydzik G. J. "Giant enhancement in luminescence intensity in Er-doped Si/SiO_2 resonant cavities" *Appl. Phys. Lett.* **61**, 1381 (1992b)

Schubert E. F. *Doping in III–V Semiconductors* p. 512 (Cambridge University Press, Cambridge UK, 1993)

Schubert E. F., Hunt N. E. J., Micovic M., Malik R. J., Sivco D. L., Cho A. Y., and Zydzik G. J. *Science* **265**, 943 (1994)

Schubert E. F., Hunt N. E. J., Malik R. J., Micovic M., and Miller D. L. "Temperature and modulation characteristics of resonant cavity light-emitting diodes" *IEEE J. Lightwave Technol.* **14**, 1721 (1996)

Streubel K., Linder N., Wirth R., and Jaeger A. "High brightness AlGaInP light-emitting diodes" *IEEE J. Sel. Top. Quantum Electron.* **8**, 321 (2002)

Suzuki M., Yokoyama H., Brorson S. D., and Ippen E. P. "Observation of spontaneous emission lifetime change of dye-containing Langmuir–Blodgett films in optical microcavities" *Appl. Phys. Lett.* **58**, 998 (1991)

Vredenberg A. M., Hunt N. E. J., Schubert E. F., Jacobson D. C., Poate J. M., and Zydzik G. J. *Phys. Rev. Lett.* **71**, 517 (1993)

Yablonovitch E., Gmitter T. J., and Bhat R. "Inhibited and enhanced spontaneous emission from optically thin AlGaAs/GaAs double heterostructures" *Phys. Rev. Lett.* **61**, 2546 (1988)

Yablonovitch E. personal communication (1994)

Yariv A. *Theory and Applications of Quantum Mechanics* (John Wiley and Sons, New York, 1982) p. 143

Yokoyama H., Nishi K., Anan T., Yamada H., Boorson S.D., and Ippen E. P. "Enhanced spontaneous emission from GaAs quantum wells in monolithic microcavities" *Appl. Phys. Lett.* **57**, 2814 (1990)

Yokoyama H. "Physics and device applications of optical microcavities" *Science* **256**, 66 (1992)

15

Resonant-cavity light-emitting diodes

15.1 Introduction and history

The resonant-cavity light-emitting diode (RCLED) is a light-emitting diode that has a light-emitting region inside an optical cavity. The optical cavity has a thickness of typically one-half or one times the wavelength of the light emitted by the LED, i.e. a fraction of a micrometer for devices emitting in the visible or in the infrared. The resonance wavelength of the cavity coincides or is in resonance with the emission wavelength of the light-emitting active region of the LED. Thus the cavity is a *resonant cavity*. The spontaneous emission properties from a light-emitting region located inside the resonant cavity are enhanced by the resonant-cavity effect. The RCLED is the first practical device making use of spontaneous emission enhancement occurring in microcavities.

The placement of an active region inside a resonant cavity results in multiple improvements of the device characteristics. Firstly, the light intensity emitted from the RCLED along the axis of the cavity, i.e. normal to the semiconductor surface, is higher compared with conventional LEDs. The enhancement factor is typically a factor of 2–10. Secondly, the emission spectrum of the RCLED has a *higher spectral purity* compared with conventional LEDs. In conventional LEDs, the spectral emission linewidth is determined by the thermal energy kT. However, in RCLEDs, the emission linewidth is determined by the quality factor (Q factor) of the optical cavity. As a result, the spectral emission width of the RCLED is a factor of 2–5 narrower compared with conventional LEDs. For the same reason, the wavelength shift with temperature is determined by the temperature coefficient of the optical cavity and not by the energy gap of the active material. This results in a significantly higher temperature stability of the RCLED emission wavelength compared with conventional LEDs. Thirdly, the emission far-field pattern of the RCLED is more *directed* compared with conventional LEDs. In conventional LEDs, the emission pattern is lambertian (i.e. cosine-function-like). In an RCLED, the emission pattern is directed mostly along the optical axis of the cavity.

These characteristics of RCLEDs are desirable for local-area, medium bit rate optical communication systems. LEDs play an important role in local-area (< 5 km) medium bit rate (< 1 Gbit/s) optical communication networks. In particular, plastic optical fibers are increasingly used for optical communication over short distances. The higher emission intensity and the more directed emission pattern afforded by the RCLED increase the power coupled into the optical fiber. As a result, the RCLED can transmit data over longer distances. Furthermore, the higher spectral purity of RCLEDs results in less chromatic dispersion allowing for higher bit rates.

Light-emitting diodes are the transmitter device of choice for medium bit rate optical communication over distances less than 5 km. Compared with lasers, LEDs are less expensive, more reliable, and less temperature sensitive. The RCLED has improved characteristics compared with conventional LEDs while maintaining the inherent advantages of LEDs. The reflectivity of the RCLED reflectors is lower compared with vertical-cavity surface-emitting lasers (VCSELs), thereby allowing for a lower RCLED manufacturing cost compared with VCSELs. At 650 nm, the preferred communication wavelength for plastic optical fibers, VCSELs are difficult to manufacture due to the lack of high-reflectivity reflectors.

RCLEDs are also used for high-brightness applications (Streubel et al., 1998; Wirth et al., 2001, 2002). In these devices, the resonance wavelength is designed to be at the long-wavelength end of the spontaneous emission spectrum of the semiconductor. This ensures that the emission intensity, integrated over all spatial directions, is maximized.

The enhanced spontaneous emission occurring in resonant-cavity structures can be beneficially employed in semiconductor and polymer LEDs. Resonant-cavity light-emitting diodes were first realized in 1992 (Schubert et al., 1992a) in the GaAs material system. About a year later, RCLEDs were demonstrated in organic materials (Nakayama et al., 1993).

Resonant-cavity structures with enhanced spontaneous emission also include Er-doped microcavities (Schubert et al., 1992b). Owing to the inherently narrow luminescence line of intra-atomic Er radiative transitions, there is a very good overlap between the cavity optical mode and the Er luminescence line. At the present time, no Er-doped current-injection devices exist. However, the great potential of Er-doped resonant cavities makes the realization of Er-doped RCLEDs likely in the future.

15.2 RCLED design rules

The basic structure of an RCLED is shown in Fig. 15.1 and comprises two mirrors with reflectivity R_1 and R_2. The reflectivity of the two mirrors is chosen to be unequal so that the light exits the cavity predominantly through one of the mirrors. This mirror is called the **light-exit mirror**. Here we designate the mirror with reflectivity R_1 as the light-exit mirror. An active region is located between the mirrors, preferably at the antinode location of the standing optical wave of the cavity, as shown in Fig. 15.1. Metal mirrors are assumed in Fig. 15.1 so that the wave amplitude is zero at the location of the mirrors.

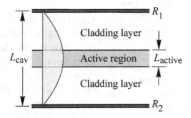

Fig. 15.1. Schematic illustration of a resonant cavity consisting of two metal mirrors with reflectivity R_1 and R_2. The active region has a thickness L_{active} and an absorption coefficient α. Also shown is the standing optical wave. The cavity length L_{cav} is equal to $\lambda/2$.

Next, we summarize several design rules intended to maximize the spontaneous emission enhancement in resonant-cavity structures (Schubert *et al.*, 1994, 1996; Hunt *et al.*, 1995a, 1995b). These rules will provide further insight into the fundamental operating principles of RCLEDs and the differences of these devices with respect to VCSELs.

The *first design criterion* for RCLEDs is that the reflectivity of the light-exit reflector, R_1, should be much lower than the reflectivity of the back reflector, i.e.

$$R_1 \ll R_2 . \tag{15.1}$$

This condition ensures that light exits the device mainly through the reflector with reflectivity R_1. Equation (15.1) applies to the design of communication RCLEDs, where light is emitted into the small core of a multimode fiber, and display RCLEDs, where light should be emitted towards the observer.

The *second design criterion* calls for the shortest possible cavity length L_{cav}. In order to derive this criterion, the integrated enhancement, discussed in a preceding section, can be rewritten by using the expressions for the cavity finesse F and cavity quality factor Q. One obtains

$$G_{\text{int}} = \frac{\xi}{2} \frac{2}{\pi} \frac{1 - R_1}{1 - \sqrt{R_1 R_2}} \sqrt{\pi \ln 2} \frac{\lambda}{\Delta \lambda_n} \frac{\lambda_{\text{cav}}}{L_{\text{cav}}} \frac{\tau_{\text{cav}}}{\tau} \tag{15.2}$$

where λ and λ_{cav} are the active region emission wavelengths in vacuum and inside the cavity, respectively. Since the emission wavelength λ and the natural linewidth of the active medium, $\Delta\lambda_n$, are given quantities, Eq. (15.2) shows that minimization of the cavity length L_{cav} maximizes the integrated intensity.

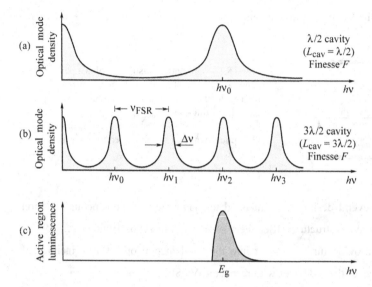

Fig. 15.2. Optical mode density for (a) a short and (b) a long cavity with the same finesse F. (c) Spontaneous free-space emission spectrum of an LED active region. The spontaneous emission spectrum has a better overlap with the short-cavity mode spectrum compared with the long-cavity mode spectrum.

The importance of a short cavity length is elucidated by Fig. 15.2. The optical mode densities of two different cavities, namely a short and a long cavity, are shown in Figs. 15.2 (a) and (b), respectively. Both cavities have the same mirror reflectivities and finesse. The natural emission spectrum of the active region is shown in Fig. 15.2 (c). The best overlap between the resonant optical mode and the active region emission spectrum is obtained for the shortest cavity.

The largest enhancements are achieved with the shortest cavities, which in turn are obtained if the *fundamental* cavity mode is in resonance with the emission from the active medium. The cavity length is also reduced by using a DBR with a short penetration depth, i.e. a DBR consisting of two materials with a large difference in the refractive index.

The **third design criterion** is the minimization of self-absorption in the active region. This criterion can be stated as follows: the reabsorption probability of photons emitted from the active region into the cavity mode should be much smaller than the escape probability of photons through one of the reflectors. Assuming $R_2 \approx 1$, this criterion can be written as

$$2\xi\alpha L_{active} < (1 - R_1) \tag{15.3}$$

where α and L_{active} are the absorption coefficient and the thickness of the active region, respectively. If the criterion of Eq. (15.3) were not fulfilled, photons would most likely be reabsorbed by the active region. Subsequently, re-emission will, with a certain probability, occur along the lateral direction (waveguided modes), i.e. not into the cavity mode. Another possibility is that the electron–hole pairs generated by reabsorption recombine non-radiatively. In either case, reabsorption processes occurring in high-finesse cavities *reduce the cavity mode emission out of the cavity*. Thus, if the condition of Eq. (15.3) is not fulfilled, the emission intensity of resonant cavities is lowered rather than enhanced.

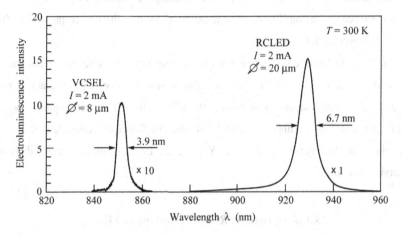

Fig. 15.3. Spontaneous electroluminescence spectrum of a vertical-cavity surface-emitting laser (VCSEL) emitting at 850 nm and of a resonant-cavity light-emitting diode (RCLED) emitting at 930 nm. The drive current for both devices is 2 mA. The VCSEL spectrum is multiplied by a factor of 10. The threshold current of the VCSEL is 7 mA (after Schubert *et al.*, 1996).

Whereas the condition of Eq. (15.3) is fulfilled in RCLEDs, it is clearly not fulfilled in VCSELs. The spontaneous emission intensities of RCLEDs and VCSELs were compared by Schubert *et al.* (1996). In this comparison, the VCSEL and the RCLED were driven by an injection current of 2 mA, which is below the threshold current of the VCSEL of I_{th} = 7 mA. The spontaneous emission spectra of an RCLED and a VCSEL are shown in Fig. 15.3. The VCSEL has an AlGaAs/GaAs quantum well active region emitting at 850 nm. Both reflectors of the VCSEL are AlGaAs/AlAs DBRs. Figure 15.3 reveals that the emission intensity of the VCSEL in the *spontaneous* regime is more than a factor of 15 lower than the emission intensity from the RCLED.

Because the magnitude of the maximum gain in semiconductors is always lower than the magnitude of the absorption coefficient in an unpumped semiconductor ($|g| < |\alpha|$), VCSELs could not lase if the condition of Eq. (15.3) were met. Thus, the spontaneous emission intensity in VCSELs is low and *must be low* in order to enable the device to lase. Figure 15.3 also reveals that the emission spectral linewidth of VCSELs is narrower than that of RCLEDs. The higher spectral purity is due to the higher values of R_1 and R_2 as required for VCSELs.

The fulfillment of Eq. (15.3) by RCLEDs also implies that these devices *cannot lase*. As stated above, it is always $|g| < |\alpha|$. Consequently, the mirror loss $(1 - R_1)$ is always larger than the maximum achievable round-trip gain ($2\xi g L_{active}$). The fundamental inability of RCLEDs to lase has been verified experimentally by pulsed injecting currents of large magnitude without finding any evidence for lasing. These considerations show that the device physics of RCLEDs and VCSELs is fundamentally different.

The arguments used above imply that the *spontaneous* emission into the fundamental cavity mode in VCSEL structures is very low due to reabsorption of photons by the active region. A reduction of the threshold current by increasing the reflectivity will be accompanied by a further decrease of the *spontaneous* emission below threshold. We therefore conclude that the so-called zero-threshold laser (Kobayashi *et al.* 1982; Yokoyama, 1992) cannot be realized by a planar resonant-cavity structure.

15.3 GaInAs/GaAs RCLEDs emitting at 930 nm

The structure of an RCLED with a GaInAs active region is shown in Fig. 15.4 (a). The cavity is defined by one distributed Bragg reflector (DBR) and one metallic reflector. Also included are two confinement regions and a four-quantum-well active region. The heavily doped n-type substrate is coated with a ZrO_2 anti-reflection layer (Schubert *et al.*, 1994). A picture of the first RCLED is shown in Fig. 15.4 (b).

The motivation for the metal reflector is twofold. Firstly, the metallic Ag reflector serves as a non-alloyed ohmic contact to the heavily doped p-type ($N_A \approx 5 \times 10^{19}$ cm^{-3}) GaAs top layer, thus effectively confining the pumped region to the area below the contact. Secondly, it was shown in the preceding section that the cavity length must be kept as short as possible for maximizing the emission enhancement. Owing to the lack of a penetration depth, metal reflectors allow for a short cavity length. Cavities with two metallic reflectors have been reported (Wilkinson *et al.*, 1995). However, optical absorption losses in the light-exit mirror can be large in a double metal mirror structure, unless very thin metallic reflectors are used (Tu *et al.*, 1990). The lack of a p-type DBR also avoids the well-known problem of high resistance in p-type DBRs (Schubert *et* al., 1992c; Lear and Schneider, 1996). It has been shown that parabolic grading yields the lowest ohmic resistance in DBRs. Such parabolic grading is suited to eliminating heterojunction band discontinuities (Schubert et al., 1992c).

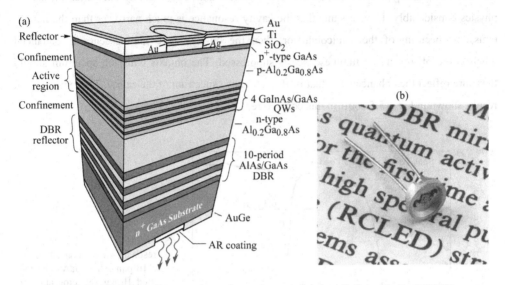

Fig. 15.4. (a) Schematic structure of a substrate-emitting GaInAs/GaAs RCLED consisting of a metal top reflector and a bottom distributed Bragg reflector (DBR). The RCLED emits at 930 nm. The reflectors are an AlAs/GaAs DBR and a Ag top reflector. (b) Picture of the first RCLED (after Schubert et al., 1994).

The magnitude of the reflectivity of the DBR needs to be consistent with Eqs. (15.1) and (15.3). The Ag back mirror has a reflectivity of approximately 96%. According to Eq. (15.1), the DBR reflectivity must be < 96%. The second criterion of Eq. (15.3) requires that $2\xi\alpha L_{active} < 1 - R_1$. Assuming $\xi = 1.3$, $\alpha = 10^4$ cm^{-1}, $L_{active} = 400$ Å, one obtains the condition $R_1 < 90\%$. Thus the mirror reflectivity of RCLEDs must be much lower than that of VCSELs. A high reflectivity would increase self-absorption and decrease the light output of the device as discussed earlier. De Neve et al. (1995) used an extensive theoretical model to calculate the mirror reflectivity. The maximum efficiency was calculated at a reflectivity of $R_1 = 50$–60%.

The reflection and emission properties of the RCLED are shown in Figs. 15.5 (a) and (b). The reflection spectrum of the RCLED (Fig. 15.5 (a)) exhibits a highly reflective band for wavelengths > 900 nm and a dip in the reflectivity at the cavity resonance. The spectral width of the cavity resonance is 6.3 nm. The emission spectrum of an electrically pumped device, shown in Fig. 15.5 (b), has nearly the same shape and width as the cavity resonance.

15 Resonant cavity light-emitting diodes

In conventional LEDs, the spectral characteristics of the devices reflect the thermal distribution of electrons and holes in the conduction and valence bands. The spectral characteristics of light emission from microcavities are as intriguing as they are complex. However, restricting our considerations to the optical axis of the cavity simplifies the cavity physics considerably. If we assume that the cavity resonance is much narrower than the natural emission spectrum of the semiconductor, then the on-resonance luminescence is enhanced whereas the off-resonance luminescence is suppressed. The on-axis emission spectrum should therefore reflect the enhancement, that is, the resonance spectrum of the cavity. The experimental results shown in Fig. 15.5 confirm this conjecture.

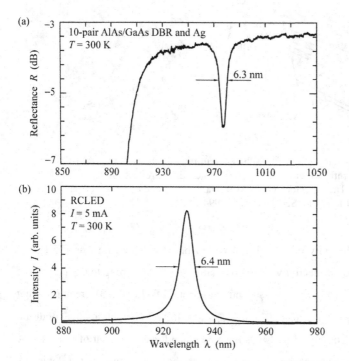

Fig. 15.5. (a) Reflectance of a resonant cavity consisting of a 10-pair AlAs/GaAs distributed Bragg reflector and an Ag reflector. (b) Emission spectrum of an RCLED consisting of a 10-pair AlAs/GaAs distributed Bragg reflector and an Ag reflector (after Schubert et al., 1994).

Owing to the cavity, the emission spectrum of an RCLED is much narrower than the emission spectrum of regular LEDs (Schubert et al. 1992a; Hunt et al., 1992, 1993). The spectral width of the RCLED emitted into a certain direction is given by the optical characteristics of the cavity. In contrast, the spectral width of a regular LED is about $1.8\,kT$, a value that is much wider than the RCLED emission spectrum. A comparison of a regular GaAs LED and a GaInAs

RCLED emission spectrum is shown in Fig. 15.6. Comparison of the spectra shows that the RCLED emission spectrum is a factor of about 10 narrower than the spectrum of the GaAs LED.

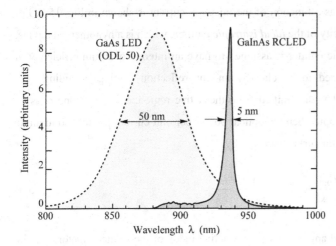

Fig. 15.6. Comparison of the emission spectra of a GaAs LED emitting at 870 nm (AT&T ODL 50 product) and a GaInAs RCLED emitting at 930 nm (after Hunt et al., 1993).

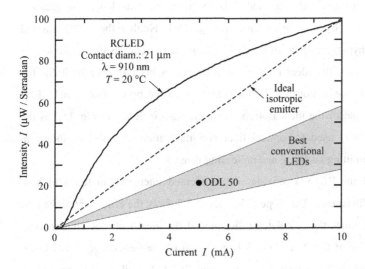

Fig. 15.7. Light-versus-current curves of a GaInAs/GaAs RCLED and of the *ideal isotropic emitter*. The ideal isotropic emitter is a hypothetial device emitting light isotropically with a quantum efficiency of 100%. The shaded region shows the intensity of the best conventional LEDs. The ODL 50 is a commercial LED product (after Schubert et al., 1994).

A regular LED has little or no angle dependence of the emission spectrum. However, the reflective properties of DBRs and of cavities consisting of DBRs are angle-dependent. Consequently, the emission from an RCLED along a certain direction is narrower than that of a regular LED. Integrated over all directions, the RCLED has a broad emission spectrum.

A figure of merit for LEDs used in optical fiber communication systems is the photon flux density emitted from the diode at a given current, which, for a given wavelength, can be characterized in terms of the unit microwatts per steradian. The optical power coupled into a fiber is directly proportional to the photon flux density.

The intensity of an RCLED as a function of the injection current is shown in Fig. 15.7. For comparison, the calculated intensity of the **ideal isotropic emitter**, which is a hypothetical device, is also shown. The ideal isotropic emitter is assumed to have an internal quantum efficiency of 100% and the device is assumed to be clad by an anti-reflection coating providing zero reflectivity ($R = 0$) for all wavelengths emitted from the active region. If the photon emission inside the semiconductor is isotropic, then the optical power per unit current per unit solid angle normal to the planar semiconductor surface is given by

$$\frac{P_{\text{optical}}}{\Omega} = \frac{1}{4\pi \bar{n}^2} \frac{hc}{e\lambda} \tag{15.4}$$

where Ω represents the unit solid angle, $\bar{n}$ is the refractive index of the semiconductor, c is the velocity of light, e is the electronic charge, and λ is the emission wavelength in vacuum. Equation (15.4) is represented by the dashed line in Fig. 15.7. Neither the 100% internal quantum efficiency nor the hypothetical anti-reflection coating can be reproduced in practice for fundamental reasons. Therefore, the ideal isotropic emitter represents an upper limit for the intensity attainable with any conventional LED. Of course, even the best conventional LEDs have intensities lower than that of the ideal isotropic emitter. Also included in Fig. 15.7 is the ODL 50 GaAs LED, frequently used for optical fiber communication. All devices shown in Fig. 15.7 have planar light-emitting surfaces, and no lensing is used.

Figure 15.7 reveals that the RCLED provides unprecedented intensities in terms of both absolute values and slope efficiencies. The slope efficiency is 7.3 times the efficiency of the best conventional LEDs and 3.1 times the calculated efficiency of the ideal isotropic emitter. At a current of 5 mA, the intensity of the RCLED is 3.3 times that of the best conventional LEDs including the ODL 50. The high efficiencies make the RCLED well suited for optical interconnect and communication systems.

The higher spectral purity of RCLEDs reduces chromatic dispersion in optical fiber communications (Hunt et al., 1993). The chromatic dispersion is directly proportional to the linewidth of the source. Since RCLEDs have linewidths 5–10 times narrower than conventional LEDs, chromatic dispersion effects, which dominate at wavelengths of 800–900 nm, are reduced as well. Hunt et al. (1993) showed that the bandwidth of RCLEDs is a factor of 5–10 higher than that of conventional LEDs. An RCLED-versus-LED comparison in a transmission experiment is shown in Fig. 15.8. The results show the received signal after transmission lengths of 5 m and 3.4 km for the two devices. The fiber used is a graded-index multimode fiber with a core diameter of 62.5 µm. After a transmission length of 5 m, no marked difference is found for the two devices. However, a substantial difference is found after a transmission length of 3.4 km. Inspection of Fig. 15.8 reveals that the RCLED exhibits much less pulse broadening as compared to the conventional LED. This difference is due to reduced material dispersion for the RCLED.

Schubert et al. (1996) demonstrated the high-speed modulation capability of RCLEDs. Eye diagram measurements with a random bit pattern generator revealed wide-open eyes at 622 Mbit/s. Due to the small size of the current injected region, the parasitic capacitances of communication RCLEDs are small. It is expected that RCLEDs will be suitable for modulation frequencies exceeding 1 Gbit/s.

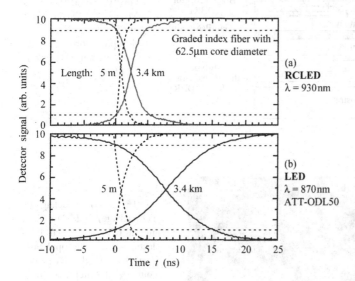

Fig. 15.8. Signal detected at the receiver end of a graded-index multimode fiber with a core diameter of 62.5 µm using (a) a GaInAs RCLED and (b) a GaAs LED source. After a short transmission distance of 5 m, no marked difference is found for the two sources. After a trans-mission distance of 3.4 km, the RCLED exhibits much less pulse broadening than the LED (after Hunt et al., 1993).

15.4 AlGaInP/GaAs RCLEDs emitting at 650 nm

RCLEDs have also been demonstrated in the visible wavelength range using the AlGaInP material system (Streubel et al., 1998; Whitaker, 1999; Wirth et al., 2001, 2002). The AlGaInP material system is commonly used for high-brightness red, orange, and yellow emitters and can be grown lattice matched on GaAs substrates. The active region of RCLEDs is an AlGaInP/GaInP multiple-quantum well structure emitting at 650 nm. The RCLEDs are suited for use in communication systems using plastic optical fibers. It is difficult to fabricate VCSELs in this wavelength range due to the unavailability of lattice-matched and transparent DBR materials with a large refractive-index contrast.

The basic structure of a top-emitting AlGaInP RCLED emitting at 650 nm is shown in Fig. 15.9. The device consists of an AlGaInP/GaInP MQW active region and AlGaInP cladding layers. The DBRs consist of AlAs/AlGaAs layers. The Al content in the AlGaAs layers of the DBR is chosen to be sufficiently high to make the DBR transparent to the light emitted by the active region. As a result, the index contrast of the AlAs/AlGaAs DBR layers is somewhat low.

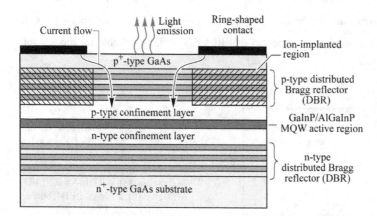

Fig. 15.9. Structure of a GaInP/AlGaInP/GaAs MQW RCLED emitting at 650 nm used for plastic optical fiber applications (after Whitaker, 1999)

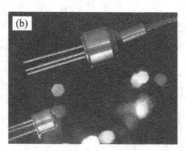

Fig. 15.10. (a) Packaged (TO package) RCLED emitting at 650 nm suited for plastic optical fiber applications. (b) Pig-tailed RCLED (courtesy of Mitel Corporation, Sweden, 1999).

The RCLED has a ring-shaped top contact configuration. The current is funneled into the center region of the ring using ion implantation to create an insulating region under the ring-shaped metalization. Hydrogen and, more frequently, oxygen implantation is used to render the semiconductor highly resistive. Note that the ion-implanted region is located in the p-type region only and does not extend into the active region, thereby avoiding the creation of defects in the active region where they would act as luminescence killers.

Packaged RCLEDs in a lensed TO package and in a pig-tailed package are shown in Figs. 15.10 (a) and (b) (Mitel Corporation, 1999), respectively. The lens is used for beam collimation, thereby enhancing the coupling efficiency to fibers.

Three RCLEDs under current injection conditions are shown in Fig. 15.11 (Osram Opto Semiconductors Corporation). The picture shows that the emission pattern is directed towards the surface normal of the devices. The emission wavelength is 650 nm.

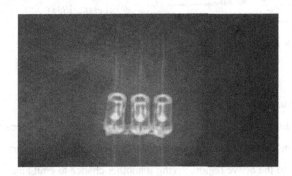

Fig. 15.11. AlGaInP/GaAs RCLEDs emitting at 650 nm. Note the forward-directed emission pattern similar to that of a semiconductor laser (courtesy of Osram Opto Semiconductors Corporation, Germany, 1999).

Optical spectra of a 650 nm RCLED and of a conventional LED injected at different current levels are shown in Fig. 15.12. The spectra are measured after the light is coupled into a plastic optical fiber. Thus, the magnitude of the spectra is a direct measure of the device efficiency *and* of the coupling efficiency. Inspection of the figure reveals several features. Firstly, the RCLED has a higher coupled peak power as well as integrated power than the LED. Secondly, the RCLED has a higher spectral purity than the LED.

Streubel *et al.* (1998) also reported that the emission spectrum was, at room temperature, intentionally blue-shifted with respect to the cavity resonance in order to improve the temperature stability of the RCLED output power. This cavity tuning results in a heart-shaped (double-lobed) emission pattern at room temperature, because the resonance wavelength of the cavity decreases for off-normal emission directions. As the temperature increases, the natural emission spectrum from the active region red-shifts, so that the cavity resonance (along the normal direction) has a better overlap with the natural emission spectrum. As a result, the temperature sensitivity of the RCLED is reduced. The natural emission spectrum red-shifts at a rate of about 0.5 nm/°C, whereas the cavity resonance shifts at only about one-tenth of that rate.

15 Resonant cavity light-emitting diodes

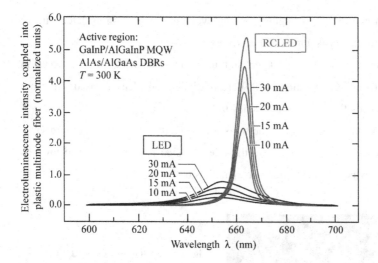

Fig. 15.12. Spectra of light coupled into a plastic optical fiber from a GaInP/AlGaInP MQW RCLED and a conventional GaInP/AlGaInP LED at different drive currents. Note the narrower spectrum and higher coupled power of the RCLED (after Streubel *et al.*, 1998).

15.5 Large-area photon recycling LEDs

Instead of devising ways to *reduce the lateral emission*, one could devise ways to *recycle* lateral emission, thereby redirecting the emission towards the top of the device. By reabsorbing the lateral emission, one can recapture the energy into the active region, giving it another chance to emit along the desired direction. Below, two examples of such devices are discussed.

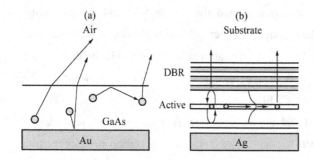

Fig. 15.13. Two approaches to photon recycling LEDs. (a) Bulk epilayer placed on top of gold. Most spontaneous emission that does not escape into air is reabsorbed and has a chance to emit again. (b) Microcavity designed with a waveguiding active region. Waveguided light is reabsorbed after some tens of micrometers, and has a chance to re-emit out of the top of the device.

The first example of such a device is an optically pumped semiconductor structure reported by Schnitzer *et al.* (1993). Consider the structure shown in Fig. 15.13 (a). The backside of a thin semiconductor layer is coated with gold. If one optically pumps the sample at low intensities, the active region remains absorbing. Most of the emission (about 95%) that hits the semiconductor–air interface will totally internally reflect, and stay within the semiconductor. Gold is a good reflector at infrared wavelengths. The semiconductor forming the active region is absorbing at the emission wavelength, so after some average absorption length, L_{abs}, the trapped light has a chance to re-emit.

This structure exhibited 72% external efficiency, which required a 99.7% internal quantum efficiency. The fabrication of the structure required the active epilayer to be etched and floated off its substrate. Note that the structure had no electrical contacts. The extraction efficiency of contactless structures is always higher compared with current-injection devices with contacts. An electrical device fabricated in this manner would probably have reliability problems due to the fragile nature of unsupported epitaxial films. A similar but more practical photon-recycling device was made by Blondelle et al. (1995).

The concept of photon recycling was used by De Neve et al. (1995) in a GaInAs/GaAs/AlGaAs RCLED. A simplified diagram of their structure is shown in Fig. 15.13 (b). If the graded-composition carrier confinement region is thick enough around the quantum wells, it also acts as a graded-index waveguide. De Neve et al. calculated that 30% of the light emitted by the active region goes into this waveguide mode. This is reabsorbed after some tens of micrometers, allowing the photons another chance to re-emit out of the top of the structure. In this way, about a 30% increase in external quantum efficiency was attained. The active region of quantum well RCLEDs is not thick enough to support a strongly guided optical mode. One might think that a strong waveguiding mode would take power away from the normal emission, but this does not appear to be the case. The energy tends to be at the expense of other high-angle modes instead, meaning that just modifying the waveguiding mode does little to change the external efficiency. Making use of this high-angle light by achieving photon recycling of the waveguiding emission is an attractive option.

One drawback to photon recycling is that it requires a device with large enough diameter that multiple reabsorption events can occur within the emitting area of the device. This makes such devices less attractive for fiber-optic communications, where small diameters couple better to fibers, especially when coupling lenses are used. Another drawback is that self-absorption necessarily increases the lifetime of the spontaneous emission, thereby slowing down the maximum modulation rate of the devices.

The device reported by De Neve et al. (1995) was shown to work with highest efficiency at large diameters. The authors were not targeting this RCLED for fiber communications. At high current densities, the carrier confinement in the quantum wells is reduced, and the efficiency decreases, so larger devices tend to be more efficient. Also, their device was resonant to the long-wavelength side of the natural emission peak, rather than at the peak. This shifts the maximum intensity of the main emission lobe to an off-normal angle, rather than being on axis, but it maximizes the total amount of emission into the main lobe. The Blondelle et al. (1995) and De

Neve et al. (1995) device is therefore designed for maximization of total emission from the top of the device, rather than for the specific needs of fiber coupling. Display devices and free-space communication devices benefit from this approach. The authors achieved an external quantum efficiency of 16%, compared with the theoretical 2% for an ideal planar emitter of the same index.

15.6 Thresholdless lasers

There have been a number of papers describing a thresholdless laser as a possible communications device (De Martini et al., 1987). This would essentially be a light-emitting device that emitted most or all of its light into the fundamental spatial cavity mode. In this way, one can achieve the lasing condition of there being one photon in the cavity almost without gain. Let us denote the probability that light is emitted into the fundamental spatial cavity mode as β. If the β of the LED were almost 1, then the light-versus-current curve of the device would be linear, and *indistinguishable* from a laser with no threshold. A comparison of the light-versus-current for a conventional laser, a high-β laser, and a truly thresholdless laser is given in Fig. 15.14. If one envisions a thresholdless device as fabricated from a resonant-cavity pillar, there are a number of problems besides those of fabrication. Simply having a good sub-threshold intensity, giving only a small kink in the L–I curve between sub-threshold and lasing, is not good enough for a thresholdless laser. The output modulation of such a device would be extremely slow below threshold compared to the lasing regime. To enable lasing, there must be some gain, because the β will never be 1. For there to be gain, we must pump the active region past transparency first, since it is absorbing under no-pump conditions. This gain condition requires a certain carrier density, which, in fact, determines the threshold.

One can imagine a device such that all optical modes other than the fundamental are highly suppressed, and the internal quantum efficiency is nearly perfect. In this case, with even a small current, the carrier densities within the device would slowly build until the active region became transparent, and some light could get out through the fundamental optical mode. Such a device would have a very low threshold, but one would have to be careful not to ever drive the device below its threshold even momentarily, because of the long time required to build the necessary space charge inside the device for transparency. Also, if the fundamental emission was suppressed at low carrier densities, it is unlikely that the other modes would be suppressed by an even higher degree, which is what would be required not to lose the injected carriers. In fact, it is the difficulty in suppressing any side emission that makes semiconductor thresholdless lasers difficult if not impractical.

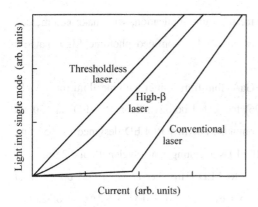

Fig. 15.14. Light-power-versus-current curves for single spatial-mode emission from (*i*) a conventional laser, (*ii*) a high β-factor laser, and (*iii*) a thresholdless laser. The conventional laser has a distinct current threshold. The high β-factor laser has a less distinct threshold. It would be noticeable in the spectrum and device modulation speed, however. A hypothetical thresholdless laser would have a β close to 1, and would somehow suppress all other lossy emission until the carrier density required for gain (or at least transparency) was achieved.

One could, of course, make a laser with a very small threshold by making the laser very small. The problem is that such a device would be capable of only a very small output power. For a hypothetical four-quantum-well single-mode LED, which does not rely on gain, but achieves a β close to 1, the current densities cannot be more than about 50 µA/µm² or 5 kA/cm². Using the formulas for β for a dielectric pillar (β > 0.5), and this current density, one can find that less than 2 µW could be emitted into a single mode for an RCLED pillar structure. This is clearly too small for high-speed communications in the 100–650 Mbit/s regime, where 10 µW is considered a minimum for an emitter. An RCLED with a very short carrier lifetime owing to extreme enhancement of the desired emission could have higher pump currents, but no semiconductor design can do this. If the device is a laser, higher powers could be drawn because of the reduced carrier lifetime, but the device sizes are incredibly small for real fabrication, or for any sort of heat dissipation.

In conclusion, semiconductor RCLEDs should remain devices larger than about 5 or 10 µm in diameter, and should remain multimode emitters. Semiconductor vertical-cavity lasers are best fabricated at sizes that make sense technologically, without regard to the β factor of the spontaneous emission. A high-β laser may not be desirable anyway, since the spontaneous emission will introduce excess noise into the emission.

15.7 Other RCLED devices

It is likely that the future of resonant-cavity LEDs and confined-photon emitters will involve new materials and applications from those currently used in communications and display systems. Spontaneously emitting confined-photon devices will always have competition from lasers, conventional LEDs, and other forms of devices. It is only a matter of time, however, before the right materials combination at the right wavelengths, for the right application, will make commercial devices a reality. It is therefore worthwhile to mention other confined photonic LED material systems.

Wilkinson et al. (1995) fabricated an AlGaAs/GaAs thin-film emitter with metal mirrors on a Si substrate, emitting at 880 nm. Pavesi et al. (1996) fabricated porous silicon RCLEDs at 750 nm wavelength. Fisher et al., (1995) investigated a conjugate polymer RCLED designed for 650 nm. Hadji et al. (1995) have realized CdHgTe/HgTe RCLEDs operating at a wavelength of 3.2 μm.

A structure of particular note is a GaAs/Al_xO_y RCLED (Huffaker et al., 1995) operating at 950 nm, where the aluminum oxide in the output mirror was produced from AlAs by oxidation. This composition of output mirror allows the effective cavity length to remain small, maximizing the output enhancement in the RCLED. The back mirror of the structure was Ag. Another particularly interesting device is the broadly tunable RCLED by Larson and Harris (1995). The top mirror is a deformable membrane which can be moved by electrostatic forces. Tunable emission was shown from 938 to 970 nm.

15.8 Other novel confined-photon emitters

A complete discussion of other light-emitting structures that confine photons is beyond the scope of this chapter. It is useful, however, to discuss the properties of some confined-photon emitters. **Photonic crystal** or **photonic bandgap structures** or involve two- or three-dimensional photon confinement achieved by periodic patterning of the light-emitting active region or the material adjoining the active region. Examples of photonic crystal structures were given by Joannopoulos et al. (1995), Baba and Matsuzaki (1996), and Fan et al. (1997). Erchak et al. (2001) reported very encouraging results on photonic crystal LEDs, namely a six-fold enhancement of light extraction along the surface-normal direction.

Photonic crystal structures can consist of a series of rods or holes arranged in a regular pattern, such as a hexagonal close-packed array. The periodicity of the array can create an optical bandgap for lateral emission at certain emission energies and one or both polarizations. By suppressing the lateral emission, a structure consisting of rods will have a large bandgap for TM emission and a smaller bandgap for TE emission, but not at the same emission energies. However, if the emitting region had a dipole oriented mainly along the rods (such as in quantum well electron-to-light-hole recombination), the lateral emission could be efficiently suppressed. A structure consisting of patterned holes will have a smaller bandgap than that of the rod structure, but it has the advantage that it can have a true optical bandgap for both polarizations of light. Photonic crystal structures either by themselves, or combined with a planar resonant cavity, enable a strong longitudinal emission enhancement.

Another confined-photon emitter is the ***microdisk laser*** (McCall *et al.*, 1992), which is fabricated as a thin dielectric disk that couples light out the edges of the disk. Lasing modes can be described by a mode number M, where $\exp(iM\phi)$ is the form of the electric field around the cylindrical disk. Because waves can propagate both ways, M can be positive or negative. The disk can be fabricated with a thickness such that the emission perpendicular to the disk is suppressed. Small disks will only support a few modes, and therefore can have a high spontaneous emission factor β. The Q of these modes are also high enough to achieve lasing. One attractive aspect of such disks is that the lasing emission occurs in the plane of the sample, from a very small device. This could be useful for integration of many photonic devices on a single wafer. However, the output is difficult to efficiently couple into waveguides and fibers, as it only couples evanescently. Advances have been made in improving the longevity, operating temperature range, and active-region passivation of such devices (Mohideen *et al.*, 1993). Room-temperature cw electrical pumping is still a problem, however.

References

Baba T. and Matsuzaki T., "GaInAsP/InP 2-dimensional photonic crystals" in *Microcavities and Photonic Bandgaps* edited by J. Rarity and C. Weisbuch, p. 193 (Kluwer Academic Publishers, Netherlands, 1996)

Blondelle J., De Neve H., Demeester P., Van Daele P., Borghs G., and Baets R. "16% external quantum efficiency from planar microcavity LED's at 940 nm by precise matching of the cavity wavelength" *Electron. Lett.* **31**, 1286 (1995)

De Martini F., Innocenti G., Jacobovitz G. R., and Mataloni P. "Anomalous spontaneous emission time in a microscopic optical cavity" *Phys. Rev. Lett.* **59**, 2955 (1987)

De Neve H., Blondelle J., Baets R., Demeester P., Van Daele P., and Borghs G. "High efficiency planar microcavity LEDs: Comparison of design and experiment" *IEEE Photonics Technol. Lett.* **7**, 287 (1995)

Erchak A. A., Ripin D. J., Fan S., Rakich P., Joannopoulos J. D., Ippen E. P., Petrich G. S., and Kolodziejski L. A. "Enhanced coupling to vertical radiation using a two-dimensional photonic crystal in a semiconductor light-emitting diode" *Appl. Phys. Lett.* **78**, 563 (2001)

Fan S., Villeneuve P. R., Joannopolous J. D., and Schubert E. F. "High extraction efficiency of spontaneous emission from slabs of photonic crystals" *Phys. Rev. Lett.* **78**, 3294 (1997)

Fisher T. A., Lidzey D. G., Pate M. A., Weaver M. S., Whittaker D. M., Skolnick M. S., and Bradley D. D. C. "Electroluminescence from a conjugated polymer microcavity structure" *Appl. Phys. Lett.* **67**, 1355 (1995)

Hadji E., Bleuse J., Magnea N., and Pautrat J. L. "3.2 μm infrared resonant cavity light emitting diode" *Appl. Phys. Lett.* **67**, 2591 (1995)

Huffaker D. L., Lin C. C., Shin J., and Deppe D. G. "Resonant cavity light emitting diode with an Al_xO_y/GaAs reflector" *Appl. Phys. Lett.* **66**, 3096 (1995)

Hunt N. E. J., Schubert E. F., Logan R. A., and Zydzik G. J. "Enhanced spectral power density and reduced linewidth at 1.3 μm in an InGaAsP quantum well resonant-cavity light-emitting diode" *Appl. Phys. Lett.* **61**, 2287 (1992)

Hunt N. E. J., Schubert E. F., Kopf R. F., Sivco D. L., Cho A. Y., and Zydzik G. J. "Increased fiber communications bandwidth from a resonant cavity light-emitting diode emitting at $\lambda = 940$ nm" *Appl. Phys. Lett.* **63**, 2600 (1993)

Hunt N. E. J., Schubert E. F., Sivco D. L., Cho A. Y., Kopf R. F., Logan R. A., and Zydzik G. J. "High efficiency, narrow spectrum resonant cavity light-emitting diodes" in *Confined Electrons and Photons* edited by E. Burstein and C. Weisbuch (Plenum Press, New York, 1995a)

Hunt N. E. J., Vredenberg A. M., Schubert E. F., Becker P. C., Jacobson D. C., Poate J. M., and Zydzik G. J. "Spontaneous emission control of Er^{3+} in Si/SiO_2 microcavities" in *Confined Electrons and Photons* edited by E. Burstein and C. Weisbuch (Plenum Press, New York, 1995b)

Joannopoulos J. D., Meade R. D., and Winn J. N. *Photonic Crystals* (Princeton University Press, Princeton NJ, 1995)

Kobayashi T., Segawa T., Morimoto A., and Sueta T., paper presented at the 43rd fall meeting of the Japanese Society of Applied Physics, Tokyo, Sept. (1982)

Larson M. C. and Harris Jr. J. S. "Broadly tunable resonant-cavity light emission" *Appl. Phys. Lett.* **67**, 590 (1995)

Lear K. L. and Schneider Jr. R. P. "Uniparabolic mirror grading for vertical cavity surface emitting lasers" *Appl. Phys. Lett.* **68**, 605 (1996)

McCall S. L., Levi A. F. J., Slusher R. E., Pearton S. J., and Logan R. A. "Whispering-gallery mode microdisk laser" *Appl. Phys. Lett.* **60**, 289 (1992)

Mitel Corporation, Sweden. Photograph of RCLED is gratefully acknowledged (1999)

Mohideen U., Hobson W. S., Pearton J., Ren F., and Slusher R. E. "GaAs/AlGaAs microdisk lasers" *Appl. Phys. Lett.* **64**, 1911 (1993)

Nakayama T., Itoh Y., and Kakuta A. "Organic photo- and electroluminescent devices with double mirrors" *Appl. Phys. Lett.* **63**, 594 (1993)

Osram Opto Semiconductors Corp., Germany. RCLED photograph is gratefully acknowledged (1999)

Pavesi L., Guardini R., and Mazzoleni C. "Porous silicon resonant cavity light emitting diodes" *Solid State Comm.* **97**, 1051 (1996)

Schnitzer I., Yablonovitch E., Caneau C., and Gmitter T. J. "Ultra-high spontaneous emission quantum efficiency, 99.7% internally and 72% externally, from AlGaAs/GaAs/AlGaAs double heterostructures" *Appl. Phys. Lett.* **62**, 131 (1993)

Schubert E. F., Wang Y.-H., Cho A. Y., Tu L.-W., and Zydzik G. J. "Resonant cavity light-emitting diode" *Appl. Phys. Lett.* **60**, 921 (1992a)

Schubert E. F., Vredenberg A. M., Hunt N. E. J., Wong Y. H., Becker P. C., Poate J. M., Jacobson D. C., Feldman L. C., and Zydzik G. J. "Giant enhancement of luminescence intensity in Er-doped Si/SiO_2 resonant cavities" *Appl. Phys. Lett.* **61**, 1381 (1992b)

Schubert E. F., Tu L. W., Zydzik G. J., Kopf R. F., Benvenuti A., and Pinto M. R. "Elimination of heterojunction band discontinuities by modulation doping" *Appl. Phys. Lett.* **60**, 466 (1992c)

Schubert E. F., Hunt N. E. J., Micovic M., Malik R. J., Sivco D. L., Cho A. Y., and Zydzik G. J. "Highly efficient light-emitting diodes with microcavities" *Science* **265**, 943 (1994)

Schubert E. F., Hunt N. E. J., Malik R. J., Micovic M., and Miller D. L. "Temperature and modulation characteristics of resonant cavity light-emitting diodes" *IEEE J. Lightwave Technol.* **14**, 1721 (1996)

Streubel K., Helin U., Oskarsson V., Backlin E., and Johanson A. "High-brightness visible (660 nm) resonant-cavity light-emitting diode" *IEEE Photonics Technol. Lett.* **10**, 1685 (1998)

Tu L. W., Schubert E. F., Zydzik G. J., Kopf R. F., Hong M., Chu S. N. G., and Mannaerts J. P. "Vertical cavity surface emitting lasers with semitransparent metallic mirrors and high quantum efficiencies" *Appl. Phys. Lett.* **57**, 2045 (1990)

Whitaker T. "Resonant cavity LEDs" *Compound Semiconductors* **5**, 32 (1999)

Wilkinson S. T., Jokerst N. M., and Leavitt R. P. "Resonant-cavity-enhanced thin-film AlGaAs/GaAs/AlGaAs LED's with metal mirrors" *Appl. Opt.* **34**, 8298 (1995)

Wirth R., Karnutsch C., Kugler S., and Streubel K. "High-efficiency resonant-cavity LEDs emitting at 650 nm" *IEEE Photonics Technol. Lett.* **13**, 421 (2001)

Wirth R., Huber W., Karnutsch C., and Streubel K. "Resonators provide LEDs with laser-like performance" *Compound Semiconductors* **8**, 49 (2002)

Yokoyama H. "Physics and device applications of optical microcavities" *Science* **256**, 66 (1992)

16

Human eye sensitivity and photometric quantities

The recipient of the light emitted by most visible-spectrum LEDs is the human eye. In this chapter, the characteristics of human vision and of the human eye and are summarized, in particular as these characteristics relate to human eye sensitivity and photometric quantities.

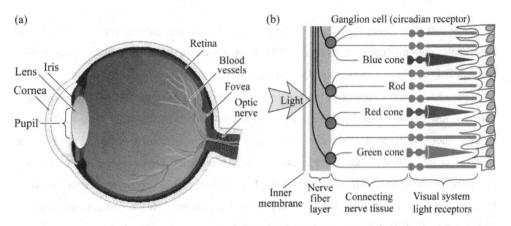

Fig. 16.1. (a) Cross section through a human eye. (b) Schematic view of the retina including rod and cone light receptors (adapted from Encyclopedia Britannica, 1994).

16.1 Light receptors of the human eye

Figure 16.1 (a) shows a schematic illustration of the human eye (Encyclopedia Britannica, 1994). The inside of the eyeball is clad by the retina, which is the light-sensitive part of the eye. The illustration also shows the fovea, a cone-rich central region of the retina which affords the high acuteness of central vision. Figure 16.1 (b) shows the cell structure of the retina including the light-sensitive *rod cells* and *cone cells*. Also shown are the ganglion cells and nerve fibers that transmit the visual information to the brain. Rod cells are more abundant and more light sensitive than cone cells. Rods are sensitive over the entire visible spectrum. There are three types of cone

16 Human eye sensitivity and photometric quantities

cells, namely cone cells sensitive in the red, green, and blue spectral range. The cone cells are therefore denoted as the red-sensitive, green-sensitive, and blue-sensitive cones, or simply as the red, green, and blue cones.

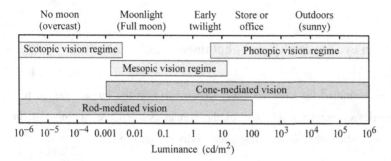

Fig. 16.2. Approximate ranges of vision regimes and receptor regimes (after Osram Sylvania, 2000).

Three different vision regimes are shown in Fig. 16.2 along with the receptors relevant to each of the regimes (Osram Sylvania, 2000). **Photopic vision** relates to human vision at high ambient light levels (e.g. during daylight conditions) when vision is mediated by the cones. The photopic vision regime applies to luminance levels > 3 cd/m^2. **Scotopic vision** relates to human vision at low ambient light levels (e.g. at night) when vision is mediated by rods. Rods have a much higher sensitivity than the cones. However, the sense of color is essentially lost in the scotopic vision regime. At low light levels such as in a moonless night, objects lose their colors and only appear to have different gray levels. The scotopic vision regime applies to luminance levels < 0.003 cd/m^2. **Mesopic vision** relates to light levels between the photopic and scotopic vision regime (0.003 cd/m^2 < mesopic luminance < 3 cd/m^2).

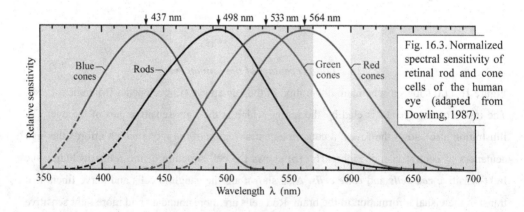

Fig. 16.3. Normalized spectral sensitivity of retinal rod and cone cells of the human eye (adapted from Dowling, 1987).

276

The approximate spectral sensitivity functions of the rods and three types or cones are shown in Fig. 16.3 (Dowling, 1987). Inspection of the figure reveals that night-time vision (scotopic vision) is weaker in the red spectral range and thus stronger in the blue spectral range as compared to day-time vision (photopic vision). The following discussion mostly relates to the photopic vision regime.

16.2 Basic radiometric and photometric units

The physical properties of electromagnetic radiation are characterized by **radiometric units**. Using radiometric units, we can characterize light in terms of physical quantities; for example, the number of photons, photon energy, and **optical power** (in the lighting community frequently called the **radiant flux**). However, the radiometric units are irrelevant when it comes to light perception by a human being. For example, infrared radiation causes no luminous sensation in the eye. To characterize the light and color sensation by the human eye, different types of units are needed. These units are called **photometric units**.

The **luminous intensity**, which is a photometric quantity, represents the light intensity of an optical source, as perceived by the human eye. The luminous intensity is measured in units of **candela** (cd), which is a base unit of the International System of Units (SI unit). The present definition of luminous intensity is as follows: *a monochromatic light source emitting an optical power of (1/683) watt at 555 nm into the solid angle of 1 steradian (sr) has a luminous intensity of 1 candela (cd).*

The unit *candela* has great historical significance. All light intensity measurements can be traced back to the candela. It evolved from an older unit, the **candlepower**, or simply, the **candle**. The original, now obsolete, definition of one candela was the light intensity emitted by a plumber's candle, as shown in Fig. 16.4, which had a specified construction and dimensions:

> one standardized candle emits a luminous intensity of 1.0 cd

.

Fig. 16.4. Plumber's candle, as used by plumbers in the nineteenth century to melt lead solder when joining water pipes.

The luminous intensity of a light source can thus be characterized by giving the number of standardized candles that, when combined, would emit the same luminous intensity. Note that *candlepower* and *candle* are non-SI units that are no longer current and rarely used at the present time.

The **luminous flux**, which is also a photometric quantity, represents the light power of a source as perceived by the human eye. The unit of luminous flux is the **lumen** (lm). It is defined as follows: *a monochromatic light source emitting an optical power of (1/683) watt at 555 nm has a luminous flux of 1 lumen (lm)*. The lumen is an SI unit.

A comparison of the definitions for the candela and lumen reveals that 1 candela equals 1 lumen per steradian or cd = lm/sr. Thus, an isotropically emitting light source with luminous intensity of 1 cd has a luminous flux of 4π lm = 12.57 lm.

The **illuminance** is the luminous flux incident per unit area. The illuminance measured in **lux** (lux = lm/m^2). It is an SI unit used when characterizing illumination conditions. Table 16.1 gives typical values of the illuminance in different environments.

Table 16.1. Typical illuminance in different environments.

Illumination condition	Illuminance
Full moon	1 lux
Street lighting	10 lux
Home lighting	30 to 300 lux
Office desk lighting	100 to 1 000 lux
Surgery lighting	10 000 lux
Direct sunlight	100 000 lux

The **luminance** of a **surface source** (i.e. a source with a non-zero light-emitting surface area such as a display or an LED) is the ratio of the luminous intensity emitted in a certain direction (measured in cd) divided by the *projected surface area* in that direction (measured in m^2). The luminance is measured in units of cd/m^2. In most cases, the direction of interest is normal to the chip surface. In this case, the luminance is the luminous intensity emitted along the chip-normal direction divided by the chip area.

The *projected surface area* mentioned above follows a cosine law, i.e. the projected area is given by $A_{\text{projected}} = A_{\text{surface}} \cos \Theta$, where Θ is the angle between the direction considered and the surface normal. The light-emitting surface area and the projected area are shown in Fig. 16.5. The luminous intensity of LEDs with lambertian emission pattern also depends on the angle Θ

according to a cosine law. Thus the luminance of lambertian LEDs is a constant, independent of angle.

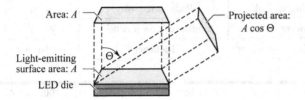

Fig. 16.5. Area of LED, A, and projected area, $A \cos \Theta$, used for the definition of the luminance of an LED.

For LEDs, it is desirable to maximize luminous intensity and luminous flux while keeping the LED chip area minimal. Thus the luminance is a measure of how efficiently the valuable semiconductor wafer area is used to attain, at a given injection current, a certain luminous intensity.

There are several units that are used to characterize the luminance of a source. The names of these common units are given in Table 16.2.

Typical luminances of displays, organic LEDs, and inorganic LEDs are given in Table 16.3. The table reveals that displays require a comparatively low luminance because the observer directly views the display from a close distance. This is not the case for high-power inorganic LEDs used for example in traffic light and illumination applications.

Photometric and the corresponding radiometric units are summarized in Table 16.4.

Table 16.2. Conversion between common SI and non-SI units for luminance.

Unit	Common name	Unit	Common name
1 cd/cm²	1 stilb	$(1/\pi)$ cd/m²	1 apostilb
$(1/\pi)$ cd/cm²	1 lambert	$(1/\pi)$ cd/ft²	1 foot-lambert
1 cd/m²	1 nit		

Table 16.3. Typical values for the luminance of displays, LEDs fabricated from organic materials, and inorganic LEDs.

Device	Luminance (cd/m²)		Device	Luminance (cd/m²)
Display	100	(operation)	Organic LED	100–10 000
Display	250–750	(max. value)	III–V LED	1 000 000–10 000 000

Table 16.4. Photometric and corresponding radiometric units.

Photometric unit	Dimension	Radiometric unit	Dimension
Luminous flux	lm	Radiant flux (optical power)	W
Luminous intensity	lm / sr = cd	Radiant intensity	W / sr
Illuminance	lm / m² = lux	Irradiance (power density)	W / m²
Luminance	lm / (sr m²) = cd / m²	Radiance	W / (sr m²)

Exercise: *Photometric units*. A 60 W incandescent light bulb has a luminous flux of 1000 lm. Assume that light is emitted isotropically from the bulb.
(a) What is the luminous efficiency (i.e. the number of lumens emitted per watt of electrical input power) of the light bulb?
(b) What number of standardized candles emit the same luminous intensity?
(c) What is the illuminance, E_{lum}, in units of lux, on a desk located 1.5 m below the bulb?
(d) Is the illuminance level obtained under (c) sufficiently high for reading?
(e) What is the luminous intensity, I_{lum}, in units of candela, of the light bulb?
(f) Derive the relationship between the illuminance at a distance r from the light bulb, measured in *lux*, and the luminous intensity, measured in *candela*.
(g) Derive the relationship between the illuminance at a distance r from the light bulb, measured in *lux*, and the luminous flux, measured in *lumen*.
(h) The definition of the cd involves the optical power of (1/683) W. What, do you suppose, is the origin of this particular power level?

Solution: (a) 16.7 lm/W. (b) 80 candles. (c) $E_{\text{lum}} = 35.4 \text{ lm/m}^2 = 35.4$ lux. (d) Yes.
(e) 79.6 lm/sr = 79.6 cd. (f) $E_{\text{lum}} \, r^2 = I_{\text{lum}}$. (g) $E_{\text{lum}} \, 4\pi r^2 = \Phi_{\text{lum}}$.
(h) Originally, the unit of luminous intensity had been defined as the intensity emitted by a real candle. Subsequently the unit was defined as the intensity of a light source with specified wavelength and optical power. When the power of that light source is (1/683) W, it has the same intensity as the candle. Thus this particular power level has a historical origin and results from the effort to maintain continuity.

16.3 Eye sensitivity function

The conversion between radiometric and photometric units is provided by the **luminous efficiency function** or **eye sensitivity function**, $V(\lambda)$. In 1924, the CIE introduced the photopic eye sensitivity function $V(\lambda)$ for point-like light sources where the viewer angle is 2° (CIE, 1931). This function is referred to as the **CIE 1931 $V(\lambda)$ function**. It is the current photometric standard in the United States.

A *modified* $V(\lambda)$ function was introduced by Judd and Vos in 1978 (Vos, 1978; Wyszecki and Stiles, 1982, 2000) and this modified function is here referred to as the **CIE 1978 $V(\lambda)$ function**. The modification was motivated by the underestimation of the human eye sensitivity in the blue and violet spectral region by the CIE 1931 $V(\lambda)$ function. The modified function $V(\lambda)$ has higher values in the spectral region below 460 nm. The CIE has endorsed the CIE 1978 $V(\lambda)$

function by stating "the spectral luminous efficiency function for a point source may be adequately represented by the Judd modified $V(\lambda)$ function" (CIE, 1988) and "the Judd modified $V(\lambda)$ function would be the preferred function in those conditions where luminance measurements of short wavelengths consistent with color normal observers is desired" (CIE, 1990).

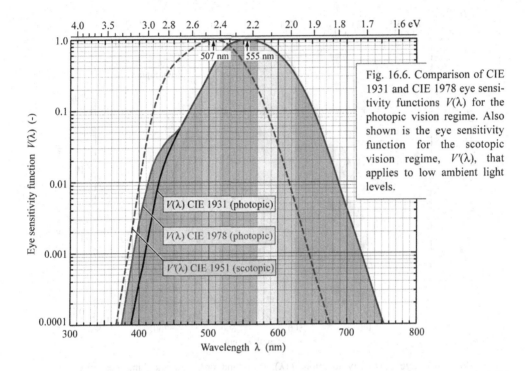

Fig. 16.6. Comparison of CIE 1931 and CIE 1978 eye sensitivity functions $V(\lambda)$ for the photopic vision regime. Also shown is the eye sensitivity function for the scotopic vision regime, $V'(\lambda)$, that applies to low ambient light levels.

The CIE 1931 $V(\lambda)$ function and the CIE 1978 $V(\lambda)$ function are shown in Fig. 16.6. The photopic eye sensitivity function has maximum sensitivity in the green spectral range at 555 nm, where $V(\lambda)$ has a value of unity, i.e. $V(555 \text{ nm}) = 1$. Inspection of the figure also reveals that the CIE 1931 $V(\lambda)$ function underestimated the eye sensitivity in the blue spectral range ($\lambda < 460$ nm). Numerical values of the CIE 1931 and CIE 1978 $V(\lambda)$ function are tabulated in Appendix 16.1.

Also shown in Fig. 16.6 is the scotopic eye sensitivity function $V'(\lambda)$. The peak sensitivity in the scotopic vision regime occurs at 507 nm. This value is markedly shorter than the peak sensitivity in the photopic vision regime. Numerical values of the CIE 1951 $V'(\lambda)$ function are tabulated in Appendix 16.2.

Note that even though the CIE 1978 $V(\lambda)$ function is preferable, it is not the standard, mostly for practical reasons such as possible ambiguities created by changing standards. Wyszecki and Stiles (2000) note that even though the CIE 1978 $V(\lambda)$ function is not a standard, it has been used in several visual studies. The CIE 1978 $V(\lambda)$ function, which can be considered the most accurate description of the eye sensitivity in the photopic vision regime, is shown in Fig. 16.7.

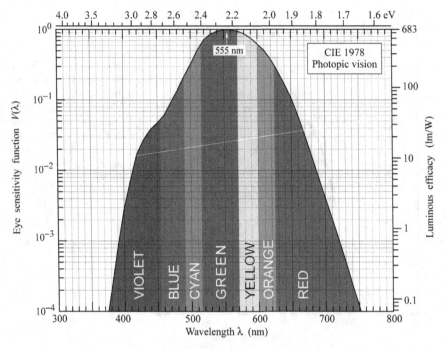

Fig. 16.7. Eye sensitivity function, $V(\lambda)$, (left-hand ordinate) and luminous efficacy, measured in lumens per watt of optical power (right-hand ordinate). $V(\lambda)$ is maximum at 555 nm (after 1978 CIE data).

The eye sensitivity function has been determined by the ***minimum flicker method***, which is the classic method for luminance comparison and for the determination of $V(\lambda)$. The stimulus is a light-emitting small circular area, alternatingly illuminated (with a frequency of 15 Hz) with the standard color and the comparison color. Since the hue-fusion frequency is lower than 15 Hz, the hues fuse. However, the brightness-fusion frequency is higher than 15 Hz and thus if the two colors differ in brightness, then there will be visible flicker. The human subject's task is to adjust the target color until the flicker is minimal.

Any desired chromaticity can be obtained with an infinite variety of spectral power distributions $P(\lambda)$. One of these distributions has the greatest possible luminous efficacy. This limit can be obtained in only one way, namely by the mixture of suitable intensities emitted by two monochromatic sources (MacAdam, 1950). The maximum attainable luminous efficacy obtained with a single monochromatic pair of emitters is shown in Fig. 16.8. The maximum luminous efficacy of *white* light depends on the color temperature; it is about 420 lm/W for a color temperature of 6500 K and can exceed 500 lm/W for lower color temperatures. The exact value depends on the exact location within the white area of the chromaticity diagram.

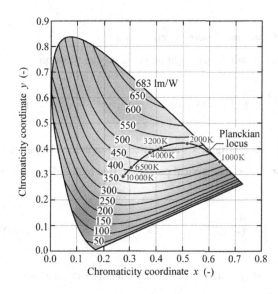

Fig. 16.8. Relation of maximum possible luminous efficacy (lumens per optical watt) and chromaticity in the CIE 1931 x, y chromaticity diagram (adapted from MacAdam, 1950).

16.4 Colors of near-monochromatic emitters

For wavelengths ranging from 390 to 720 nm, the eye sensitivity function $V(\lambda)$ is greater than 10^{-3}. Although the human eye is sensitive to light with wavelengths < 390 nm and > 720 nm, the sensitivity at these wavelengths is extremely low. Therefore, the wavelength range 390 nm $\leq \lambda \leq$ 720 nm can be considered the *visible wavelength range*. The relationship between color and wavelength within the visible wavelength range is given in Table 16.5. This relationship is valid for monochromatic or near-monochromatic light sources such as LEDs. Note that color is, to some extent, a subjective quantity. Also note that the transition between different colors is continuous.

16 Human eye sensitivity and photometric quantities

Table 16.5. Colors and associated typical LED peak wavelength ranges

Color	Wavelength	Color	Wavelength
Ultraviolet	< 390 nm	Yellow	570–600 nm
Violet	390–455 nm	Amber	590–600 nm
Blue	455–490 nm	Orange	600–625 nm
Cyan	490–515 nm	Red	625–720 nm
Green	515–570 nm	Infrared	> 720 nm

16.5 Luminous efficacy and luminous efficiency

The *luminous flux*, Φ_{lum}, is obtained from the radiometric light power using the equation

$$\Phi_{lum} = 683 \frac{lm}{W} \int_\lambda V(\lambda) P(\lambda) d\lambda \tag{16.1}$$

where $P(\lambda)$ is the power spectral density, i.e. the light power emitted per unit wavelength, and the prefactor 683 lm/W is a normalization factor. The optical power emitted by a light source is then given by

$$P = \int_\lambda P(\lambda) d\lambda . \tag{16.2}$$

High-performance single-chip visible-spectrum LEDs can have a luminous flux of about 10–100 lm at an injection current of 100–1 000 mA.

The *luminous efficacy of optical radiation* (also called the *luminosity function*), measured in units of lumens per watt of optical power, is the conversion efficiency from optical power to luminous flux. The luminous efficacy is defined as

$$\text{Luminous efficacy} = \frac{\Phi_{lum}}{P} = \left[683 \frac{lm}{W} \int_\lambda V(\lambda) P(\lambda) d\lambda \right] \Big/ \left[\int_\lambda P(\lambda) d\lambda \right] . \tag{16.3}$$

For strictly monochromatic light sources ($\Delta\lambda \to 0$), the luminous efficacy is equal to the eye sensitivity function $V(\lambda)$ multiplied by 683 lm/W. However, for multicolor light sources and especially for white light sources, the luminous efficacy needs to be calculated by integration over all wavelengths. The luminous efficacy is shown on the right-hand ordinate of Fig. 16.4.

The *luminous efficiency of a light source*, also measured in units of lm/W, is the luminous flux of the light source divided by the electrical input power.

$$\text{Luminous efficiency} = \Phi_{\text{lum}} / (IV) \qquad (16.4)$$

where the product (IV) is the electrical input power of the device. Note that in the lighting community, luminous efficiency is often referred to as *luminous efficacy of the source*.

Inspection of Eqs. (16.3) and (16.4) reveals that the luminous efficiency is the product of the luminous efficacy and the electrical-to-optical power conversion efficiency. The luminous efficiency of common light sources is given in Table 16.6.

Table 16.6. Luminous efficiencies of different light sources. (a) Incandescent sources. (b) Fluorescent sources. (c) High-intensity discharge (HID) sources.

Light source		Luminous efficiency
Edison's first light bulb (with C filament)	(a)	1.4 lm/W
Tungsten filament light bulbs	(a)	15–20 lm/W
Quartz halogen light bulbs	(a)	20–25 lm/W
Fluorescent light tubes and compact bulbs	(b)	50–80 lm/W
Mercury vapor light bulbs	(c)	50–60 lm/W
Metal halide light bulbs	(c)	80–125 lm/W
High-pressure sodium vapor light bulbs	(c)	100–140 lm/W

The luminous efficiency is a highly relevant figure of merit for visible-spectrum LEDs. It is a measure of the perceived light power normalized to the electrical power expended to operate the LED. For light sources with a perfect electrical-power-to-optical-power conversion, the luminous source efficiency is equal to the luminous efficacy of radiation.

Exercise: Luminous efficacy and luminous efficiency of LEDs. Consider a red and an amber LED emitting at 625 and 590 nm, respectively. For simplicity, assume that the emission spectra are monochromatic ($\Delta\lambda \to 0$). What is the luminous efficacy of the two light sources? Calculate the luminous efficiency of the LEDs, assuming that the red and amber LEDs have an external quantum efficiency of 50%. Assume that the LED voltage is given by $V = E_g / e = h\nu / e$.

Assume next that the LED spectra are thermally broadened and have a gaussian lineshape with a linewidth of $1.8 kT$. Again calculate the luminous efficacy and luminous efficiency of the two light sources. How accurate are the results obtained with the approximation of monochromaticity?

Some LED structures attain excellent power efficiency by using small light-emitting areas (current injection in a small area of chip) and advanced light-output-coupling structures (see, for example, Schmid et al., 2002). However, such devices have low luminance because only a small fraction of the chip area is injected with current. Table 16.7 summarizes frequently used figures of merit for light-emitting diodes.

Table 16.7. Summary of photometric, radiometric, and quantum performance measures for LEDs.

Figure of merit	Explanation	Unit
Luminous efficacy	Luminous flux per optical unit power	lm/W
Luminous efficiency	Luminous flux per input electrical unit power	lm/W
Luminous intensity efficiency	Luminous flux per sr per input electrical unit power	cd/W
Luminance	Luminous flux per sr per chip unit area	cd/m^2
Power efficiency	Optical output power per input electrical unit power	%
Internal quantum efficiency	Photons emitted in active region per electron injected	%
External quantum efficiency	Photons emitted from LED per electron injected	%
Extraction efficiency	Escape probability of photons emitted in active region	%

16.6 Brightness and linearity of human vision

Although the term **brightness** is frequently used, it lacks a standardized scientific definition. The frequent usage is due to the fact that the general public can more easily relate to the term *brightness* than to photometric terms such as *luminance* or *luminous intensity*. Brightness is an attribute of visual perception and is frequently used as synonym for *luminance* and (incorrectly) for the radiometric term *radiance*.

To quantify the brightness of a source, it is useful to differentiate between point and surface area sources. For *point sources*, brightness (in the photopic vision regime) can be approximated by the luminous intensity (measured in cd). For *surface sources*, brightness (in the photopic vision regime) can be approximated by the luminance (measured in cd/m^2). However, due to the lack of a formal standardized definition of the term brightness, it is frequently avoided in technical publications.

Standard CIE photometry assumes human vision to be **linear** within the photopic regime. It is clear that an isotropically emitting blue point source and an isotropically emitting red point source each having a luminous flux of, e.g., 5 lm, have the same luminous intensity. Assuming *linearity* of photopic vision, both sources still have the same luminous intensity as the luminous fluxes of the sources are increased from 5 to, e.g., 5000 lm.

However, if the luminous fluxes of the two sources are reduced so that the mesopic or scotopic vision regime is entered, the blue source will appear brighter than the red source due to the shift of the eye sensitivity function to shorter wavelengths in the scotopic regime.

It is important to keep in mind that the linearity of human vision within the photopic regime is an *approximation*. Linearity clearly simplifies photometry. However, human subjects may feel discrepancies between the experience of brightness and measured luminance of a light source, especially for colored light sources if the luminous flux is changed over orders of magnitude.

16.7 Circadian rhythm and circadian sensitivity

The human wake-sleep rhythm has a period of approximately 24 hours and the rhythm therefore is referred to as the **circadian rhythm** or **circadian cycle**, with the name being derived from the Latin words *circa* and *dies* (and its declination *diem*), meaning *approximately* and *day*, respectively. Light has been known for a long time to be the synchronizing clock (*zeitgeber*) of the human circadian rhythm. For reviews on the development of the understanding of the circadian rhythm including the identification of light as the dominant trigger for the endogenous *zeitgeber*, see Pittendrigh (1993) and Sehgal (2004).

The wake-sleep rhythm of humans is synchronized by the intensity and spectral composition of light. Sunlight is the natural *zeitgeber*. During mid-day hours sunlight has high intensity, a high color temperature, and a high content of blue light. During evening hours, intensity, color temperature, and blue content of sunlight strongly decrease. Humans have adapted to this variation and the circadian rhythm is most likely synchronized by the following three factors: intensity, color temperature, and blue content.

Exposure to inappropriately high intensities of light in the late afternoon or evening can upset the regular wake-seep rhythm and lead to sleeplessness and even serious illnesses such as cancer (Brainard et al., 2001; Blask et al, 2003). It is therefore highly advisable to limit exposure to high intensity light in the late afternoon and evening hours, to not be counterproductive to the natural circadian rhythm (Schubert, 1997).

It was believed for a long time that rod cells and the three types of cone cells are the only optically sensitive cells in the human eye. However, Brainard *et al.* (2001) postulated that an unknown photoreceptor in the human eye would control the circadian rhythm. Evidence presented by Berson *et al.* (2002) and Hattar *et al.*, (2002) indicates that retinal ganglion cells have an optical sensitivity as well. For a schematic illustration of ganglion cells, see Fig. 16.1. The spectral sensitivity of mammalian ganglion cells was measured and the responsivity curve is shown in Fig. 16.9. Inspection of the figure reveals a ganglion-cell peak-sensitivity at 484 nm, i.e. in the blue spectral range.

Berson et al. (2002) presented evidence that the photosensitive ganglion cells are instrumental in the control of the circadian rhythm. Due to their sensitivity in the blue spectral range, it can be hypothesized that the blue sky occurring near mid-day is a strong factor in synchronizing the endogenous circadian rhythm. The photosensitive ganglion cells have therefore been referred to as *blue-sky receptors*.

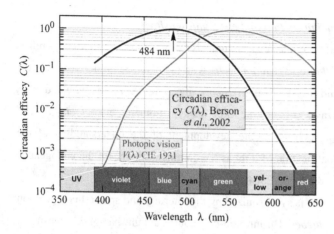

Fig. 16.9. Circadian efficacy curve derived from retinal ganglion cell photoresponse measurements. The ganglion cells on which the measurements were performed originated from mammals. The figure reveals the significant difference between circadian and visual sensitivity (adapted from Berson et al., 2002).

Inspection of the spectral sensitivity of the ganglion cells shown in Fig. 16.9 reveals the huge difference of red light and blue light for circadian efficacy: The efficacy of blue light in synchronizing the circadian rhythm can be three orders of magnitude greater than the efficacy of red light. This particular role of blue light should be taken into account in lighting design and the use of artificial lighting by consumers.

References

Berson D. M., Dunn F. A., and Takao M. "Phototransduction by retinal ganglion cells that set the circadian clock" *Science* **295**, 1070 (2002)

Brainard G. C., Hanifin J. P., Greeson J. M., Byrne B., Glickman G., Gerner E., and Rollag M. D. "Action spectrum for melatonin regulation in humans: Evidence for a novel circadian photoreceptor" *J. Neuroscience* **21**, 6405 (2001)

Blask D. E., Dauchy R. T., Sauer L. A., Krause J. A., Brainard G. C. "Growth and fatty acid metabolism of human breast cancer (MCF-7) xenografts in nude rats: Impact of constant light-induced nocturnal melatonin suppression" *Breast Cancer Research and Treatment* **79**, 313 (2003)

CIE *Commission Internationale de l'Eclairage Proceedings* (Cambridge University Press, Cambridge, 1931)

CIE *Proceedings* **1**, Sec. 4; **3**, p. 37; Bureau Central de la CIE, Paris (1951)

CIE data of 1931 and 1978 available at http://cvision.ucsd.edu and http://www.cvrl.org (1978). The CIE 1931 $V(\lambda)$ data were modified by D. B. Judd and J. J. Vos in 1978. The Judd–Vos-modified eye-sensitivity function is frequently referred to as $V_M(\lambda)$; see J. J. Vos "Colorimetric and photometric properties of a 2-deg fundamental observer" *Color Res. Appl.* **3**, 125 (1978)

CIE publication 75-1988 *Spectral Luminous Efficiency Functions Based Upon Brightness Matching for Monochromatic Point Sources with 2° and 10° Fields* ISBN 3900734119 (1988)

CIE publication 86-1990 *CIE 1988 2° Spectral Luminous Efficiency Function for Photopic Vision* ISBN 3900734232 (1990)

Dowling J. E. *The retina: An Approachable Part of the Brain* (Harvard University Press, Cambridge, Massachusetts, 1987)

Encyclopedia Britannica, Inc. Illustration of human eye adopted from 1994 edition of the encyclopedia (1994)

Hattar S., Liao H.-W., Takao M., Berson D. M., and Yau K.-W. "Melanopsin-containing retinal ganglion cells: Architecture, projections, and intrinsic photosensitivity" *Science* **295**, 1065 (2002)

MacAdam D. L. "Maximum attainable luminous efficiency of various chromaticities" *J. Opt. Soc. Am.* **40**, 120 (1950)

Osram Sylvania Corporation *Lumens and mesopic vision* Application Note FAQ0016-0297 (2000)

Pittendrigh C. S. "Temporal organization: Reflections of a Darwinian clock-watcher" *Ann. Rev. Physiol.* **55**, 17 (1993)

Schmid W., Scherer M., Karnutsch C., Plobl A., Wegleiter W., Schad S., Neubert B., and Streubel K. "High-efficiency red and infrared light-emitting diodes using radial outcoupling taper" *IEEE J. Sel. Top. Quantum Electron.* **8**, 256 (2002)

Schubert E. F. The author of this book noticed in 1997 that working after 8 PM under bright illumination conditions in the office allowed him to fall asleep only very late, typically after midnight. The origin of sleeplessness was traced back to high-intensity office lighting conditions. Once the high intensity of the office lighting was reduced, the sleeplessness vanished (1997)

Sehgal A., editor *Molecular Biology of Circadian Rhythms* (John Wiley and Sons, New York, 2004)

Vos J. J. "Colorimetric and photometric properties of a 2-deg fundamental observer" *Color Res. Appl.* **3**, 125 (1978)

Wyszecki G. and Stiles W. S. *Color Science – Concepts and Methods, Quantitative Data and Formulae* 2nd edition (John Wiley and Sons, New York, 1982)

Wyszecki G. and Stiles W. S. *Color Science – Concepts and Methods, Quantitative Data and Formulae* 2nd edition (John Wiley and Sons, New York, 2000)

Appendix 16.1

Tabulated values of the 2° degree CIE 1931 photopic eye sensitivity function and the CIE 1978 Judd–Vos-modified photopic eye sensitivity function for point sources (after CIE, 1931 and CIE, 1978).

λ (nm)	CIE 1931 $V(\lambda)$	CIE 1978 $V(\lambda)$	λ (nm)	CIE 1931 $V(\lambda)$	CIE 1978 $V(\lambda)$
			590	0.75700	0.75700
			595	0.69490	0.69483
360	3.9170 E–6	0.0000E–4	600	0.63100	0.63100
365	6.9650 E–6	0.0000E–4	605	0.56680	0.56654
370	1.2390 E–5	0.0000E–4	610	0.50300	0.50300
375	2.2020 E–5	0.0000E–4	615	0.44120	0.44172
380	3.9000 E–5	2.0000E–4	620	0.38100	0.38100
385	6.4000 E–5	3.9556E–4	625	0.32100	0.32052
390	1.2000 E–4	8.0000E–4	630	0.26500	0.26500
395	2.1700 E–4	1.5457E–3	635	0.21700	0.21702
400	3.9600 E–4	2.8000E–3	640	0.17500	0.17500
405	6.4000 E–4	4.6562E–3	645	0.13820	0.13812
410	1.2100 E–3	7.4000E–3	650	0.10700	0.1.0700
415	2.1800 E–3	1.1779E–2	655	8.1600 E–2	8.1652E–2
420	4.0000 E–3	1.7500E–2	660	6.1000 E–2	6.1000E–2
425	7.3000 E–3	2.2678E–2	665	4.4580 E–2	4.4327E–2
430	1.1600 E–2	2.7300E–2	670	3.2000 E–2	3.2000E–2
435	1.6840 E–2	3.2584E–2	675	2.3200 E–2	2.3454E–2
440	2.3000 E–2	3.7900E–2	680	1.7000 E–2	1.7000E–2
445	2.9800 E–2	4.2391E–2	685	1.1920 E–2	1.1872E–2
450	3.8000 E–2	4.6800E–2	690	8.2100 E–3	8.2100E–3
455	4.8000 E–2	5.2122E–2	695	5.7230 E–3	5.7723E–3
460	6.0000 E–2	6.0000E–2	700	4.1020 E–3	4.1020E–3
465	7.3900 E–2	7.2942E–2	705	2.9290 E–3	2.9291E–3
470	9.0980 E–2	9.0980E–2	710	2.0910 E–3	2.0910E–3
475	0.11260	0.11284	715	1.4840 E–3	1.4822E–3
480	0.13902	0.13902	720	1.0470 E–3	1.0470E–3
485	0.16930	0.16987	725	7.4000 E–4	7.4015E–4
490	0.20802	0.20802	730	5.2000 E–4	5.2000E–4
495	0.25860	0.25808	735	3.6110 E–4	3.6093E–4
500	0.32300	0.32300	740	2.4920 E–4	2.4920E–4
505	0.40730	0.40540	745	1.7190 E–4	1.7231E–4
510	0.50300	0.50300	750	1.2000 E–4	1.2000E–4
515	0.60820	0.60811	755	8.4800 E–5	8.4620E–5
520	0.71000	0.71000	760	6.0000 E–5	6.0000E–5
525	0.79320	0.79510	765	4.2400 E–5	4.2446E–5
530	0.86200	0.86200	770	3.0000 E–5	3.0000E–5
535	0.91485	0.91505	775	2.1200 E–5	2.1210E–5
540	0.95400	0.95400	780	1.4990 E–5	1.4989E–5
545	0.98030	0.98004	785	1.0600 E–5	1.0584E–5
550	0.99495	0.99495	790	7.4657 E–6	7.4656E–6
555	1.00000	1.00000	795	5.2578 E–6	5.2592E–6
560	0.99500	0.99500	800	3.7029 E–6	3.7028E–6
565	0.97860	0.97875	805	2.6078 E–6	2.6076E–6
570	0.95200	0.95200	810	1.8366 E–6	1.8365E–6
575	0.91540	0.91558	815	1.2934 E–6	1.2950E–6
580	0.87000	0.87000	820	9.1093 E–7	9.1092E–7
585	0.81630	0.81623	825	6.4153 E–7	6.3564E–7

Appendix 16.2

Tabulated values of the CIE 1951 eye sensitivity function of the scotopic vision regime, $V'(\lambda)$ (after CIE, 1951).

λ (nm)	CIE 1951 $V'(\lambda)$	λ (nm)	CIE 1951 $V'(\lambda)$
380	5.890e-004	585	8.990e-002
385	1.108e-003	590	6.550e-002
390	2.209e-003	595	4.690e-002
395	4.530e-003	600	3.315e-002
400	9.290e-003	605	2.312e-002
405	1.852e-002	610	1.593e-002
410	3.484e-002	615	1.088e-002
415	6.040e-002	620	7.370e-003
420	9.660e-002	625	4.970e-003
425	1.436e-001	630	3.335e-003
430	1.998e-001	635	2.235e-003
435	2.625e-001	640	1.497e-003
440	3.281e-001	645	1.005e-003
445	3.931e-001	650	6.770e-004
450	4.550e-001	655	4.590e-004
455	5.130e-001	660	3.129e-004
460	5.670e-001	665	2.146e-004
465	6.200e-001	670	1.480e-004
470	6.760e-001	675	1.026e-004
475	7.340e-001	680	7.150e-005
480	7.930e-001	685	5.010e-005
485	8.510e-001	690	3.533e-005
490	9.040e-001	695	2.501e-005
495	9.490e-001	700	1.780e-005
500	9.820e-001	705	1.273e-005
505	9.980e-001	710	9.140e-006
510	9.970e-001	715	6.600e-006
515	9.750e-001	720	4.780e-006
520	9.350e-001	725	3.482e-006
525	8.800e-001	730	2.546e-006
530	8.110e-001	735	1.870e-006
535	7.330e-001	740	1.379e-006
540	6.500e-001	745	1.022e-006
545	5.640e-001	750	7.600e-007
550	4.810e-001	755	5.670e-007
555	4.020e-001	760	4.250e-007
560	3.288e-001	765	3.196e-007
565	2.639e-001	770	2.413e-007
570	2.076e-001	775	1.829e-007
575	1.602e-001	780	1.390e-007
580	1.212e-001		

17

Colorimetry

The assessment and quantification of color is referred to as *colorimetry* or the "science of color". Colorimetry is closely associated with human color vision. Both colorimetry and human vision have attracted a great deal of interest that spans many centuries. For a thorough and entertaining review of the history of colorimetry including early attempts to understand color, we recommended the collection of historical reprints complied by MacAdam (1993).

The human sense of vision is very different from the human sense of hearing. If we hear two frequencies simultaneously, e.g. two frequencies generated by a musical instrument, we will be able to recognize the musical tone as having two distinct frequencies. This is not the case for optical signals and the sense of vision. Mixing two monochromatic optical signals will appear to us as one color and we are unable to recognize the original dichromatic composition of that color.

17.1 Color-matching functions and chromaticity diagram

Light causes different levels of excitation of the red, green, and blue cones. However, the sensation of color and luminous flux caused a particular light source varies slightly among different individuals. Furthermore, the sensation of color is, to some extent, a subjective quantity. For these reasons, *The International Commission for Illumination* (*Commission Internationale de l'Eclairage*, CIE) has *standardized the measurement of color* by means of **color-matching functions** and the **chromaticity diagram** (CIE, 1931).

How are color-matching functions obtained? Consider two lights lying side by side: One being monochromatic and the other one being a mixture of three primary lights with color red, green, and blue, as shown in Fig. 17.1. A human subject will be able to make the two lights appear identical (i.e. "match" them) by adjusting the relative intensities of the red, green, and blue light. The three color-matching functions are obtained from a series of such matches, in which the subject sets the intensities of the three primary lights required to match a series of monochromatic lights across the visible spectrum.

17.1 Color-matching functions and chromaticity diagram

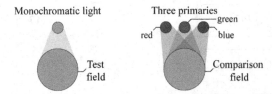

Fig. 17.1. Principle of color matching: A monochromatic test light (imaged on the "test field") is color-matched by mixing three adjustable primary lights, red, green, and blue (imaged on the "comparison field").

Subsequently, the measured set of color-matching functions is mathematically transformed into a new set of color-matching functions for which the green color-matching function, $\bar{y}(\lambda)$, is chosen to be identical to the eye sensitivity function, $V(\lambda)$, i.e.

$$\bar{y}(\lambda) = V(\lambda) \quad . \tag{17.1}$$

The CIE 1931 and CIE 1978 color-matching functions $\bar{x}(\lambda)$, $\bar{y}(\lambda)$, and $\bar{z}(\lambda)$ are shown in Fig. 17.2. The numerical values of these color-matching functions are tabulated in Appendices 17.1 and 17.2, respectively. The three color-matching functions reflect the fact that human color vision possesses **trichromacy**, that is, the color of any light source can be described by just three variables. Note that $\bar{x}(\lambda)$, $\bar{y}(\lambda)$, and $\bar{z}(\lambda)$ are dimensionless quantities. Also note that neither the color-matching functions nor the chromaticity diagram is unique (see, for example, Judd, 1951 or Vos, 1978). In fact there have been several different versions of the color-matching functions and of the chromaticity diagram.

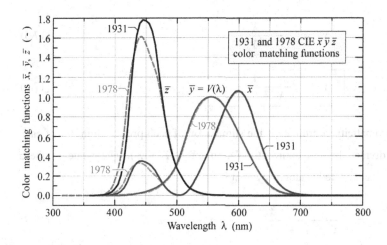

Fig. 17.2. CIE (1931) and CIE (1978) $\bar{x}\bar{y}\bar{z}$ color-matching functions. The $\bar{y}$ color-matching function is identical to the eye sensitivity function $V(\lambda)$. Note that the CIE 1931 color-matching functions are the currently valid official standard in the United States.

For a given power-spectral density $P(\lambda)$, the degree of stimulation required to match the color of $P(\lambda)$ is given by

293

17 Colorimetry

$$X = \int_\lambda \bar{x}(\lambda) \, P(\lambda) \, d\lambda \qquad (17.2)$$

$$Y = \int_\lambda \bar{y}(\lambda) \, P(\lambda) \, d\lambda \qquad (17.3)$$

$$Z = \int_\lambda \bar{z}(\lambda) \, P(\lambda) \, d\lambda \qquad (17.4)$$

where X, Y, and Z are the **tristimulus values** that give the stimulation (i.e. power) of each of the three primary red, green, and blue lights needed to match the color of $P(\lambda)$. Large values of X, Y, and Z indicate red, green, and blue colors of the spectrum $P(\lambda)$, respectively.

Because of the distinct similarity of the three retinal-cone-sensitivity functions on one hand, and the color-matching functions on the other hand (both groups of functions have three peaks), each tristimulus value represents the *approximate* (but not *exact*) degree of stimulation that each type of retinal cone experiences when illuminated by a light source with spectrum $P(\lambda)$.

Inspection of Eqs. (17.2)–(17.4) suggests that the unit of the tristimulus values is "watt". However, the tristimulus values are usually given as dimensionless quantities. The prefactor "watt^{-1}" in front of the integral can be included so that the tristimulus values become dimensionless. If only *ratios* of tristimulus values are employed, as below, the prefactors and units cancel and thus become irrelevant.

The **chromaticity coordinates** x and y are calculated from the tristimulus values according to

$$x = \frac{X}{X + Y + Z} \qquad (17.5)$$

$$y = \frac{Y}{X + Y + Z} \qquad (17.6)$$

Thus, the value of a chromaticity coordinate is the stimulation of each primary light (or of each type of retinal cone) divided by the entire stimulation ($X + Y + Z$). The value of the z chromaticity coordinate is calculated analogously, that is

$$z = \frac{Z}{X + Y + Z} = 1 - x - y \, . \qquad (17.7)$$

Note that the z chromaticity value can be obtained from x and y, so that there is no new information provided by the z chromaticity coordinate. Therefore, the z coordinate is redundant and, for this reason, does not need to be used.

The (x, y) chromaticity diagram is shown in Fig. 17.3. Reddish and greenish colors are found for large values of x and y, respectively. Bluish colors are found for large values of z, which is, according to Eq. (17.7), for low values of x and y, or near the origin of the chromaticity diagram.

The chromaticity diagram of Fig. 17.4 shows a detailed attribution of colors to their locations in the chromaticity diagram. The assignment of colors was given by Gage *et al.* (1977).

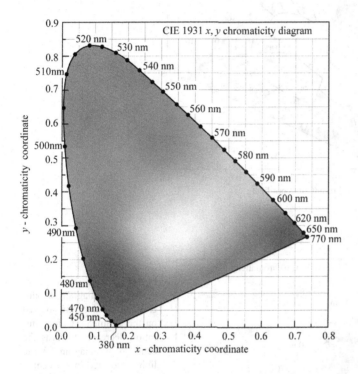

Fig. 17.3. CIE 1931 (x, y) chromaticity diagram. Monochromatic colors are located on the perimeter and white light is located in the center of the diagram.

An assignment of common colors in the chromaticity diagram is given in Fig. 17.5. The figure also shows the **equal-energy point** located in the center of the chromaticity diagram at $(x, y, z) = (1/3, 1/3, 1/3)$. The optical spectrum corresponding to the equal-energy point has a constant spectral distribution, i.e. the optical energy per wavelength interval $d\lambda$ is constant across the visible spectrum. Such a spectrum also results in equal tristimulus values, i.e. $X = Y = Z$.

Monochromatic or pure colors are found on the perimeter of the chromaticity diagram. White light is found in the center of the chromaticity diagram. All colors can be characterized in terms of their location in the chromaticity diagram.

17 Colorimetry

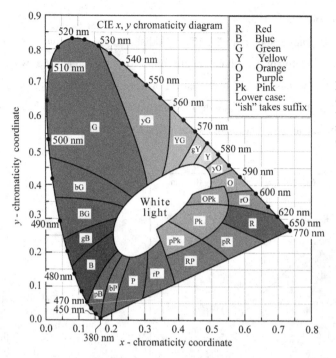

Fig. 17.4. 1931 CIE chromaticity diagram with areas attributed to distinct colors (after Gage et al., 1977).

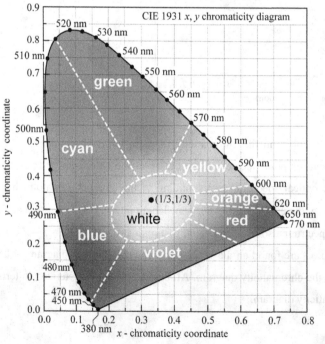

Fig. 17.5. CIE 1931 (x, y) chromaticity diagram. Monochromatic colors are located on the perimeter. Color saturation decreases towards the center of the diagram. White light is located in the center. Also shown are the regions of distinct colors. The equal-energy point is located at the center and has the coordinates $(x, y) = (1/3, 1/3)$.

MacAdam (1943) analyzed the color differences of closely spaced points in the chromaticity diagram. The author found that two chromaticity points must have a *minimum geometrical distance* to yield a perceptible difference in color. Colors within a certain small region in the chromaticity diagram appear identical to human subjects. MacAdam showed that these regions have the shape of an ellipses. Such ellipses, now known as the ***MacAdam ellipses***, are shown in Fig. 17.6 (MacAdam, 1943, 1993; Wright, 1943). Inspection of the figure shows that ellipses in the blue and green regions are very different in size. Thus, the geometric distance between two points in the (x, y) chromaticity diagram does *not* scale linearly with the color difference.

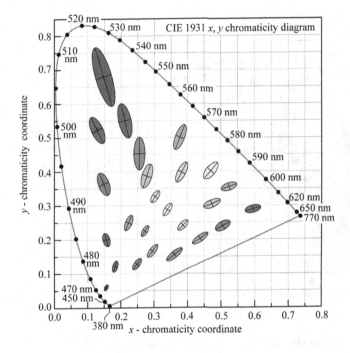

Fig. 17.6. MacAdam ellipses plotted in the CIE 1931 (x, y) chromaticity diagram. The axes of the ellipses are ten times their actual lengths (after MacAdam, 1943; Wright, 1943; MacAdam, 1993).

The total number of differentiable chromaticities can be obtained by dividing the area of the chromaticity diagram through the average area of the MacAdam ellipses. This calculation yields the result that humans can discern approximately 50 000 distinct chromaticities. If possible variations in luminance (brightness) are taken into account, the number of differentiable colors increases to a value greater than 10^6.

In the chromaticity diagram, it is very desirable for the color difference to be proportional to the geometric difference. This has motivated the *uniform* chromaticity diagram. In 1960, the CIE introduced the (u, v) and in 1976 the (u', v') ***uniform chromaticity coordinates*** (Wyszecki and

Stiles, 2000). These coordinates form the **uniform chromaticity diagram**. The uniform chromaticity coordinates are calculated from the tristimulus values according to

$$u = \frac{4X}{X + 15Y + 3Z} \qquad v = \frac{6Y}{X + 15Y + 3Z} \qquad \text{(CIE, 1960)} \qquad (17.8)$$

and

$$u' = \frac{4X}{X + 15Y + 3Z} \qquad v' = \frac{9Y}{X + 15Y + 3Z} \qquad \text{(CIE, 1976)}. \qquad (17.9)$$

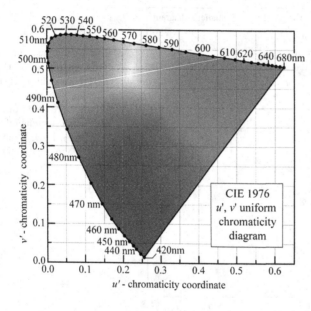

Fig. 17.7. CIE 1976 (u', v') uniform chromaticity diagram calculated using the CIE 1931 2° standard observer.

The CIE 1976 (u', v') uniform chromaticity diagram is shown in Fig. 17.7. The (u, v) and (u', v') uniform chromaticity coordinates can be calculated from the (x, y) chromaticity coordinates according to

$$u = u' = \frac{4x}{-2x + 12y + 3} \qquad (17.10)$$

and

$$v = \frac{6y}{-2x + 12y + 3} \qquad v' = \frac{9y}{-2x + 12y + 3}. \qquad (17.11)$$

Conversely, one obtains

$$x = \frac{9u'}{6u' - 16v' + 12} \qquad y = \frac{2v'}{3u' - 8v' + 6} \qquad (17.12)$$

and

$$x = \frac{3u}{2u - 8v + 4} \qquad y = \frac{2v}{2u - 8v + 4}. \qquad (17.13)$$

The color differences between two points in the (x, y) chromaticity diagram are spatially very non-uniform, that is, the color changes much more rapidly in one direction, e.g. the x-direction, compared with the other direction, e.g. the y-direction. This deficiency of the (x, y) chromaticity diagram is strongly reduced, although not eliminated, in the (u, v) and (u', v') uniform chromaticity diagrams. As a result, the *color difference* between two locations in the uniform chromaticity diagram is (approximately) directly proportional to the *geometrical distance* between these points.

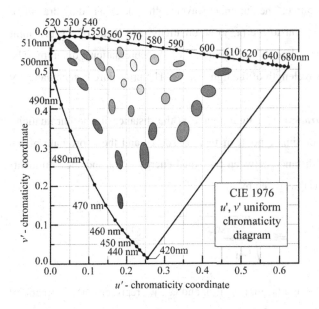

Fig. 17.8. MacAdam ellipses transformed to uniform CIE 1976 (u', v') chromaticity coordinates. For clarity, the axes of the transformed ellipses are ten times their actual lengths. Transformed ellipses are not ellipses in a strict mathematical sense, but their shapes closely resemble those of ellipses. The areas of the transformed ellipses in the (u', v') diagram are much more similar than the MacAdam ellipses in the (x, y) diagram.

The 1943 MacAdam ellipses in the (x, y) chromaticity diagram can be transformed to the uniform (u', v') chromaticity diagram. This transformation is shown in Fig. 17.8, which reveals that the areas of non-discernable colors are much more uniform in shape and area than the

MacAdam ellipses in the (x, y) chromaticity diagram.

Note that the transformation from the (x, y) to the (u', v') coordinate system is mathematically *non-linear*, and thus ellipses in the (x, y) chromaticity diagram do *not* transform into ellipses in the (u', v') coordinate system. However, if the ellipses in the (x, y) chromaticity diagram are sufficiently small in size, non-linear distortions are small as well, so that the transformed regions are very close to ellipses.

17.2 Color purity

Monochromatic sources ($\Delta\lambda \to 0$) are located on the perimeter of the chromaticity diagram. However, as the spectral linewidth of a light source gets broader, the color location in the chromaticity diagram moves towards the center of the chromaticity diagram. If the spectral width of a light source becomes comparable to the entire visible range, the light source is *white* and thus located near the center of the chromaticity diagram.

The **dominant wavelength** of a test light source is defined as the wavelength (i.e. monochromatic color) located on the perimeter of the chromaticity diagram that appears to be closest to the color of the test light source. The dominant wavelength is determined by drawing a straight line from the equal-energy point to the (x, y) chromaticity coordinate of the test light source, and by extending the straight line to the perimeter of the chromaticity diagram. The intersection point is the dominant wavelength of the light source. The procedure is schematically shown in Fig. 17.9.

The **color purity** or **color saturation** of a light source is the distance in the chromaticity diagram between the (x, y) color-coordinate point of the test source and the coordinate of the equal-energy point divided by the distance between the equal-energy point and the dominant-wavelength point. The color purity is thus given by

$$\text{color purity} = \frac{a}{a + b} = \frac{\sqrt{(x - x_{ee})^2 + (y - y_{ee})^2}}{\sqrt{(x_d - x_{ee})^2 + (y_d - y_{ee})^2}} \qquad (17.14)$$

where a and b are shown in Fig. 17.9 and (x, y), (x_{ee}, y_{ee}), and (x_d, y_d) represent the chromaticity coordinates of the light source under test, of the equal-energy reference illuminant, and of the dominant-wavelength point, respectively. Thus the color purity is the relative distance of a light source under test from the center of the chromaticity diagram. Generally, the color purity is 100% for monochromatic light sources ($\Delta\lambda \to 0$) located on the perimeter of the chromaticity diagram and near 0% for white illuminants.

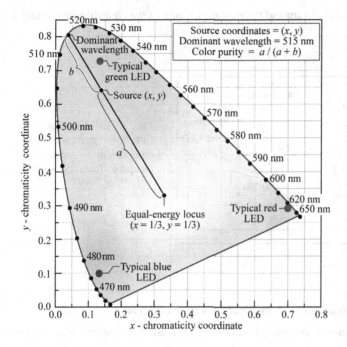

Fig. 17.9. Chromaticity diagram showing the determination of the *dominant wavelength* and *color purity* of a light source with chromaticity coordinates (x, y) using the equal-energy locus $(x = 1/3, y = 1/3)$ as the white-light reference. Also shown are typical locations of blue, green, and red LEDs.

Note that the dominant wavelength and color purity are an alternative way to uniquely characterize the location of an emitter on the chromaticity diagram. Dominant wavelength and color purity are quite intuitive quantities (more so than the numerical (x, y) chromaticity coordinates) and they are therefore frequently preferred.

17.3 LEDs in the chromaticity diagram

Monochromatic light sources ($\Delta\lambda \to 0$) are located on the perimeter of the chromaticity diagram. Light emission from LEDs is monochromatic (single color) to the eye but LEDs are not monochromatic in the strict physical sense since LEDs have a spectral linewidth of about $1.8\ kT$. Owing to the finite spectral linewidth of LEDs, they are not located on the very perimeter of the chromaticity diagram but are located *close* to the perimeter. When a source emits light distributed over a range of wavelengths, then the chromaticity location moves towards the center of the diagram.

The location of different LEDs on the chromaticity diagram is shown in Fig. 17.10. Inspection of the figure reveals that the location of red and blue LEDs is on the perimeter of the chromaticity diagram. That is, their color purity is very high, close to 100%. However, blue–green and green LEDs are located off the perimeter closer to the center of the diagram due to the finite linewidth of the emission spectrum and the strong curvature of the chromaticity diagram in the green wavelength range.

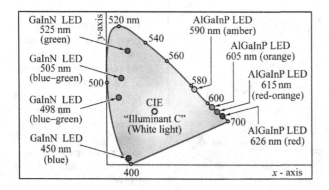

Fig. 17.10. Location of LED light emission on the chromaticity diagram (adapted from Schubert and Miller, 1999).

17.4 Relationship between chromaticity and color

Having completed the discussion of *chromaticity* allows us to revisit the question: What is *color*? One could certainly define a color by its location in the chromaticity diagram (i.e. by its chromaticity). However, the CIE (1986) adopted a more general definition of the term "color" that goes beyond the location in the chromaticity diagram. The CIE's broader definition of color includes **chromaticity** as well as **brightness**. That is, we can keep the chromaticity of a light source the same but change its luminous intensity (brightness); this, according to the CIE definition, changes the color of the source. If, on the other hand, we restrict our considerations to a single brightness level, the terms "chromaticity" and "color" can be used synonymously.

The location of a chromaticity point can be expressed by the *dominant wavelength* and *saturation* (or, alternatively, by *hue* and *saturation*). Thus, the CIE definition of the *color of light* is given by the *dominant wavelength, saturation, and* **brightness**. The *color of an object* is given by the *dominant wavelength, saturation, and* **lightness**.

References
CIE *Commission Internationale de l'Eclairage Proceedings* (Cambridge University Press, Cambridge, 1931)
CIE data of 1931 and 1978 available at http://cvision.ucsd.edu and http://www.cvrl.org (1978)
CIE data of 1960 relating to the (u, v) chromaticity coordinates can be found in CIE, 1986
CIE data of 1976 relating to the (u', v') chromaticity coordinates can be found in CIE, 1986

References

CIE publication No. 15 (E.1.3.1) 1971: *Colorimetry*; this publication was updated in 1986 to CIE Publication 15.2 *Colorimetry* (CIE, Vienna, Austria, 1986)

Gage S., Evans D., Hodapp M. W., and Sorensen H. *Optoelectronics Applications Manual* 1st edition (McGraw Hill, New York, 1977)

Judd D. B. "Report of US Secretariat Committee on Colorimetry and Artificial Daylight" in *Proceedings of the 12th Session of the CIE* **1**, p. 11 (Bureau Central de la CIE, Paris, 1951)

MacAdam D. L. "Specification of small chromaticity differences" *J. Opt. Soc. Am.* **33**, 18 (1943)

MacAdam D. L. (Editor) *Colorimetry – Fundamentals* (SPIE Optical Engineering Press, Bellingham, Washington, 1993)

Schubert E. F. and Miller J. N "Light-emitting diodes - An introduction" *Encyclopedia of Electrical Engineering*, edited by John G. Webster, Vol. **11**, p. 326 (John Wiley and Sons, New York, March 1999)

Vos J. J. "Colorimetric and photometric properties of a 2-degree fundamental observer" *Color Res. Appl.* **3**, 125 (1978)

Wyszecki G. and Stiles W. S. *Color Science – Concepts and Methods, Quantitative Data and Formulae* 2nd edition (John Wiley and Sons, New York, 2000)

Wright W. D. "The graphical representation of small color differences" *J. Opt. Soc. Am.* **33**, 632 (1943)

Appendix 17.1

Tabulated values of the CIE 1931 two-degree color-matching functions and eye sensitivity function for point sources (after CIE, 1931).

λ (nm)	$\bar{x}(\lambda)$ red	$\bar{y} = V(\lambda)$ green	$\bar{z}(\lambda)$ blue	λ	$\bar{x}(\lambda)$	$\bar{y} = V(\lambda)$	$\bar{z}(\lambda)$
360	1.2990 E–4	3.9170 E–6	6.0610 E–4	590	1.02630	0.75700	1.1000 E–3
365	2.3210 E–4	6.9650 E–6	1.0860 E–3	595	1.05670	0.69490	1.0000 E–3
370	4.1490 E–4	1.2390 E–5	1.9460 E–3	600	1.06220	0.63100	8.0000 E–4
375	7.4160 E–4	2.2020 E–5	3.4860 E–3	605	1.04560	0.56680	6.0000 E–4
380	1.3680 E–3	3.9000 E–5	6.4500 E–3	610	1.00260	0.50300	3.4000 E–4
385	2.2360 E–3	6.4000 E–5	1.0550 E–2	615	0.93840	0.44120	2.4000 E–4
390	4.2430 E–3	1.2000 E–4	2.0050 E–2	620	0.85445	0.38100	1.9000 E–4
395	7.6500 E–3	2.1700 E–4	3.6210 E–2	625	0.75140	0.32100	1.0000 E–4
400	1.4310 E–2	3.9600 E–4	6.7850 E–2	630	0.64240	0.26500	5.0000 E–5
405	2.3190 E–2	6.4000 E–4	0.11020	635	0.54190	0.21700	3.0000 E–5
410	4.3510 E–2	1.2100 E–3	0.20740	640	0.44790	0.17500	2.0000 E–5
415	7.7630 E–2	2.1800 E–3	0.37130	645	0.36080	0.13820	1.0000 E–5
420	0.13438	4.0000 E–3	0.64560	650	0.28350	0.10700	0.0000 E–5
425	0.21477	7.3000 E–3	1.03905	655	0.21870	8.1600 E–2	0.0000 E–5
430	0.28390	1.1600 E–2	1.38560	660	0.16490	6.1000 E–2	0.0000 E–5
435	0.32850	1.6840 E–2	1.62296	665	0.12120	4.4580 E–2	0.0000 E–5
440	0.34828	2.3000 E–2	1.74706	670	8.7400 E–2	3.2000 E–2	0.0000 E–5
445	0.34806	2.9800 E–2	1.78260	675	6.3600 E–2	2.3200 E–2	0.0000 E–5
450	0.33620	3.8000 E–2	1.77211	680	4.6770 E–2	1.7000 E–2	0.0000 E–5
455	0.31870	4.8000 E–2	1.74410	685	3.2900 E–2	1.1920 E–2	0.0000 E–5
460	0.29080	6.0000 E–2	1.66920	690	2.2700 E–2	8.2100 E–3	0.0000 E–5
465	0.25110	7.3900 E–2	1.52810	695	1.5840 E–2	5.7230 E–3	0.0000 E–5
470	0.19536	9.0980 E–2	1.28764	700	1.1359 E–2	4.1020 E–3	0.0000 E–5
475	0.14210	0.11260	1.04190	705	8.1109 E–3	2.9290 E–3	0.0000 E–5
480	9.5640 E–2	0.13902	0.81295	710	5.7903 E–3	2.0910 E–3	0.0000 E–5
485	5.7950 E–2	0.16930	0.61620	715	4.1065 E–3	1.4840 E–3	0.0000 E–5
490	3.2010 E–2	0.20802	0.46518	720	2.8993 E–3	1.0470 E–3	0.0000 E–5
495	1.4700 E–2	0.25860	0.35330	725	2.0492 E–3	7.4000 E–4	0.0000 E–5
500	4.9000 E–3	0.32300	0.27200	730	1.4400 E–3	5.2000 E–4	0.0000 E–5
505	2.4000 E–3	0.40730	0.21230	735	9.9995 E–4	3.6110 E–4	0.0000 E–5
510	9.3000 E–3	0.50300	0.15820	740	6.9008 E–4	2.4920 E–4	0.0000 E–5
515	2.9100 E–2	0.60820	0.11170	745	4.7602 E–4	1.7190 E–4	0.0000 E–5
520	6.3270 E–2	0.71000	7.8250 E–2	750	3.3230 E–4	1.2000 E–4	0.0000 E–5
525	0.10960	0.79320	5.7250 E–2	755	2.3483 E–4	8.4800 E–5	0.0000 E–5
530	0.16550	0.86200	4.2160 E–2	760	1.6615 E–4	6.0000 E–5	0.0000 E–5
535	0.22575	0.91485	2.9840 E–2	765	1.1741 E–4	4.2400 E–5	0.0000 E–5
540	0.29040	0.95400	2.0300 E–2	770	8.3075 E–5	3.0000 E–5	0.0000 E–5
545	0.35970	0.98030	1.3400 E–2	775	5.8707 E–5	2.1200 E–5	0.0000 E–5
550	0.43345	0.99495	8.7500 E–3	780	4.1510 E–5	1.4990 E–5	0.0000 E–5
555	0.51205	1.00000	5.7500 E–3	785	2.9353 E–5	1.0600 E–5	0.0000 E–5
560	0.59450	0.99500	3.9000 E–3	790	2.0674 E–5	7.4657 E–6	0.0000 E–5
565	0.67840	0.97860	2.7500 E–3	795	1.4560 E–5	5.2578 E–6	0.0000 E–5
570	0.76210	0.95200	2.1000 E–3	800	1.0254 E–5	3.7029 E–6	0.0000 E–5
575	0.84250	0.91540	1.8000 E–3	805	7.2215 E–6	2.6078 E–6	0.0000 E–5
580	0.91630	0.87000	1.6500 E–3	810	5.0859 E–6	1.8366 E–6	0.0000 E–5
585	0.97860	0.81630	1.4000 E–3	815	3.5817 E–6	1.2934 E–6	0.0000 E–5
				820	2.5225 E–6	9.1093 E–7	0.0000 E–5
				825	1.7765 E–6	6.4153 E–7	0.0000 E–5

Appendix 17.2

Tabulated values of the CIE 1978 two-degree color-matching functions and eye sensitivity function for point sources (after CIE, 1978). The functions are also called the Judd–Vos-modified color-matching functions.

λ (nm)	$\bar{x}(\lambda)$ red	$\bar{y} = V(\lambda)$ green	$\bar{z}(\lambda)$ blue
380	2.6899E–3	2.0000E–4	1.2260E–2
385	5.3105E–3	3.9556E–4	2.4222E–2
390	1.0781E–2	8.0000E–4	4.9250E–2
395	2.0792E–2	1.5457E–3	9.5135E–2
400	3.7981E–2	2.8000E–3	1.7409E–1
405	6.3157E–2	4.6562E–3	2.9013E–1
410	9.9941E–2	7.4000E–3	4.6053E–1
415	1.5824E–1	1.1779E–2	7.3166E–1
420	2.2948E–1	1.7500E–2	1.0658
425	2.8108E–1	2.2678E–2	1.3146
430	3.1095E–1	2.7300E–2	1.4672
435	3.3072E–1	3.2584E–2	1.5796
440	3.3336E–1	3.7900E–2	1.6166
445	3.1672E–1	4.2391E–2	1.5682
450	2.8882E–1	4.6800E–2	1.4717
455	2.5969E–1	5.2122E–2	1.3740
460	2.3276E–1	6.0000E–2	1.2917
465	2.0999E–1	7.2942E–2	1.2356
470	1.7476E–1	9.0980E–2	1.1138
475	1.3287E–1	1.1284E–1	9.4220E–1
480	9.1944E–2	1.3902E–1	7.5596E–1
485	5.6985E–2	1.6987E–1	5.8640E–1
490	3.1731E–2	2.0802E–1	4.4669E–1
495	1.4613E–2	2.5808E–1	3.4116E–1
500	4.8491E–3	3.2300E–1	2.6437E–1
505	2.3215E–3	4.0540E–1	2.0594E–1
510	9.2899E–3	5.0300E–1	1.5445E–1
515	2.9278E–2	6.0811E–1	1.0918E–1
520	6.3791E–2	7.1000E–1	7.6585E–2
525	1.1081E–1	7.9510E–1	5.6227E–2
530	1.6692E–1	8.6200E–1	4.1366E–2
535	2.2768E–1	9.1505E–1	2.9353E–2
540	2.9269E–1	9.5400E–1	2.0042E–2
545	3.6225E–1	9.8004E–1	1.3312E–2
550	4.3635E–1	9.9495E–1	8.7823E–3
555	5.1513E–1	1.0000	5.8573E–3
560	5.9748E–1	9.9500E–1	4.0493E–3
565	6.8121E–1	9.7875E–1	2.9217E–3
570	7.6425E–1	9.5200E–1	2.2771E–3
575	8.4394E–1	9.1558E–1	1.9706E–3
580	9.1635E–1	8.7000E–1	1.8066E–3
585	9.7703E–1	8.1623E–1	1.5449E–3
590	1.0230	7.5700E–1	1.2348E–3
595	1.0513	6.9483E–1	1.1177E–3
600	1.0550	6.3100E–1	9.0564E–4
605	1.0362	5.6654E–1	6.9467E–4
610	9.9239E–1	5.0300E–1	4.2885E–4
615	9.2861E–1	4.4172E–1	3.1817E–4
620	8.4346E–1	3.8100E–1	2.5598E–4
625	7.3983E–1	3.2052E–1	1.5679E–4
630	6.3289E–1	2.6500E–1	9.7694E–5
635	5.3351E–1	2.1702E–1	6.8944E–5
640	4.4062E–1	1.7500E–1	5.1165E–5
645	3.5453E–1	1.3812E–1	3.6016E–5
650	2.7862E–1	1.0700E–1	2.4238E–5
655	2.1485E–1	8.1652E–2	1.6915E–5
660	1.6161E–1	6.1000E–2	1.1906E–5
665	1.1820E–1	4.4327E–2	8.1489E–6
670	8.5753E–2	3.2000E–2	5.6006E–6
675	6.3077E–2	2.3454E–2	3.9544E–6
680	4.5834E–2	1.7000E–2	2.7912E–6
685	3.2057E–2	1.1872E–2	1.9176E–6
690	2.2187E–2	8.2100E–3	1.3135E–6
695	1.5612E–2	5.7723E–3	9.1519E–7
700	1.1098E–2	4.1020E–3	6.4767E–7
705	7.9233E–3	2.9291E–3	4.6352E–7
710	5.6531E–3	2.0910E–3	3.3304E–7
715	4.0039E–3	1.4822E–3	2.3823E–7
720	2.8253E–3	1.0470E–3	1.7026E–7
725	1.9947E–3	7.4015E–4	1.2207E–7
730	1.3994E–3	5.2000E–4	8.7107E–8
735	9.6980E–4	3.6093E–4	6.1455E–8
740	6.6847E–4	2.4920E–4	4.3162E–8
745	4.6141E–4	1.7231E–4	3.0379E–8
750	3.2073E–4	1.2000E–4	2.1554E–8
755	2.2573E–4	8.4620E–5	1.5493E–8
760	1.5973E–4	6.0000E–5	1.1204E–8
765	1.1275E–4	4.2446E–5	8.0873E–9
770	7.9513E–5	3.0000E–5	5.8340E–9
775	5.6087E–5	2.1210E–5	4.2110E–9
780	3.9541E–5	1.4989E–5	3.0383E–9
785	2.7852E–5	1.0584E–5	2.1907E–9
790	1.9597E–5	7.4656E–6	1.5778E–9
795	1.3770E–5	5.2592E–6	1.1348E–9
800	9.6700E–6	3.7028E–6	8.1565E–10
805	6.7918E–6	2.6076E–6	5.8626E–10
810	4.7706E–6	1.8365E–6	4.2138E–10
815	3.3550E–6	1.2950E–6	3.0319E–10
820	2.3534E–6	9.1092E–7	2.1753E–10
825	1.6377E–6	6.3564E–7	1.5476E–10

18

Planckian sources and color temperature

White light is a unique color. There are a very large number of optical spectra that can be used to generate white light. Among these spectra, the planckian black-body radiation spectrum forms a unique and very useful standard because it allows us to describe the spectrum with only one parameter, namely the color temperature. Furthermore, natural daylight closely resembles the planckian spectrum.

18.1 The solar spectrum

White light usually has a broad spectrum extending over the entire visible range. An instructive model for white light is sunlight. The sun's optical spectrum is shown in Fig. 18.1, including the spectrum at sea level with the sun at zenith, incident above the earth's atmosphere, and at sunset and sunrise (Jackson, 1975). The spectrum of sunlight extends over the entire visible region. However, the sun's spectrum depends on the time of day, season, altitude, weather, and other factors.

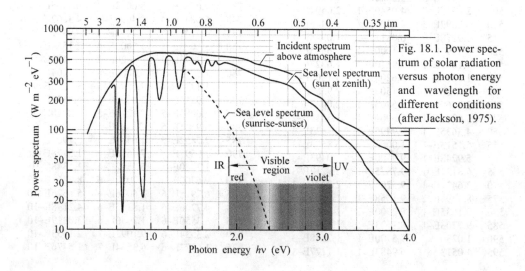

Fig. 18.1. Power spectrum of solar radiation versus photon energy and wavelength for different conditions (after Jackson, 1975).

306

Exact replication of the solar spectrum for white-light illumination sources would not yield an efficient source due to the large infrared (IR) and ultraviolet (UV) components of the solar spectrum. Thus, "mother nature" does not serve as a good example for an efficient white source. Even if the IR and UV components of the spectrum were to be eliminated, the solar spectrum still would not be optimum, due to the high intensity at the visible-IR and visible-UV boundaries.

18.2 The planckian spectrum

It is desirable to define an independent standard for white light. The *planckian black-body radiation spectrum* is used as one such standard. The black-body spectrum is characterized by only one parameter, the temperature of the body. The black-body spectrum was first derived by Max Planck (1900) and is given by

$$I(\lambda) = \frac{2hc^2}{\lambda^5 \left[\exp\left(\dfrac{hc}{\lambda kT}\right) - 1 \right]}. \tag{18.1}$$

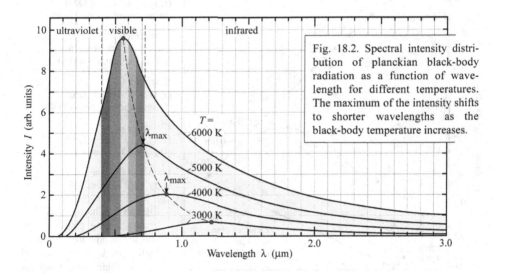

Fig. 18.2. Spectral intensity distribution of planckian black-body radiation as a function of wavelength for different temperatures. The maximum of the intensity shifts to shorter wavelengths as the black-body temperature increases.

The planckian spectrum is shown for different black-body temperatures in Fig. 18.2. The maximum intensity of radiation emanating from a black body of temperature T occurs at a specific wavelength which is given by **Wien's law**

18 Planckian sources and color temperature

$$\lambda_{max} = \frac{2880 \ \mu m \ K}{T}. \tag{18.2}$$

At "low" black-body temperatures, e.g. 3 000 K, the radiation occurs mostly in the infrared. As the temperature increases, the maximum of the radiation shifts into the visible wavelength range.

The location of the black-body radiation in the (x, y) chromaticity diagram (called planckian locus) is shown in Fig. 18.3. As the temperature of the black body increases, the chromaticity location moves from the red wavelength range towards the center of the diagram. Typical black-body temperatures in the white region of the chromaticity diagram range between 2 500 and 10 000 K. Also shown in Fig. 18.3 are the locations of several illuminants standardized by the CIE. These standard illuminants include Illuminants A, B, C, D_{65}, and E. The planckian locus and the locations of the black-body temperatures in the (u', v') uniform chromaticity diagram are shown in Fig. 18.4. The (x, y) and in (u', v') chromaticity coordinates of the planckian radiator are tabulated in Appendix 18.1.

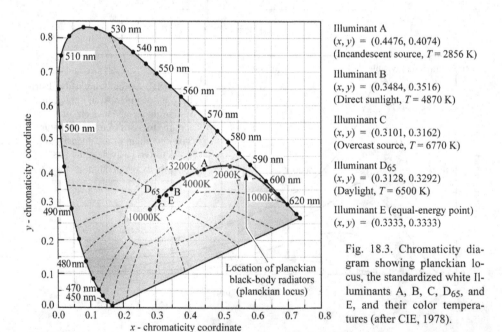

Illuminant A
$(x, y) = (0.4476, 0.4074)$
(Incandescent source, $T = 2856$ K)

Illuminant B
$(x, y) = (0.3484, 0.3516)$
(Direct sunlight, $T = 4870$ K)

Illuminant C
$(x, y) = (0.3101, 0.3162)$
(Overcast source, $T = 6770$ K)

Illuminant D_{65}
$(x, y) = (0.3128, 0.3292)$
(Daylight, $T = 6500$ K)

Illuminant E (equal-energy point)
$(x, y) = (0.3333, 0.3333)$

Fig. 18.3. Chromaticity diagram showing planckian locus, the standardized white Illuminants A, B, C, D_{65}, and E, and their color temperatures (after CIE, 1978).

In both the (x, y) and the (u', v') chromaticity diagrams, the planckian locus starts out in the red, then moves through the orange and yellow, to finally the white region. This sequence of colors is reminiscent of the colors of a real object (e.g. a piece of metal) heated to high temperatures, indicating that real objects closely follow the chromaticity of Planck's idealized black bodies.

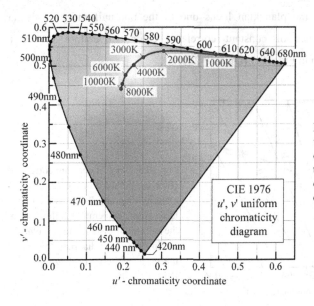

Fig. 18.4. CIE 1976 (u', v') uniform chromaticity diagram calculated using the CIE 1931 2° standard observer and planckian locus.

18.3 Color temperature and correlated color temperature

Color temperature may appear to be a somewhat surprising quantity as *color* and *temperature* don't seem to have a direct relationship with each other. However, the relationship is derived from Planck's black-body radiator. With increasing temperatures, it glows in the red, orange, yellowish white, white, and ultimately bluish white. The **color temperature** (CT) of a white light source, given in units of kelvin, is the temperature of a planckian black-body radiator that has the same chromaticity location as the white light source considered.

If the color of a white light source does not fall on the planckian locus, the **correlated color temperature** (CCT), also given in units of kelvin, is used. The correlated color temperature of a white light source is defined as the temperature of a planckian black-body radiator whose color is closest to the color of the white light source.

The correlated color temperature of a light source is determined as follows. On the (u', v') uniform chromaticity diagram, the point on the planckian locus that is *closest* to the chromaticity location of the light source is determined (i.e. shortest geometrical distance). The correlated color temperature is the temperature of the planckian black-body radiator at that point. The determination of the correlated color temperature was discussed in CIE publication No. 17.4 (1987) and by Robertson (1968).

On the (x, y) chromaticity diagram, the correlated color temperature *cannot* be determined by using the shortest distance to the planckian locus due to the non-uniformity of the (x, y) chromaticity diagram. The lines of constant correlated color temperature in the (x, y) chromaticity diagram are shown in Fig. 18.5 (Duggal, 2005).

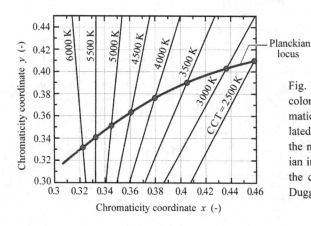

Fig. 18.5. Lines of constant correlated color temperature in the (x, y) chromaticity diagram. Whereas the correlated color temperature follows from the minimum distance to the planckian in the (u', v') diagram, this is not the case in the (x, y) diagram (after Duggal, 2005).

Table 18.1. Correlated color temperatures of common artificial and natural light sources.

Light source	Correlated color temperature (K)
Wax candle flame / CIE standard candle flame	1 500 to 2 000 / 2 000
W filament household lamp: 60 W / 100 W	2 800 / 2 850
W filament halogen lamp	2 800 to 3 200
"Warm white" fluorescent tube	3 000
"Cool daylight white" fluorescent tube	4 300
"True daylight" color match fluorescent tube	6 500
Carbon arc white flame	5 000
Xenon arc (unfiltered)	6 000
Summer sunlight (before 9.00 or after 15.00 h)	4 900 to 5 600
Summer sunlight (9.00 to 15.00 h)	5 400 to 5 700
Direct sun	5 700 to 6 500
Overcast daylight	6 500 to 7 200
Clear blue sky	8 000 to 27 000

The chromaticity locations of incandescent light sources are very close to, although not exactly on the planckian locus (Ohno, 2001). For such incandescent light sources, the *color temperature* can be specified. Standard incandescents have color temperatures ranging from 2 000 to 2 900 K. The common warm incandescent light source has a color temperature of 2 800 K. Quartz halogen incandescent lamps have a color temperature ranging from 2 800 to 3 200 K (Ohno, 1997). Other light sources, such as metal-halide sources, are further removed from the planckian locus. For such light sources, the *correlated color temperature* should be used. Bluish white lights have a correlated color temperature of about 8 000 K. Color temperatures and correlated color temperatures of common artificial and natural light sources are given in Table 18.1.

References

CIE publication No. 17.4 *International Lighting Vocabulary* see http://www.cie.co.at (CIE, Vienna, Austria, 1987)

Duggal A. R. "Organic electroluminescent devices for solid-state lighting" in *Organic Electroluminescence* edited by Z. H. Kafafi (Taylor & Francis Group, Boca Raton, Florida, 2005)

Jackson J. D. *Classical Electrodynamics* (John Wiley and Sons, New York, 1975)

Ohno Y. "Photometric standards" Chapter 3 in *OSA/AIP Handbook of Applied Photometry*, 55 (Optical Society of America, Washington DC, 1997)

Ohno Y. "Photometry and radiometry" Chapter 14 in *OSA Handbook of Optics, Volume III Review for Vision Optics, Part 2, Vision Optics* (McGraw-Hill, New York, 2001)

Planck M. "On the theory of the law on energy distribution in the normal spectrum (translated from German)" *Verhandlungen der Deutschen Physikalischen Gesellschaft* **2**, 237 (1900)

Robertson R. "Computation of correlated color temperature and distribution temperature" *J. Opt. Soc. Am.* **58**, 1528 (1968)

Appendix 18.1
Color temperature T and (x, y) and (u', v') chromaticity coordinates of a planckian emitter.

T	x	y	u'	v'
1 000 K	0.649	0.347	0.443	0.533
1 200 K	0.623	0.370	0.402	0.538
1 400 K	0.597	0.389	0.369	0.541
1 600 K	0.572	0.402	0.342	0.542
1 800 K	0.549	0.412	0.321	0.542
2 000 K	0.527	0.417	0.303	0.540
2 200 K	0.506	0.420	0.288	0.538
2 400 K	0.487	0.419	0.276	0.535
2 600 K	0.470	0.417	0.266	0.531
2 800 K	0.454	0.414	0.257	0.528
3 000 K	0.439	0.409	0.250	0.524
3 200 K	0.425	0.404	0.243	0.520
3 400 K	0.413	0.399	0.237	0.516
3 600 K	0.402	0.393	0.233	0.512
3 800 K	0.392	0.388	0.228	0.508
4 000 K	0.383	0.382	0.225	0.504
4 200 K	0.374	0.376	0.221	0.500
4 400 K	0.367	0.371	0.218	0.497
4 600 K	0.360	0.366	0.216	0.494
4 800 K	0.353	0.361	0.213	0.490
5 000 K	0.347	0.356	0.211	0.487
5 200 K	0.342	0.351	0.209	0.484
5 400 K	0.337	0.347	0.208	0.481
5 600 K	0.332	0.343	0.206	0.479
5 800 K	0.328	0.339	0.205	0.476
6 000 K	0.324	0.335	0.203	0.473
6 200 K	0.321	0.332	0.202	0.471
6 400 K	0.317	0.328	0.201	0.469
6 500 K	0.315	0.327	0.201	0.468
6 600 K	0.314	0.325	0.200	0.466
6 800 K	0.311	0.322	0.199	0.464
7 000 K	0.308	0.319	0.198	0.462
7 200 K	0.306	0.317	0.198	0.460
7 400 K	0.303	0.314	0.197	0.459
7 600 K	0.301	0.312	0.196	0.457
7 800 K	0.299	0.309	0.196	0.455
8 000 K	0.297	0.307	0.195	0.454
8 500 K	0.292	0.301	0.194	0.450
9 000 K	0.289	0.297	0.193	0.447
9 500 K	0.285	0.293	0.192	0.444
10 000 K	0.282	0.290	0.191	0.441

19

Color mixing and color rendering

19.1 Additive color mixing

The *combination* or *additive mixing* of two or more light sources is employed in a number of applications. In LED displays, three different types of LEDs, usually emitting in the red, green, and blue, are used. The three colors are mixed so that the observer can experience a wide range of colors. Another useful application of color mixing is the generation of white light by two, three, or more complementary colors. A schematic of additive color mixing and a corresponding experiment are shown in Fig. 19.1.

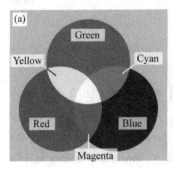
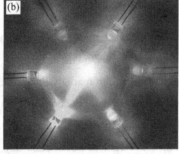

Fig. 19.1. (a) Schematic of additive color mixing of three primary colors. (b) Additive color mixing using LEDs.

Next, we determine the chromaticity coordinates of the mixture of three discrete emission bands. Assume that the three emission bands have spectral power densities $P_1(\lambda)$, $P_2(\lambda)$, and $P_3(\lambda)$ with peak wavelengths of λ_1, λ_2, λ_3, respectively. We assume that each emission band is much narrower than any of the three color-matching functions. We further assume that the three light sources have the chromaticity coordinates (x_1, y_1), (x_2, y_2), and (x_3, y_3). Then the tristimulus values are given by

$$X = \int_\lambda \bar{x}(\lambda) P_1(\lambda) \, d\lambda + \int_\lambda \bar{x}(\lambda) P_2(\lambda) \, d\lambda + \int_\lambda \bar{x}(\lambda) P_3(\lambda) \, d\lambda \approx \bar{x}(\lambda_1) P_1 + \bar{x}(\lambda_2) P_2 + \bar{x}(\lambda_3) P_3 \quad (19.1)$$

19 Color mixing and color rendering

$$Y = \int_\lambda \bar{y}(\lambda) P_1(\lambda) d\lambda + \int_\lambda \bar{y}(\lambda) P_2(\lambda) d\lambda + \int_\lambda \bar{y}(\lambda) P_3(\lambda) d\lambda \approx \bar{y}(\lambda_1) P_1 + \bar{y}(\lambda_2) P_2 + \bar{y}(\lambda_3) P_3 \quad (19.2)$$

$$Z = \int_\lambda \bar{z}(\lambda) P_1(\lambda) d\lambda + \int_\lambda \bar{z}(\lambda) P_2(\lambda) d\lambda + \int_\lambda \bar{z}(\lambda) P_3(\lambda) d\lambda \approx \bar{z}(\lambda_1) P_1 + \bar{z}(\lambda_2) P_2 + \bar{z}(\lambda_3) P_3 \quad (19.3)$$

where P_1, P_2, and P_3 are the optical powers emitted by the three sources. Using the abbreviations

$$L_1 = \bar{x}(\lambda_1) P_1 + \bar{y}(\lambda_1) P_1 + \bar{z}(\lambda_1) P_1 \quad (19.4)$$

$$L_2 = \bar{x}(\lambda_2) P_2 + \bar{y}(\lambda_2) P_2 + \bar{z}(\lambda_2) P_2 \quad (19.5)$$

$$L_3 = \bar{x}(\lambda_3) P_3 + \bar{y}(\lambda_3) P_3 + \bar{z}(\lambda_3) P_3 \quad (19.6)$$

the chromaticity coordinates of the mixed light can be calculated from the tristimulus values to yield

$$x = \frac{x_1 L_1 + x_2 L_2 + x_3 L_3}{L_1 + L_2 + L_3} \quad (19.7)$$

$$y = \frac{y_1 L_1 + y_2 L_2 + y_3 L_3}{L_1 + L_2 + L_3}. \quad (19.8)$$

Thus, the chromaticity coordinate of the multi-component light *is a linear combination of the individual chromaticity coordinates* weighted by the L_i factors.

The principle of color mixing in the chromaticity diagram is shown in Fig. 19.2. The figure shows the mixing of *two* colors with chromaticity coordinates (x_1, y_1) and (x_2, y_2). For the case of two colors, $L_3 = P_3 = 0$. The mixed color will be located on the straight line connecting the chromaticity coordinates of the two light sources. Thus *any* color (including white) located between the two chromaticity points can be created by mixing the two colors.

Figure 19.2 also shows the mixing of *three* colors, located in the red, green, and blue regions of the chromaticity diagram. The three chromaticity points, connected by a dashed line, are typical points for red, green, and blue LEDs. The area located within the dashed line, called the **color gamut**, represents all colors that can be created by mixing the three primary colors red, green, and blue. The ability to create a great variety of colors is an important quality for displays. It is desirable that the color gamut provided by the three light sources is as large as possible to create displays able to show brilliant, saturated colors.

The color gamut represents the entire range of colors that can be created from a set of primary sources. Color gamuts are polygons positioned within the perimeter of the chromaticity diagram. For the case of *three* primary colors, the color gamut is a *triangle*, as shown in Fig. 19.2. All colors created by additive mixtures of the vertex points (primary colors) of a gamut, are necessarily located inside the gamut.

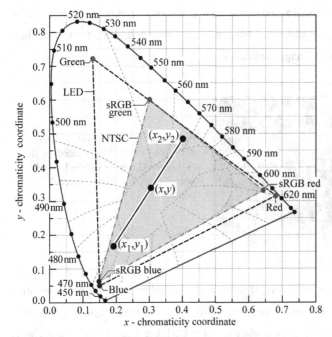

Fig. 19.2. Principle of color mixing illustrated with two light sources with chromaticity coordinates (x_1, y_1) and (x_2, y_2). The resulting color has the coordinates (x, y). Also shown is the triangular area of the chromaticity diagram (color gamut) accessible by additive mixing of a red, green, and blue LED. The locations of the red, green, and blue phosphors of the sRGB display standard ($x_r = 0.64$, $y_r = 0.33$, $x_g = 0.30$, $y_g = 0.60$, $x_b = 0.15$, $y_b = 0.06$) are also shown. The sRGB standard is similar to the NTSC standard.

The insight now gained on color mixing allows one to understand the location of different LEDs in the chromaticity diagram. The perimeter of the chromaticity diagram in the red spectral region is approximately a *straight line*, so that red LEDs, despite their thermal broadening, are located directly on the perimeter of the chromaticity diagram. In contrast, the perimeter is strongly curved in the green region, so that green LEDs, due to their spectral broadening, are displaced from the perimeter towards the center of the chromaticity diagram.

19.2 Color rendering

Another important characteristic of a white light source is its ability to show (i.e. render) the true colors of physical objects, e.g. fruits, plants, or toys, that are being illuminated by the source. The ability to render the colors of an object is measured in terms of the **color-rendering index** or **CRI** (Wyszecki and Stiles, 1982, 2000; MacAdam, 1993; Berger-Schunn, 1994; CIE, 1995). It is

315

a measure of the ability of the *illuminant* (i.e. a white illumination source) to faithfully render the colors of physical objects illuminated by the source.

Figure 19.3 shows an example of a physical object (here a painting by the impressionist Auguste Renoir) under illumination with a high-CRI source and with a low-CRI source. Colors appear richer and more vivid under illumination with a high-CRI source. Whereas high color rendering is important in locations such as museums, homes, and offices, it is less so in locations such as streets and parking lots. Finally, the color-rendering index is irrelevant for white light sources used in indicator lamp and signage applications.

Fig. 19.3. Artwork entitled "Fleurs dans un vase" illuminated with (a) high-CRI source and (b) low-CRI source (Auguste Renoir, French impressionist, 1841–1919).

The color-rendering ability of a **test light source** (to be abbreviated as **test source**) is evaluated by comparing it with the color-rendering ability of a **reference light source** (to be abbreviated as **reference source**). For the calculation of the CRI, the reference source is chosen as follows (CIE, 1995): (*i*) If the chromaticity point of the test source is located *on* the planckian locus, the reference source is a planckian black-body radiator with the same color temperature as the test source. (*ii*) If the chromaticity point of the test source is located *off* the planckian locus, the reference source is a planckian black-body radiator with the same *correlated* color temperature as the test source. (*iii*) Alternatively, one of the standardized CIE illuminants (e.g. Illuminant D_{65}) can be used as a reference source (CIE, 1995). Ideally, the test source and the reference source have the same chromaticity coordinates and luminous flux.

By convention, the planckian black-body reference source is assumed to have perfect color-rendering properties and thus its color-rendering index is CRI = 100. This convention was agreed upon because natural daylight closely resembles a planckian black-body source and thus rightfully deserves to be established as the standard reference source. Illuminants other than the reference source necessarily have a color-rendering index lower than 100. Because the CRI depends sensitively on the choice of the reference source, the selection of the reference source is of critical importance when calculating the CRIs of test sources.

Because the emission spectrum of an incandescent lamp closely follows that of a planckian black-body radiator, such lamps have the highest possible CRI and thus the best color-rendering properties of all artificial light sources. Incandescent quartz-halogen lamps are used in locations where color rendering is of prime importance, such as in museums, art galleries, and clothing shops. The drawback of quartz-halogen lamps is high power consumption.

In addition to the test source and the reference source, *test-color samples* are instrumental in determining the CRI of a test source. Test-color samples could be derived from real objects, e.g. fruit, flowers, wood, furniture, and clothes. However, in the interest of international standardization, a specific set of 14 test-color samples has been agreed upon for the purpose of determining the CRI. These 14 test-color samples are a subset of a larger collection of test-color samples initially introduced by Albert H. Munsell, a professor who taught at Rochester Institute of Technology (Rochester, NY) in the late 1800s and early 1900s (Munsell, 1905; 2005; Billmeyer, 1987; Long and Luke, 2001). Munsell introduced a color notation – the *Munsell color system* – which is a notation for defining a very wide range of colors.

The CRI calculation has been discussed in detail by Wyszecki and Stiles (1982; 2000) and by CIE (1995). The *CIE general CRI* is an average calculated according to

$$\mathrm{CRI}_{general} = \frac{1}{8} \sum_{i=1}^{8} \mathrm{CRI}_i \qquad (19.9)$$

where the CRI_i are the *special CRIs* for a set of eight test-color samples. The special color-rendering indices are calculated according to

$$\mathrm{CRI}_i = 100 - 4.6\, \Delta E_i^* \qquad (19.10)$$

where ΔE_i^* represents the quantitative color change that occurs when a test-color sample is illuminated with, first, the reference illumination source ("reference source"), and subsequently with the test illumination source ("test source"). The special color-rendering indices are calculated in such a way that they have a value of 100 if there is no difference in color appearance. The quantitative color change, ΔE_i^*, plays a key role in the calculation of the CRI and the determination of ΔE_i^* will be discussed in detail in the two subsequent sections where we will differentiate between on-planckian-locus and off-planckian-locus test sources.

At the time Eq. (19.10) was established, the pre-factor 4.6 had been chosen in such a way

19 Color mixing and color rendering

that the general CRI equals 60 when a "standard warm white" fluorescent lamp was used as a test source and a planckian black-body radiator was used as a reference source. Current fluorescent light sources have higher CRIs, typically in the range 60–85 (Kendall and Scholand, 2001).

The test-color samples mentioned above are defined in terms of their spectral reflectivity. The reflectivity curves of eight internationally agreed-upon test-color samples are shown in Fig. 19.4. The numerical values of the reflectivity of the eight test-color samples are listed in Appendix 19.1. The general color-rendering index is calculated from these eight test-color samples ($i = 1$–8).

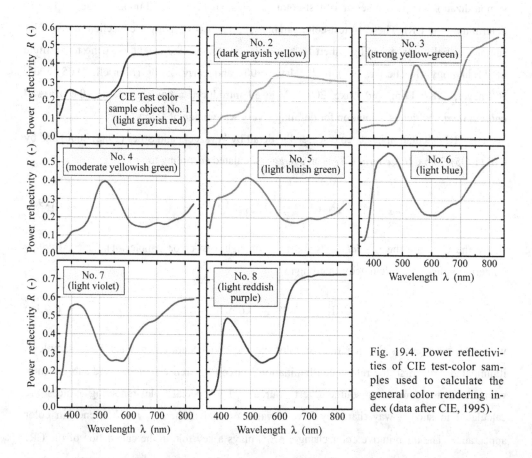

Fig. 19.4. Power reflectivities of CIE test-color samples used to calculate the general color rendering index (data after CIE, 1995).

In addition to the test-color samples (with numbers 1–8) used to calculate the *general* color rendering index, six supplemental test-color samples (with numbers 9–14) are used to further assess the color rendering capabilities of test sources. These supplemental test-color samples have the following colors: 9 – strong red; 10 – strong yellow; 11 – strong green; 12 – strong purplish blue; 13 – complexion of white person; 14 – leaf of tree. The reflectivity spectra and the numerical values of the reflectivity of the supplemental test-color samples are given in Fig. 19.5 and in Appendix 19.2, respectively. Inspection of the reflectivity curves reveals that the colors of the test-color samples 9–14 have particularly strong colors with relatively narrow peaks. CRI_9 to CRI_{14} are referred to as the **special color-rendering indices 9–14**.

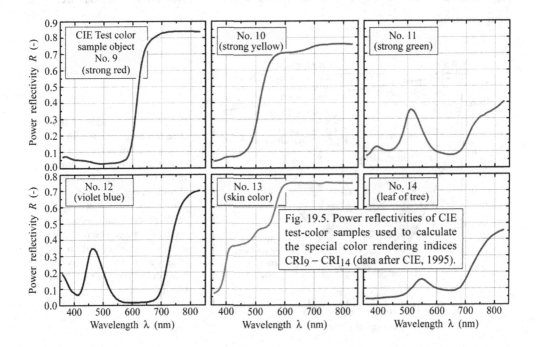

Fig. 19.5. Power reflectivities of CIE test-color samples used to calculate the special color rendering indices $CRI_9 - CRI_{14}$ (data after CIE, 1995).

The meaning of *chromaticity difference* of a test and a reference illumination source and the rendered colors of a test-color sample, when illuminated with the test and reference illumination sources, is illustrated in Fig. 19.6. In the example shown in the figure, the test source is located slightly off the planckian locus. The reference source is a planckian source with the least possible distance from the test-source chromaticity point. As a result, the color temperature of the reference source is equal to the correlated color temperature of the test source. The four chromaticity points shown in Fig. 19.6 enter the calculation of the CRI.

Note, however, that the term "color" as used by the CIE, is not equal to "chromaticity". The broad CIE definition for color includes hue, saturation, and additionally, brightness (for light) or lightness (for physical objects). Whereas hue and saturation are fully defined by location in the chromaticity coordinate system, brightness and lightness are not. To allow for a graphical representation of object lightness (or source brightness) a third axis could be added to the chromaticity diagram, as done for illustrative purposes in Fig. 19.7. The *color difference* of a physical object when illuminated with, first, a reference source, and subsequently with a test source, thus consists of the *chromaticity difference* and the *lightness difference*, as represented by the geometrical distance of the two points shown in Fig. 19.7. The reader is cautioned that the representation shown in Fig. 19.7 is for educational purposes and not a standardized CIE representation.

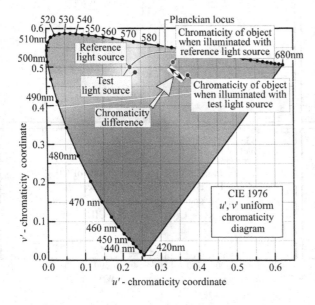

Fig. 19.6. Chromaticity difference resulting from the illumination of an object with a reference and a test light source. In the CIE 1976 u', v' uniform chromaticity diagram, the chromaticity difference is directly proportional to the geometric distance. The reference light source is located on the planckian locus at the correlated color temperature of the test light source.

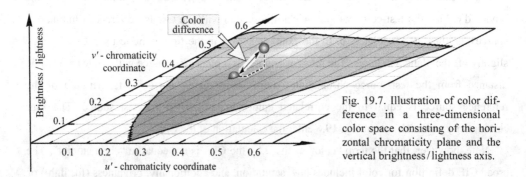

Fig. 19.7. Illustration of color difference in a three-dimensional color space consisting of the horizontal chromaticity plane and the vertical brightness/lightness axis.

A *uniform color space* (CIE, 1986) is motivated by the need for a quantitative color space that includes chromaticity and brightness/lightness. This uniform color space provides direct proportionality between color difference and geometrical distance. Thus color differences can be directly related to geometric distances between two points in the uniform color space. The CIE has introduced two three-dimensional uniform color spaces, namely the (L^*, u^*, v^*) and (L^*, a^*, b^*) spaces (CIE, 1986; Wyszecki and Stiles, 2000). For our purposes, it will be sufficient to consider the (L^*, u^*, v^*) uniform color space and we will therefore restrict our considerations to this space. The CIE (L^*, u^*, v^*) uniform color space is a three-dimensional space with cartesian coordinates, two coordinates being associated primarily with the chromaticity (u^*, v^*), and the third coordinate, L^*, representing the brightness (of a source) or lightness (of a physical object). The uniform color space is particularly suited to quantify color *differences*.

As an example and to become familiar with the CIE (L^*, u^*, v^*) color space, we consider a test-color sample i illuminated with a white *reference source*. The coordinates, L^*, u^*, and v^*, which give the color difference between object color and reference-source color, are defined as

$$L^*|_{\text{ref}} = 116(Y_{\text{ref},i}/Y_{\text{ref}})^{1/3} - 16 \qquad (19.11a)$$

$$u^*|_{\text{ref}} = 13L^*(u_{\text{ref},i} - u_{\text{ref}}) \qquad (19.11b)$$

$$v^*|_{\text{ref}} = 13L^*(v_{\text{ref},i} - v_{\text{ref}}) \qquad (19.11c)$$

where $Y_{\text{ref},i}$, $u_{\text{ref},i}$, and $v_{\text{ref},i}$ describe the color stimulus of the test-color sample i when illuminated with the *reference source*, and Y_{ref}, u_{ref}, and v_{ref} describe the color stimulus of the white reference illuminant.

Next consider the test-color sample illuminated with a white *test source*. The coordinates, L^*, u^*, and v^*, which give the color difference between object color and test-source color, are then given by

$$L^*|_{\text{test}} = 116(Y_{\text{test},i}/Y_{\text{test}})^{1/3} - 16 \qquad (19.12a)$$

$$u^*|_{\text{test}} = 13L^*(u_{\text{test},i} - u_{\text{test}}) \qquad (19.12b)$$

$$v^*|_{\text{test}} = 13L^*(v_{\text{test},i} - v_{\text{test}}) \qquad (19.12c)$$

where $Y_{test, i}$, $u_{test, i}$, and $v_{test, i}$ describe the color stimulus of the test-color sample i when illuminated with the *test source*, and Y_{test}, u_{test}, and v_{test} describe the color of the white test illuminant.

The difference in color between the two points located in the (L^*, u^*, v^*) space given by Eqs. (19.11) and (19.12) is equal to the euclidean distance between the points, that is

$$\Delta E^* = \sqrt{(\Delta L^*)^2 + (\Delta u^*)^2 + (\Delta v^*)^2} \qquad (19.13)$$

where

$$\Delta L^* = L^*|_{test} - L^*|_{ref} \qquad (19.14a)$$

$$\Delta u^* = u^*|_{test} - u^*|_{ref} \qquad (19.14b)$$

$$\Delta v^* = v^*|_{test} - v^*|_{ref} . \qquad (19.14c)$$

The calculation of the CRI will be detailed in the two subsequent sections.

Table 19.1. General color-rendering indices (CRIs) of different light sources. (a) Using sunlight as reference source. (b) Using incandescent light with the same correlated color temperature as the reference source. (c) Using Illuminant D_{65} as the reference source (some data after Kendall and Scholand, 2001).

Light source	Color-rendering index	
Sunlight	100	(a)
Quartz halogen W filament incandescent light	100	(b)
W filament incandescent light	100	(b)
Fluorescent light	60–95	(b)
Trichromatic white LED	60–95	(b, c)
Tetrachromatic white LED	70–95	(b, c)
Phosphor-based LED	55–95	(b, c)
Broadened dichromatic white LED	10–60	(b, c)
Hg vapor light coated with phosphor	50	(b)
Hg vapor light	33	(b)
Low and high-pressure Na vapor light	10 and 22	(b)
Green monochromatic light	−50	(c)

An overview of the general color-rendering indices of common light sources is given in

Table 19.1. The table includes several types of LED sources including dichromatic white LEDs, trichromatic white LEDs, and phosphor-based white LEDs. A CRI between 90 and 100 is suitable for virtually all illumination applications. A CRI between 70 and 90 is suitable for many standard illumination applications. Light sources with a CRI below 70 are considered to be of lower quality.

19.3 Color-rendering index for planckian-locus illumination sources

The following calculation gives the value of the quantity ΔE_i^*, which, according to Eq. (19.10), is needed to determine the CRI of a test source. The calculation is suited for test sources that are located on the planckian locus or extremely close to the planckian locus. The calculation described here follows Wyszecki and Stiles (1982, 2000) and CIE (1995). ΔE_i^*, which is the difference in appearance of a test-color sample when illuminated with the test and reference sources, is calculated according to

$$\Delta E_i^* = \sqrt{(\Delta L^*)^2 + (\Delta u^*)^2 + (\Delta v^*)^2} \tag{19.15}$$

where

$$\Delta L^* = L_{\text{ref}}^* - L_{\text{test}}^* = \left[116\left(\frac{Y_{\text{ref},i}}{Y_{\text{ref}}}\right)^{1/3} - 16\right] - \left[116\left(\frac{Y_{\text{test},i}}{Y_{\text{ref}}}\right)^{1/3} - 16\right] \tag{19.16}$$

$$\Delta u^* = u_{\text{ref}}^* - u_{\text{test}}^* = 13 L_{\text{ref}}^* \left(u_{\text{ref},i} - u_{\text{ref}}\right) - 13 L_{\text{test}}^* \left(u_{\text{test},i} - u_{\text{ref}}\right) \tag{19.17}$$

$$\Delta v^* = v_{\text{ref}}^* - v_{\text{test}}^* = 13 L_{\text{ref}}^* \left(v_{\text{ref},i} - v_{\text{ref}}\right) - 13 L_{\text{test}}^* \left(v_{\text{test},i} - v_{\text{ref}}\right) \tag{19.18}$$

and

$$u = \frac{4X}{X + 15Y + 3Z} \quad \text{and} \quad v = \frac{6Y}{X + 15Y + 3Z}. \tag{19.19}$$

Note that u and v are calculated from the tristimulus values of the reference source spectrum (subscript "ref"), from the reference source spectrum reflected off the test-color samples (subscript "ref, i"), and from the test source spectrum reflected off the test-color samples (subscript "test, i").

When calculating the CRI using the equations given above, the chromaticity coordinates and luminous flux of the test and reference sources should be identical in order to get the highest possible CRI for the test source. That is, the conditions $u_{\text{test}} = u_{\text{ref}}$, $v_{\text{test}} = v_{\text{ref}}$, and $Y_{\text{test}} = Y_{\text{ref}}$ should be satisfied.

The calculation of the color-rendering index in terms of Eqs. (19.9)–(19.19) illustrates that it is calculated from the ability of a test source to render the chromaticity of physical objects (taken into account by Δu^* and Δv^*) but also from the ability of the test source to render the lightnesses of the physical objects (taken into account by ΔL^*). The CRI calculation is based on the premise that the reference source renders the true chromaticity and lightness, i.e. the true color, of physical objects.

The choice of the numerical prefactors in Eqs. (19.16)–(19.19) is somewhat arbitrary. These prefactors have been determined in extensive experiments with human subjects. Evidence exists, however, that the current prefactors may not be optimal (Wyszecki and Stiles, 1982, 2000).

19.4 Color-rendering index for non-planckian-locus illumination sources

The following calculation of ΔE_i^* is suited for test sources that are located *off the planckian locus*. The calculation described here follows the procedure developed by CIE (1995) and takes into account the **adaptive color shift** that follows from the human ability of **chromatic adaptation**.

Chromatic adaptation is the well-known ability of humans to adapt to certain illumination environments without a substantial loss of color perception. For example, the yellow illumination sources used in semiconductor clean rooms provide a low-quality light. Such yellow sources do not contain short-wavelength light (violet, blue and cyan) and they are located clearly off the planckian locus. However, after having adapted to the clean-room illumination conditions (which typically takes several tens of minutes), colors appear quite natural, certainly much more natural than before the chromatic adaptation.

When calculating the CRI for off-planckian-locus sources according to the method described in the previous section, the CRI is very low. However, such low CRI values are not supported by experiments with human subjects: Due to chromatic adaptation, colors can appear vivid and natural, even for illumination sources slightly off the planckian locus. To overcome this discrepancy and to allow for a more realistic calculation of the CRI, the CIE (1995) introduced an alternative method to calculate the CRI. This alternative method takes into account the human ability of chromatic adaptation by introducing an adaptive color shift of the test source towards the planckian reference source.

19.4 Color-rendering index for non-planckian-locus illumination sources

As a rule of thumb, the pleasantness and quality of white illumination sources decreases rapidly if the chromaticity point of the illumination source deviates from the planckian locus by a distance greater than 0.01 in the x, y chromaticity system. This corresponds to the distance of about 4 MacAdam ellipses, a standard employed by the lighting industry (Duggal, 2005). Note however, that the 0.01-rule-of-thumb is a necessary but not a sufficient condition for high quality of illumination sources.

The calculation starts with the uniform chromaticity coordinates of the reference and test sources and the chromaticity coordinates of the test-color samples when illuminated with the reference and test sources, i.e. (u_{ref}, v_{ref}), (u_{test}, v_{test}), $(u_{ref,i}, v_{ref,i})$, and $(u_{test,i}, v_{test,i})$.

To account for the adaptive color shift, the (u, v) coordinates of (u_{ref}, v_{ref}), (u_{test}, v_{test}), and $(u_{test,i}, v_{test,i})$ are transformed into (c, d) coordinates using the formulae

$$c = (4 - u - 10v)/v \tag{19.20}$$

$$d = (1.708v + 0.404 - 1.481u)/v. \tag{19.21}$$

Note that these two equations correspond to six equations when transforming (u_{ref}, v_{ref}), (u_{test}, v_{test}), and $(u_{test,i}, v_{test,i})$ into (c_{ref}, d_{ref}), (c_{test}, d_{test}), and $(c_{test,i}, d_{test,i})$, respectively. Subsequently the adaptive-color-shifted chromaticity coordinates of the test-color samples are calculated according to

$$u_{test,i}^{**} = \frac{10.872 + 0.404 \dfrac{c_{ref}}{c_{test}} c_{test,i} - 4 \dfrac{d_{ref}}{d_{test}} d_{test,i}}{16.518 + 1.481 \dfrac{c_{ref}}{c_{test}} c_{test,i} - \dfrac{d_{ref}}{d_{test}} d_{test,i}} \tag{19.22}$$

$$v_{test,i}^{**} = \frac{5.520}{16.518 + 1.481 \dfrac{c_{ref}}{c_{test}} c_{test,i} - \dfrac{d_{ref}}{d_{test}} d_{test,i}}. \tag{19.23}$$

Correspondingly, the adaptive-color-shifted chromaticity coordinates of the test source are calculated according to

$$u_{test}^{**} = \frac{10.872 + 0.404 c_{ref} - 4 d_{ref}}{16.518 + 1.481 c_{ref} - d_{ref}} = u_{ref} \tag{19.24}$$

$$v_{\text{test}}^{**} = \frac{5.520}{16.518 + 1.481 c_{\text{ref}} - d_{\text{ref}}} = v_{\text{ref}}^{*}. \tag{19.25}$$

The values of u_{test}^{**} and v_{test}^{**} are the chromaticity coordinates of the light source to be tested after the adaptive color shift has been performed (note that $u_{\text{test}}^{**} = u_{\text{ref}}$ and $v_{\text{test}}^{**} = v_{\text{ref}}$). Finally the color difference is calculated in terms of the uniform color space coordinates

$$\Delta E_i^* = \sqrt{(\Delta L^{**})^2 + (\Delta u^{**})^2 + (\Delta v^{**})^2} \tag{19.26}$$

where

$$\Delta L^{**} = L_{\text{ref},i}^{**} - L_{\text{test},i}^{**} = \left[25(Y_{\text{ref},i})^{1/3} - 17\right] - \left[25(Y_{\text{test},i})^{1/3} - 17\right] \tag{19.27}$$

$$\Delta u^{**} = u_{\text{ref}}^{***} - u_{\text{test}}^{***} = 13 L_{\text{ref},i}^{**}\left(u_{\text{ref},i}^{**} - u_{\text{ref}}^{**}\right) - 13 L_{\text{test},i}^{**}\left(u_{\text{test},i}^{**} - u_{\text{test}}^{**}\right) \tag{19.28}$$

$$\Delta v^{**} = v_{\text{ref}}^{***} - v_{\text{test}}^{***} = 13 L_{\text{ref},i}^{**}\left(v_{\text{ref},i}^{**} - v_{\text{ref}}^{**}\right) - 13 L_{\text{test},i}^{**}\left(v_{\text{test},i}^{**} - v_{\text{test}}^{**}\right). \tag{19.29}$$

Note that the calculation requires that $Y_{\text{ref}} = Y_{\text{test}} = 100$ (CIE, 1995). Using the calculated values of ΔE_i^*, the general CRI is calculated using Eqs. (19.9) and (19.10). The special CRI_i for $i = 9$ to 14 may be of interest for a complete assessment of the color rendering properties of an illumination source.

Exercise: *Color rendering*. The color of a physical object, as seen by a human being, is not just a function of the object but also a function of the light source illuminating the object! In fact, the color of an object can depend very strongly on the light source illuminating the object. Some light sources do render the natural colors of an object (true color rendering) while some light sources do not (false color rendering).
(a) What is the color of a yellow banana when illuminated with a red LED?
(b) What is the color of a green banana when illuminated with a yellow LED?
(c) Could it be advantageous for a grocer to illuminate meat with red LEDs, bananas with yellow LEDs, and oranges with orange LEDs?
(d) Is it possible for two physical objects of different colors to appear to have the same color under certain illumination conditions?
(e) Why are low-pressure Na vapor lights used despite their low color-rendering index?
(f) What would be the advantage and disadvantage of using green LEDs for illumination?
Solution:
    (a) Red. (b) Yellow. (c) Yes – but his truthfulness in displaying fruit could be questioned. (d) Yes. (e) Because of their high luminous efficiency (and thus low electricity consumption). (f) High luminous efficacy would be an advantage but low color-rendering properties would be a disadvantage.

References

Berger-Schunn A. *Practical Color Measurement* (John Wiley and Sons, New York, 1994)

Billmeyer Jr. F. W. "Survey of color order systems" *Color Res. Appl.* **12**, 173 (1987)

CIE publication No. 15 (E.1.3.1) 1971: *Colorimetry* this publication was updated in 1986 to CIE Publication 15.2 *Colorimetry* (CIE, Vienna, Austria, 1986)

CIE publication No. 13.3 *Method of Measuring and Specifying Color-Rendering of Light Sources* (see also www.cie.co.at) (CIE, Vienna, Austria, 1995)

Duggal A. R. "Organic electroluminescent devices for solid-state lighting" in *Organic Electroluminescence* edited by Z. H. Kafafi (Taylor and Francis Group, Boca Raton, Florida, 2005)

Kendall M. and Scholand M. *Energy Savings Potential of Solid State Lighting in General Lighting Applications* available to the public from National Technical Information Service (NTIS), US Department of Commerce, 5285 Port Royal Road, Springfield, Virginia 22161 (2001)

Long J. and Luke J. T. *The New Munsell Student Color Set* 2nd ring-bound edition (Fairchild Books and Visuals, New York, 2001)

MacAdam D. L. (Editor) *Colorimetry – Fundamentals* (SPIE Optical Engineering Press, Bellingham, Washington, 1993)

Munsell A. H. *A Color Notation – An Illustrated System Defining All Colors and Their Relations by Measured Scales of Hue, Value, and Chroma* (G. H. Ellis, Boston, 1905)

Munsell *Munsell Book of Color, Matte Edition* is available through GretagMacbeth Corporation, gretagmacbeth.com and munsell.com (Regensdorf, Switzerland, 2005)

Wyszecki G. and Stiles W. S. *Color Science – Concepts and Methods, Quantitative Data and Formulae* 2nd edition (John Wiley and Sons, New York, 1982)

Wyszecki G. and Stiles W. S. *Color Science – Concepts and Methods, Quantitative Data and Formulae* 2nd edition "Wiley Classics Library" (John Wiley and Sons, New York, 2000)

Appendix 19.1

Spectral reflectivity $R_i(\lambda)$ of the CIE 1974 Test-Color Samples (TCS) Nos. 1–8 to be used in calculating the General Color-Rendering Index (General CRI) (after CIE, 1995)

λ (nm)	R_1 (-)	R_2 (-)	R_3 (-)	R_4 (-)	R_5 (-)	R_6 (-)	R_7 (-)	R_8 (-)
360	0.116	0.053	0.058	0.057	0.143	0.079	0.150	0.075
365	0.136	0.055	0.059	0.059	0.187	0.081	0.177	0.078
370	0.159	0.059	0.061	0.062	0.233	0.089	0.218	0.084
375	0.190	0.064	0.063	0.067	0.269	0.113	0.293	0.090
380	0.219	0.070	0.065	0.074	0.295	0.151	0.378	0.104
385	0.239	0.079	0.068	0.083	0.306	0.203	0.459	0.129
390	0.252	0.089	0.070	0.093	0.310	0.265	0.524	0.170
395	0.256	0.101	0.072	0.105	0.312	0.339	0.546	0.240
400	0.256	0.111	0.073	0.116	0.313	0.410	0.551	0.319
405	0.254	0.116	0.073	0.121	0.315	0.464	0.555	0.416
410	0.252	0.118	0.074	0.124	0.319	0.492	0.559	0.462
415	0.248	0.120	0.074	0.126	0.322	0.508	0.560	0.482
420	0.244	0.121	0.074	0.128	0.326	0.517	0.561	0.490
425	0.240	0.122	0.073	0.131	0.330	0.524	0.558	0.488
430	0.237	0.122	0.073	0.135	0.334	0.531	0.556	0.482
435	0.232	0.122	0.073	0.139	0.339	0.538	0.551	0.473
440	0.230	0.123	0.073	0.144	0.346	0.544	0.544	0.462
445	0.226	0.124	0.073	0.151	0.352	0.551	0.535	0.450
450	0.225	0.127	0.074	0.161	0.360	0.556	0.522	0.439
455	0.222	0.128	0.075	0.172	0.369	0.556	0.506	0.426
460	0.220	0.131	0.077	0.186	0.381	0.554	0.488	0.413
465	0.218	0.134	0.080	0.205	0.394	0.549	0.469	0.397
470	0.216	0.138	0.085	0.229	0.403	0.541	0.448	0.382
475	0.214	0.143	0.094	0.254	0.410	0.531	0.429	0.366
480	0.214	0.150	0.109	0.281	0.415	0.519	0.408	0.352
485	0.214	0.159	0.126	0.308	0.418	0.504	0.385	0.337
490	0.216	0.174	0.148	0.332	0.419	0.488	0.363	0.325
495	0.218	0.190	0.172	0.352	0.417	0.469	0.341	0.310
500	0.223	0.207	0.198	0.370	0.413	0.450	0.324	0.299
505	0.225	0.225	0.221	0.383	0.409	0.431	0.311	0.289
510	0.226	0.242	0.241	0.390	0.403	0.414	0.301	0.283
515	0.226	0.253	0.260	0.394	0.396	0.395	0.291	0.276
520	0.225	0.260	0.278	0.395	0.389	0.377	0.283	0.270
525	0.225	0.264	0.302	0.392	0.381	0.358	0.273	0.262
530	0.227	0.267	0.339	0.385	0.372	0.341	0.265	0.256
535	0.230	0.269	0.370	0.377	0.363	0.325	0.260	0.251
540	0.236	0.272	0.392	0.367	0.353	0.309	0.257	0.250
545	0.245	0.276	0.399	0.354	0.342	0.293	0.257	0.251
550	0.253	0.282	0.400	0.341	0.331	0.279	0.259	0.254
555	0.262	0.289	0.393	0.327	0.320	0.265	0.260	0.258
560	0.272	0.299	0.380	0.312	0.308	0.253	0.260	0.264
565	0.283	0.309	0.365	0.296	0.296	0.241	0.258	0.269
570	0.298	0.322	0.349	0.280	0.284	0.234	0.256	0.272
575	0.318	0.329	0.332	0.263	0.271	0.227	0.254	0.274
580	0.341	0.335	0.315	0.247	0.260	0.225	0.254	0.278
585	0.367	0.339	0.299	0.229	0.247	0.222	0.259	0.284
590	0.390	0.341	0.285	0.214	0.232	0.221	0.270	0.295
595	0.409	0.341	0.272	0.198	0.220	0.220	0.284	0.316

Appendix 19.1

600	0.424	0.342	0.264	0.185	0.210	0.220	0.302	0.348
605	0.435	0.342	0.257	0.175	0.200	0.220	0.324	0.384
610	0.442	0.342	0.252	0.169	0.194	0.220	0.344	0.434
615	0.448	0.341	0.247	0.164	0.189	0.220	0.362	0.482
620	0.450	0.341	0.241	0.160	0.185	0.223	0.377	0.528
625	0.451	0.339	0.235	0.156	0.183	0.227	0.389	0.568
630	0.451	0.339	0.229	0.154	0.180	0.233	0.400	0.604
635	0.451	0.338	0.224	0.152	0.177	0.239	0.410	0.629
640	0.451	0.338	0.220	0.151	0.176	0.244	0.420	0.648
645	0.451	0.337	0.217	0.149	0.175	0.251	0.429	0.663
650	0.450	0.336	0.216	0.148	0.175	0.258	0.438	0.676
655	0.450	0.335	0.216	0.148	0.175	0.263	0.445	0.685
660	0.451	0.334	0.219	0.148	0.175	0.268	0.452	0.693
665	0.451	0.332	0.224	0.149	0.177	0.273	0.457	0.700
670	0.453	0.332	0.230	0.151	0.180	0.278	0.462	0.705
675	0.454	0.331	0.238	0.154	0.183	0.281	0.466	0.709
680	0.455	0.331	0.251	0.158	0.186	0.283	0.468	0.712
685	0.457	0.330	0.269	0.162	0.189	0.286	0.470	0.715
690	0.458	0.329	0.288	0.165	0.192	0.291	0.473	0.717
695	0.460	0.328	0.312	0.168	0.195	0.296	0.477	0.719
700	0.462	0.328	0.340	0.170	0.199	0.302	0.483	0.721
705	0.463	0.327	0.366	0.171	0.200	0.313	0.489	0.720
710	0.464	0.326	0.390	0.170	0.199	0.325	0.496	0.719
715	0.465	0.325	0.412	0.168	0.198	0.338	0.503	0.722
720	0.466	0.324	0.431	0.166	0.196	0.351	0.511	0.725
725	0.466	0.324	0.447	0.164	0.195	0.364	0.518	0.727
730	0.466	0.324	0.460	0.164	0.195	0.376	0.525	0.729
735	0.466	0.323	0.472	0.165	0.196	0.389	0.532	0.730
740	0.467	0.322	0.481	0.168	0.197	0.401	0.539	0.730
745	0.467	0.321	0.488	0.172	0.200	0.413	0.546	0.730
750	0.467	0.320	0.493	0.177	0.203	0.425	0.553	0.730
755	0.467	0.318	0.497	0.181	0.205	0.436	0.559	0.730
760	0.467	0.316	0.500	0.185	0.208	0.447	0.565	0.730
765	0.467	0.315	0.502	0.189	0.212	0.458	0.570	0.730
770	0.467	0.315	0.505	0.192	0.215	0.469	0.575	0.730
775	0.467	0.314	0.510	0.194	0.217	0.477	0.578	0.730
780	0.467	0.314	0.516	0.197	0.219	0.485	0.581	0.730
785	0.467	0.313	0.520	0.200	0.222	0.493	0.583	0.730
790	0.467	0.313	0.524	0.204	0.226	0.500	0.585	0.731
795	0.466	0.312	0.527	0.210	0.231	0.506	0.587	0.731
800	0.466	0.312	0.531	0.218	0.237	0.512	0.588	0.731
805	0.466	0.311	0.535	0.225	0.243	0.517	0.589	0.731
810	0.466	0.311	0.539	0.233	0.249	0.521	0.590	0.731
815	0.466	0.311	0.544	0.243	0.257	0.525	0.590	0.731
820	0.465	0.311	0.548	0.254	0.265	0.529	0.590	0.731
825	0.464	0.311	0.552	0.264	0.273	0.532	0.591	0.731
830	0.464	0.310	0.555	0.274	0.280	0.535	0.592	0.731

Appendix 19.2

Spectral reflectivity $R_i(\lambda)$ of the CIE 1974 Test-Color Samples (TCS) Nos. 9–14 (after CIE, 1995)

λ (nm)	R_9 (-)	R_{10} (-)	R_{11} (-)	R_{12} (-)	R_{13} (-)	R_{14} (-)
360	0.069	0.042	0.074	0.189	0.071	0.036
365	0.072	0.043	0.079	0.175	0.076	0.036
370	0.073	0.045	0.086	0.158	0.082	0.036
375	0.070	0.047	0.098	0.139	0.090	0.036
380	0.066	0.050	0.111	0.120	0.104	0.036
385	0.062	0.054	0.121	0.103	0.127	0.036
390	0.058	0.059	0.127	0.090	0.161	0.037
395	0.055	0.063	0.129	0.082	0.211	0.038
400	0.052	0.066	0.127	0.076	0.264	0.039
405	0.052	0.067	0.121	0.068	0.313	0.039
410	0.051	0.068	0.116	0.064	0.341	0.040
415	0.050	0.069	0.112	0.065	0.352	0.041
420	0.050	0.069	0.108	0.075	0.359	0.042
425	0.049	0.070	0.105	0.093	0.361	0.042
430	0.048	0.072	0.104	0.123	0.364	0.043
435	0.047	0.073	0.104	0.160	0.365	0.044
440	0.046	0.076	0.105	0.207	0.367	0.044
445	0.044	0.078	0.106	0.256	0.369	0.045
450	0.042	0.083	0.110	0.300	0.372	0.045
455	0.041	0.088	0.115	0.331	0.374	0.046
460	0.038	0.095	0.123	0.346	0.376	0.047
465	0.035	0.103	0.134	0.347	0.379	0.048
470	0.033	0.113	0.148	0.341	0.384	0.050
475	0.031	0.125	0.167	0.328	0.389	0.052
480	0.030	0.142	0.192	0.307	0.397	0.055
485	0.029	0.162	0.219	0.282	0.405	0.057
490	0.028	0.189	0.252	0.257	0.416	0.062
495	0.028	0.219	0.291	0.230	0.429	0.067
500	0.028	0.262	0.325	0.204	0.443	0.075
505	0.029	0.305	0.347	0.178	0.454	0.083
510	0.030	0.365	0.356	0.154	0.461	0.092
515	0.030	0.416	0.353	0.129	0.466	0.100
520	0.031	0.465	0.346	0.109	0.469	0.108
525	0.031	0.509	0.333	0.090	0.471	0.121
530	0.032	0.546	0.314	0.075	0.474	0.133
535	0.032	0.581	0.294	0.062	0.476	0.142
540	0.033	0.610	0.271	0.051	0.483	0.150
545	0.034	0.634	0.248	0.041	0.490	0.154
550	0.035	0.653	0.227	0.035	0.506	0.155
555	0.037	0.666	0.206	0.029	0.526	0.152
560	0.041	0.678	0.188	0.025	0.553	0.147
565	0.044	0.687	0.170	0.022	0.582	0.140
570	0.048	0.693	0.153	0.019	0.618	0.133
575	0.052	0.698	0.138	0.017	0.651	0.125
580	0.060	0.701	0.125	0.017	0.680	0.118
585	0.076	0.704	0.114	0.017	0.701	0.112
590	0.102	0.705	0.106	0.016	0.717	0.106
595	0.136	0.705	0.100	0.016	0.729	0.101
600	0.190	0.706	0.096	0.016	0.736	0.098

605	0.256	0.707	0.092	0.016	0.742	0.095
610	0.336	0.707	0.090	0.016	0.745	0.093
615	0.418	0.707	0.087	0.016	0.747	0.090
620	0.505	0.708	0.085	0.016	0.748	0.089
625	0.581	0.708	0.082	0.016	0.748	0.087
630	0.641	0.710	0.080	0.018	0.748	0.086
635	0.682	0.711	0.079	0.018	0.748	0.085
640	0.717	0.712	0.078	0.018	0.748	0.084
645	0.740	0.714	0.078	0.018	0.748	0.084
650	0.758	0.716	0.078	0.019	0.748	0.084
655	0.770	0.718	0.078	0.020	0.748	0.084
660	0.781	0.720	0.081	0.023	0.747	0.085
665	0.790	0.722	0.083	0.024	0.747	0.087
670	0.797	0.725	0.088	0.026	0.747	0.092
675	0.803	0.729	0.093	0.030	0.747	0.096
680	0.809	0.731	0.102	0.035	0.747	0.102
685	0.814	0.735	0.112	0.043	0.747	0.110
690	0.819	0.739	0.125	0.056	0.747	0.123
695	0.824	0.742	0.141	0.074	0.746	0.137
700	0.828	0.746	0.161	0.097	0.746	0.152
705	0.830	0.748	0.182	0.128	0.746	0.169
710	0.831	0.749	0.203	0.166	0.745	0.188
715	0.833	0.751	0.223	0.210	0.744	0.207
720	0.835	0.753	0.242	0.257	0.743	0.226
725	0.836	0.754	0.257	0.305	0.744	0.243
730	0.836	0.755	0.270	0.354	0.745	0.260
735	0.837	0.755	0.282	0.401	0.748	0.277
740	0.838	0.755	0.292	0.446	0.750	0.294
745	0.839	0.755	0.302	0.485	0.750	0.310
750	0.839	0.756	0.310	0.520	0.749	0.325
755	0.839	0.757	0.314	0.551	0.748	0.339
760	0.839	0.758	0.317	0.577	0.748	0.353
765	0.839	0.759	0.323	0.599	0.747	0.366
770	0.839	0.759	0.330	0.618	0.747	0.379
775	0.839	0.759	0.334	0.633	0.747	0.390
780	0.839	0.759	0.338	0.645	0.747	0.399
785	0.839	0.759	0.343	0.656	0.746	0.408
790	0.839	0.759	0.348	0.666	0.746	0.416
795	0.839	0.759	0.353	0.674	0.746	0.422
800	0.839	0.759	0.359	0.680	0.746	0.428
805	0.839	0.759	0.365	0.686	0.745	0.434
810	0.838	0.758	0.372	0.691	0.745	0.439
815	0.837	0.757	0.380	0.694	0.745	0.444
820	0.837	0.757	0.388	0.697	0.745	0.448
825	0.836	0.756	0.396	0.700	0.745	0.451
830	0.836	0.756	0.403	0.702	0.745	0.454

20

White-light sources based on LEDs

As the trend of higher efficiencies in LEDs continues, the number of possible applications increases as well. A highly interesting application with a very large potential market size is *general daylight illumination* in homes and offices. The field of **solid-state lighting** (SSL) is concerned with the development of solid-state sources for illumination applications. LEDs are inherently monochromatic emitters. However, there are several ways to generate white light using LEDs. Approaches to white-light generation based on LEDs will be covered in the current chapter, whereas approaches based on LEDs and wavelength-converting materials will be discussed in the following chapter. A pivotal discussion of the promise of solid-state lighting was given by Bergh *et al.* (2001). A comprehensive introduction to lighting technology using solid-state sources was given by Zukauskas *et al.* (2002a).

In the field of general daylight illumination, devices should have the following properties (*i*) high efficiency, (*ii*) high power capability, (*iii*) good color-rendering capabilities (*iv*) high reliability, (*v*) low-cost manufacturability, and (*vi*) environmental benignity. Such properties would allow LEDs to compete with conventional illumination sources, in particular incandescent and fluorescent lamps.

20.1 Generation of white light with LEDs

Light is perceived as *white* light if the three types of cones located on the retina of the human eye are excited in a certain ratio, namely with similar intensity. For the case of white light, the tristimulus values are such that the location of the chromaticity point is near the center of the chromaticity diagram.

The generation of white light can be accomplished with a huge number of possible spectra. The creation of white light out of monochromatic visible-spectrum emitters can be based on dichromatic, trichromatic, or tetrachromatic approaches, as shown in Fig. 20.1, or on approaches of higher chromaticity. The optical sources can be classified in terms of their luminous efficacy of radiation, luminous source efficiency, and color-rendering properties.

Whereas high luminous efficacy and high luminous efficiency are always desirable properties of high-power light sources, color rendering depends strongly on the application. Generally, high-quality daylight illumination applications, e.g. illumination in museums, homes, offices, and stores, require a high color-rendering capability.

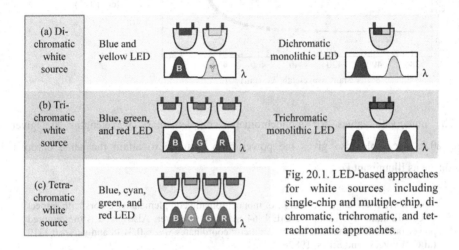

Fig. 20.1. LED-based approaches for white sources including single-chip and multiple-chip, dichromatic, trichromatic, and tetrachromatic approaches.

However, there are numerous applications where the color rendering capability is of lower priority, for example in the illumination of streets, parking garages, and stairwells. Finally, in signage applications, color rendering is irrelevant. Such signage applications include white pedestrian traffic lights, displays, and indicator lights.

There is a fundamental trade-off between the luminous efficacy of radiation and color rendering capability of a light source. Generally, dichromatic white light has the highest luminous efficacy and the poorest color-rendering capabilities. A trichromatic white source can have very acceptable color-rendering properties (CRI > 80) and luminous efficacies greater than 300 lm/W. Tetrachromatic sources can have color-rendering indices greater than 90.

20.2 Generation of white light by dichromatic sources

White light can be generated in several different ways. One way of generating white light is the use of two narrow emission bands, called *complementary wavelengths* or *complementary colors*. Two complementary colors, at a certain power ratio, result in tristimulus values that are perceived as white light. The wavelengths of complementary colors are shown in Fig. 20.2.

20 White-light sources based on LEDs

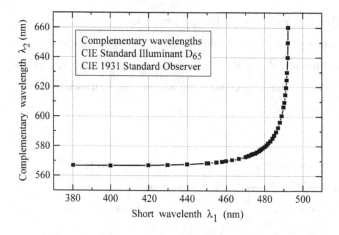

Fig. 20.2. Monochromatic complementary wavelengths resulting in the perception of white light at a certain power ratio (after Wyszecki and Stiles, 1982).

The numerical values for monochromatic complementary wavelengths are given in Table 20.1. The table also gives the power ratio required to attain the same chromaticity coordinate as Illuminant D_{65}.

Table 20.1. Wavelengths λ_1 and λ_2 of monochromatic complementary colors with respect to CIE Illuminant D_{65} and the CIE 1964 Standard Observer. Also given is the required power ratio. Illuminant D_{65} has chromaticity coordinates $x_{D65} = 0.3138$ and $y_{D65} = 0.3310$ (after Wyszecki and Stiles, 1982).

Complementary wavelengths		Power ratio	Complementary wavelengths		Power ratio
λ_1 (nm)	λ_2 (nm)	$P(\lambda_2)/P(\lambda_1)$	λ_1 (nm)	λ_2 (nm)	$P(\lambda_2)/P(\lambda_1)$
380	560.9	0.000642	460	565.9	1.53
390	560.9	0.00955	470	570.4	1.09
400	561.1	0.0785	475	575.5	0.812
410	561.3	0.356	480	584.6	0.562
420	561.7	0.891	482	591.1	0.482
430	562.2	1.42	484	602.1	0.440
440	562.9	1.79	485	611.3	0.457
450	564.0	1.79	486	629.6	0.668

Next we analyze the luminous efficacy of radiation of a source with two complementary emission lines. It is assumed that the two lines are thermally broadened to a full-width at half-maximum of ΔE. Emission lines of $\Delta E = 2\,kT$ to $10\,kT$ have been found experimentally for the GaInN system at room temperature ($kT = 25.9$ meV at 300 K). A *gaussian distribution* is assumed for the two emission lines, so that the spectral power density is given by

20.2 Generation of white light by dichromatic sources

$$P(\lambda) = P_1 \frac{1}{\sigma_1\sqrt{2\pi}} e^{-\frac{1}{2}\left(\frac{\lambda-\lambda_1}{\sigma_1}\right)^2} + P_2 \frac{1}{\sigma_2\sqrt{2\pi}} e^{-\frac{1}{2}\left(\frac{\lambda-\lambda_2}{\sigma_2}\right)^2} \quad (20.1)$$

where P_1 and P_2 are the optical powers of the two emission lines, and λ_1 and λ_2 are the peak wavelengths of the source. The gaussian standard deviation σ is related to the full-width at half-maximum of an emission spectrum, $\Delta\lambda$, by

$$\sigma = \Delta\lambda / \left[2\sqrt{2\ln 2}\right] = \Delta\lambda / 2.355 . \quad (20.2)$$

The peak emission wavelengths λ_1 and λ_2 are chosen from Table 20.1. The table also gives the required power ratio of the two light sources. Although the table applies to strictly monochromatic sources ($\Delta\lambda \to 0$), the data can be used, as an excellent approximation, for sources exhibiting moderate spectral broadening such as LEDs.

The luminous efficacy of radiation of a dichromatic source is shown in Fig. 20.3. The figure reveals that the highest luminous efficacy of 440 lm/W occurs at a primary wavelength of $\lambda_1 = 445$ nm for $\Delta E = 2kT$. The very high value of the efficacy shows the great potential of dichromatic sources.

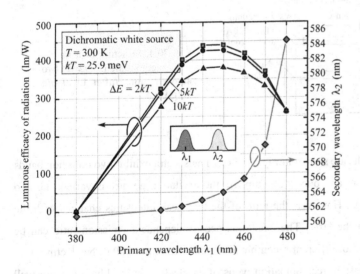

Fig. 20.3. Calculated luminous efficacy of dichromatic white-light source (with chromaticity point at D_{65} standard illuminant) for different linewidths ΔE as a function of the primary wavelength. Also shown is the complementary secondary wavelength (after Li et al., 2003).

Several approaches for the generation of white light by mixing two complementary colors have been demonstrated (Guo et al., 1999; Sheu et al., 2002; Dalmasso et al., 2002; Li et al.,

2003). One possibility uses the mixing of light emitted by two LEDs, one emitting in the blue and the other one in the yellow spectral region. Another possibility, demonstrated by Guo et al. (1999), generates white light by using a GaN-based blue LED and a second semiconductor, AlGaInP, as a wavelength converter. Sheu et al. (2002) demonstrated a codoped single active region quantum well white LED. The active region is doped with both Si and Zn. Blue light emission originates from quantum well band-to-band transitions, whereas a wide yellowish emission originates from donor-acceptor-pair (D-A) transitions. Because the D-A transition is spectrally wide, the codoped approach has the advantage of good color rendering. A dichromatic monolithic LED has also been reported by Dalmasso et al. (2002), who employed two closely spaced GaInN active regions within the pn-junction region. A strong dependence of the emission spectrum on the injection current was found.

Li et al. (2003) reported a monolithic GaInN based LED with two active regions separated by a thin GaN layer. The device was designed for emission at 465 nm and 525 nm. The device structure is shown in Fig. 20.4.

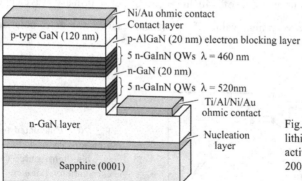

Fig. 20.4. Structure of a monolithic dichromatic LED with two active regions (after Li et al., 2003).

Photoluminescence results are shown in Fig. 20.5 (a). The spectra exhibit two emission bands one centered at about 465 nm and one at 525 nm. As the excitation density is varied, the two peak positions do not change. However, the ratio of the two peak intensities changes with the excitation power density of the laser. The excitation-density-dependent emission ratio can be explained by the competition of different recombination paths (Li et al., 2003). Neglecting non-radiative recombination, possible recombination paths of an electron in the blue quantum well (QW) are either direct radiative recombination in the blue QW or tunneling to the green QW with subsequent radiative recombination. Electrons and holes can tunnel from the blue QW to the green QW and recombine there radiatively. However, once in the green QW, carriers cannot tunnel back to the blue QW due to the higher energy of this QW.

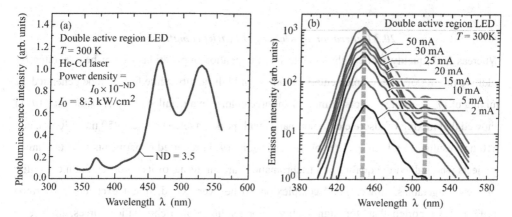

Fig. 20.5. Room temperature (a) photoluminescence and (b) electroluminescence spectra of monolithic dichromatic LED with two active regions (after Li et al., 2003).

Electroluminescence (EL) spectra of the device are shown in Fig. 20.5 (b). Two emission peaks are clearly observed with center wavelengths at about 450 nm and 520 nm. The blue peak is more intense than the green peak which can be attributed to the higher quantum efficiency of blue QWs compared with green QWs. In addition, holes are injected from the blue side (i.e. the side of the high-energy, blue-emitting QWs), whereas electrons are injected from the green side. As holes have a lower mobility and a higher effective mass, they are less likely to reach the green QWs, which can explain the higher intensity of the blue emission.

For current injection, holes are injected into the blue QWs, whereas electrons are injected into the green QWs. This is very different from optical excitation, where both types of carriers are injected from both sides of the active region. For the structure discussed here, the optical absorption length is longer than the distance between the active regions and the surface. This can explain the marked difference between the results for photoluminescence and electroluminescence (Li et al., 2003).

Room temperature I–V curves of the double-active region LED exhibit excellent forward voltages < 3.0 V at small contact diameters of 100 μm indicating high-quality ohmic contacts. The increase in forward voltage for contacts with larger diameters was attributed to an increased voltage drop in the n-type buffer layer, because, at a given current density, the current in the buffer layer scales with the area of the contact A, but the access resistance through the n-type

20 White-light sources based on LEDs

buffer layer scales with the circumference $A^{-1/2}$. In addition, the current crowding effect leads to non-uniform current injection particularly in large-diameter contacts and thus to an increased forward voltage.

20.3 Generation of white light by trichromatic sources

Whereas high-quality white light suitable for illumination applications cannot be generated by additive mixing of *two* complementary colors, such high-quality white light can be generated by mixing of *three* primary colors or more than three colors. In a detailed analysis, Thornton (1971) showed that mixing of discrete emission bands with peak wavelengths near 450 nm, 540 nm, and 610 nm resulted in a high-quality source. Thornton (1971) reported experiments with 60 human subjects who judged the quality of a trichromatic source in terms of its color rendition capability of meat, vegetables, flowers, and complexion, to be "very good, if not excellent". Thornton (1971) also reported that, for high-quality color rendition of trichromatic sources, the use of emitters near 500 nm and 580 nm should be avoided.

Although Thornton (1971) established that trichromatic sources can have high quality, the individual emission bands used in the experiments had a broad spectral width: The full-widths at half-maximum of the phosphor emitters employed in the study exceeded 50 nm. Semiconductors, with typical spectral widths < 50 nm, have much narrower emission lines than phosphors.

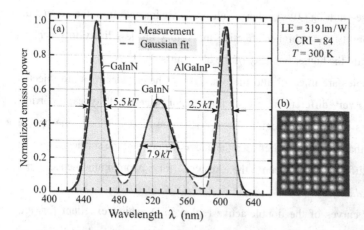

Fig. 20.6. (a) Emission spectrum of trichromatic white multi-LED source with color temperature of 6500 K (solid line) and gaussian fit (dashed line). The source has a luminous efficacy of radiation of 319 lm/W and a color rendering index of 84. (b) Photograph of source assembled from standard commercial devices (after Chhajed et al., 2005).

The trichromatic emission spectrum of a white-light source made out of three types of LEDs emitting at 455 nm, 525 nm, and 605 nm is shown in Fig. 20.6 (a). The experimentally determined full-width at half-maximum of the spectra at room temperature (20 °C) is 5.5 kT,

338

7.9 kT, and 2.5 kT for the GaInN blue, GaInN green, and AlGaInP orange emitter, respectively, where $kT = 25.25$ meV. The expression of the full-width at half-maximum in terms of kT is very useful, as it can be easily related to the theoretical full-width at half-maximum of a thermally broadened emission band of a semiconductor, which is 1.8 kT. Using the full-width at half-maximum given in the figure and

$$\Delta\lambda = \frac{\lambda^2}{hc} \Delta E = \frac{(\lambda/\text{nm})^2}{1239.8} (\Delta E/\text{eV}), \qquad (20.3)$$

the full-widths at half-maximum of the blue, green, and orange sources are 23.2 nm, 44.3 nm, and 18.6 nm, respectively. Note that the green emission line is particularly broad, which can be attributed to alloy broadening and the formation of quantum-dot-like InN-rich regions within the high-In-content GaInN.

Also shown in the figure are gaussian fits to the experimental spectra. The gaussian fits match the experimental spectra well. Note that the gaussian curves (equations were given earlier in this chapter) are symmetric in terms of wavelength. *Asymmetric gaussian distributions*, which have a more pronounced long-wavelength tail, have been employed for phosphors (Ivey, 1963). The use of such asymmetric gaussian distributions does not appear to be warranted for semiconductors as their spectral power distribution is quite symmetric when plotted versus wavelength.

A photograph of the light source assembled from a large number of LEDs is shown in Fig. 20.6 (b). The power ratio of the orange, green, and blue emitters is adjusted to match the chromaticity of a planckian radiator with color temperature 6500 K. The LED source, assembled from standard commercial devices, has a luminous efficacy of radiation of 319 lm/W, a luminous source efficiency of 32 lm/W, and a color-rendering index of 84 (Chhajed *et al.*, 2005).

There are a large number of possible wavelength combinations for trichromatic sources. To attain a high efficacy of radiation, sources near the fringes of the visible spectrum (deep red and deep violet) should be avoided. Contour plots of the luminous efficacy of radiation and of the color rendering index of a trichromatic source with color temperature of 6500 K are shown in Fig. 20.7 for a full-width at half-maximum for each emission line of 5 kT. Inspection of the figure reveals that $\lambda_1 = 455$ nm, $\lambda_2 = 530$ nm, and $\lambda_3 = 605$ nm are particularly favorable in terms of the color-rendering index. The CIE general CRI is about 85 for this wavelength combination with the luminous efficacy of radiation being 320 lm/W.

The figure also reveals that the CRI depends very sensitively on the exact peak positions. For example, changing the red peak wavelength from 605 nm to 620 nm decreases the CRI from 85 to 65. Similarly, changing the green wavelength from 530 nm to 550 nm decreases the CRI to values less than 60.

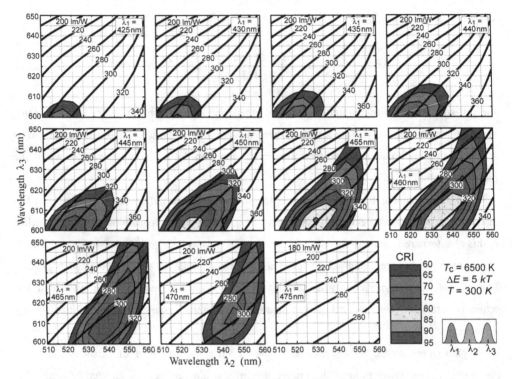

Fig. 20.7. Contour plot of luminous efficacy of radiation and CIE color-rendering index of white trichromatic LED source with color temperature 6500 K as a function of the three wavelengths for a linewidth (FWHM) of $5kT$ (after Chhajed et al., 2005).

Contour plots of the luminous efficacy of radiation and of the color-rendering index of a trichromatic source with color temperature of 6500 K are shown in Fig. 20.8 for a full-width at half-maximum for each emission line of $8kT$. A higher CRI results from the broader emission lines. A very favorable combination in terms of a high CRI is obtained for λ_1 = 450–455 nm, λ_2 = 525–535 nm, and λ_3 = 600–615 nm, where a CRI in the range 90–95 is obtained.

20.4 Temperature dependence of trichromatic LED-based white-light source

The relatively small range of wavelengths that enables a high color-rendering capability raises the question as to the stability of trichromatic sources with respect to junction and ambient temperature. It is known that emission power (P), peak wavelength (λ_{peak}), and spectral width ($\Delta\lambda$) depend on temperature, each of these quantities having a different temperature coefficient.

The optical output power of LEDs is temperature dependent in a manner that can be described by an exponential function and a characteristic temperature, T_1. The light output power of an LED is then given by

$$P = P|_{300\,K} \exp \frac{T - 300\,K}{T_1}. \qquad (20.4)$$

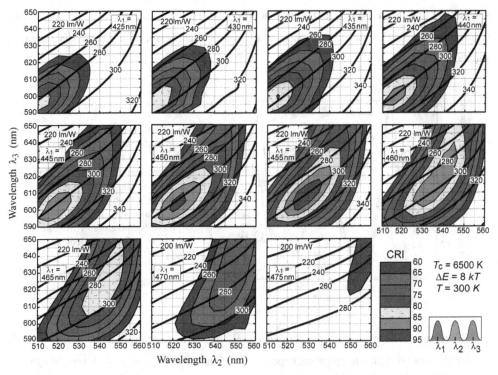

Fig. 20.8. Contour plot of luminous efficacy of radiation and CIE color-rendering index of white trichromatic LED source with color temperature 6500 K as a function of the three wavelengths for a linewidth (FWHM) of $8kT$ (after Chhajed et al., 2005).

As a result of these temperature dependences, the chromaticity point of a multi-LED white-light source changes with temperature. Consider a white-light source consisting of three types of emitters emitting in the red, green, and blue. For such LEDs, the temperature coefficients of the peak emission wavelength, spectral width, and emission power have been measured and are

20 White-light sources based on LEDs

given in Table 20.2 (Chhajed *et al.*, 2005).

Table 20.2. Experimentally determined temperature coefficients for peak wavelength, spectral width, and emission power for blue, green, and red LEDs.

	Blue	Green	Red
$d\lambda_{peak}/dT$	0.0389 nm/°C	0.0308 nm/°C	0.156 nm/°C
$d\Delta\lambda/dT$	0.0466 nm/°C	0.0625 nm/°C	0.181 nm/°C
$T_{characteristic}$	493 K	379 K	209 K

Consider further that the three currents feeding the red, green, and blue LEDs are adjusted in such a way that the resulting chromaticity point equals that of Illuminant D_{65} when the device temperature is 20 °C. The optical spectrum of such a trichromatic white source is shown in Fig. 20.9.

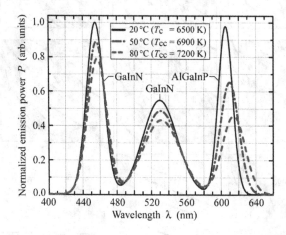

Fig. 20.9. Emission spectrum of trichromatic white LED source for different ambient temperatures (junction heating neglected). Optical power, linewidth, and peak wavelength change with temperature. As a result of these changes, the color temperature of the source increases (after Chhajed *et al.*, 2005).

However, as the device temperature increases, the chromaticity point of the trichromatic source changes due to the temperature dependences of the emission power, peak wavelength, and spectral width. This shift of the chromaticity point is shown in Fig. 20.10. Inspection of the figure reveals that the chromaticity point shifts towards higher color temperatures. This can be explained by the stronger temperature dependence of the red LED emission power. At high temperatures, the red component of the light source decreases more strongly (due to low T_1 value) than the green component, and the blue component, which is particularly stable.

Figure 20.11 shows the chromaticity shift of the trichromatic source on a magnified scale in the CIE 1931 (x, y) chromaticity coordinate system as well as in the CIE 1976 (u', v') uniform chromaticity coordinate system along with the planckian locus. At $T_j = 50\,°C$, the chromaticity point is 0.009 units away from the original point, and at 80 °C, it is shifted 0.02 units from the original point. This shift causes a clearly noticeable change in color appearance and exceeds the deviation limit of 0.01 units ("*0.01 rule*") commonly used in the lighting industry (Duggal, 2005).

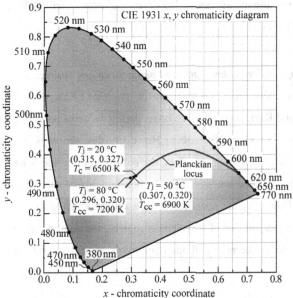

Fig. 20.10. Change in chromaticity of trichromatic white LED-based source. The source color temperature is 6500 K when devices are at room temperature. Due to the dependence of emission power, peak wavelength, and linewidth on temperature, the chromaticity point migrates off the planckian locus as the device temperature increases (after Chhajed et al., 2005).

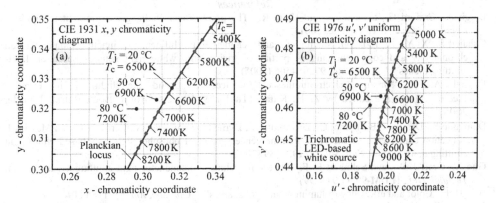

Fig. 20.11. Change in (a) x, y and (b) u', v' chromaticity of trichromatic white LED source. $T_c = 6500$ K when p-n junctions are at room temperature (after Chhajed et al., 2005).

The shift in chromaticity can be eliminated by adjusting the relative power ratio of the three LED sources. There are two possible implementations for adjusting the power ratio. In one implementation, the spectrum of the light source is constantly measured and a feedback control is used to adjust the optical power of the three components. In an alternative implementation, the device temperature is monitored and the optical power of the three components is adjusted using the known temperature dependence of the different types of emitters. The second method is easier due to the simplicity of a temperature measurement. However, the second method does not enable a compensation for device-aging effects.

20.5 Generation of white light by tetrachromatic and pentachromatic sources

Tetrachromatic and pentachromatic white sources use four and five types of LEDs, respectively (Zukauskas et al., 2002a; Schubert and Kim, 2005). The color-rendering index of polychromatic sources generally increases with the number of sources. However, the luminous efficacy generally decreases with increasing number of sources. Thus, the color-rendering index and the luminous efficacy of tetrachromatic sources are generally higher and lower than those of trichromatic sources, respectively. However, the specifics depend on the exact choice of the emission wavelengths. Due to the greater number of wavelength choices, the color temperatures of such sources can be adjusted more liberally without compromising the color-rendering capability of the source.

References

Bergh A., Craford G., Duggal A., and Haitz R. "The promise and challenges of solid-state lighting" *Physics Today* p. 42 (December 2001)

Chhajed S., Xi Y., Li Y.-L., Gessmann Th., and Schubert E. F. "Influence of junction temperature on chromaticity and color rendering properties of trichromatic white light sources based on light-emitting diodes" *J. Appl. Phys.* **97**, 054506 (2005)

Duggal A. R. "Organic electroluminescent devices for solid-state lighting" in *Organic Electroluminescence* edited by Z. H. Kafafi (Taylor & Francis Group, Boca Raton, Florida, 2005)

Dalmasso S., Damilano B., Pernot C., Dussaigne A., Byrne D., Grandjean N., Leroux M., and Massies J. "Injection dependence of the electroluminescence spectra of phosphor free GaN-based white light emitting diodes" *Phys. Stat. Sol. (a)* **192**, 139 (2002)

Guo X., Graff J. W., and Schubert E. F. "Photon-recycling semiconductor light-emitting diodes" *IEDM Technol. Dig.*, **IEDM-99**, 600 (1999)

Ivey H. F. "Color and efficiency of luminescent light sources" *J. Opt. Soc. Am.* **53**, 1185 (1963)

Li Y.-L., Gessmann Th., Schubert E. F., and Sheu J. K. "Carrier dynamics in nitride-based light-emitting p-n junction diodes with two active regions emitting at different wavelengths" *J. Appl. Phys.* **94**, 2167 (2003)

Schubert E. F. and Kim J. K. "Solid-state light sources becoming smart" *Science* **308**, 1274 (2005)

Sheu J. K., Pan C. J., Chi G. C., Kuo C. H., Wu L. W., Chen C. Chang H., S. J., and Su Y. K. "White-light emission from InGaN-GaN multiquantum-well light-emitting diodes with Si and Zn co-doped active well layer" *IEEE Photonics Technol. Lett.*, **14**, 450 (2002)

Thornton W. A. "Luminosity and color rendering capability of white light" *J. Opt. Soc. Am.* **61**, 1155 (1971)

Wyszecki G. and Stiles W. S. *Color Science – Concepts and Methods, Quantitative Data and Formulae* 2nd edition (John Wiley and Sons, New York, 1982)

Zukauskas A., Shur M. S., and Gaska R. *Introduction to Solid-State Lighting* (John Wiley and Sons, New York, 2002a)

Zukauskas A., Vaicekauskas R., Ivanauskas F., Gaska R., and Shur M. S. "Optimization of white polychromatic semiconductor lamps" *Appl. Phys. Lett.* **80**, 234 (2002b)

21

White-light sources based on wavelength converters

The generation of white light by a semiconductor LED whose light is partially or fully used to optically excite one or several phosphors, is a viable and common method to generate white light for general illumination applications. There are several different approaches to generate white light based on phosphors excited by semiconductor LEDs, which are shown in Fig. 21.1. They can be classified in dichromatic, trichromatic, and tetrachromatic approaches. These approaches use either UV-excitation sources or visible-spectrum-excitation sources (mostly blue semiconductor LEDs).

Generally, the luminous source efficiency decreases with increasing multi-chromaticity of the source. Thus, dichromatic sources have the highest luminous efficacy of radiation and also the highest potential luminous source efficiency. On the other hand, the color-rendering capability is lowest for dichromatic sources and it increases with the multi-chromaticity of the source. The color-rendering index can reach values very close to CRI = 100 for tetrachromatic sources.

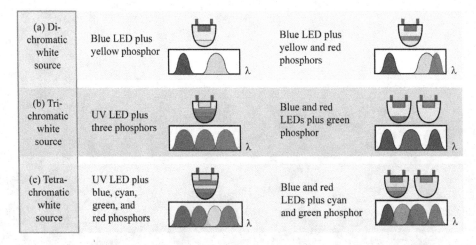

Fig. 21.1. White sources using phosphors that are optically excited by UV or blue LEDs.

21.1 Efficiency of wavelength-converter materials

The conversion efficiency of short-wavelength light to long-wavelength light by a *wavelength-converter* (λ-converter) material is determined by two distinct factors, namely (*i*) the external quantum efficiency of the λ-converter and (*ii*) the inherent quantum-mechanical-energy loss incurred in wavelength conversion.

The *external quantum efficiency* of the converter material, η_{ext}, is given by

$$\eta_{ext} = \frac{\text{number of photons emitted into free space by } \lambda\text{-converter per second}}{\text{number of photons absorbed by } \lambda\text{-converter per second}} . \quad (21.1)$$

The external efficiency originates in the *internal efficiency* and the *extraction efficiency* of the converter material according to $\eta_{ext} = \eta_{internal} \, \eta_{extraction}$. Note that the internal quantum efficiency depends on the inherent efficiency of the material whereas the extraction efficiency depends on the spatial distribution of the λ-converter material. Generally, thin films have high extraction efficiencies whereas lumpy aggregations of converter materials have lower extraction efficiency due to reabsorption. It is therefore desirable to employ λ-converter materials in the form of thin layers.

The inherent *wavelength-conversion loss* (sometimes called *quantum deficit* or *Stokes shift*) incurred when converting a photon with wavelength λ_1 to a photon with wavelength λ_2 ($\lambda_1 < \lambda_2$) is given by

$$\Delta E = h\nu_1 - h\nu_2 = \frac{hc}{\lambda_1} - \frac{hc}{\lambda_2} . \quad (21.2)$$

Thus the wavelength-conversion efficiency is given by

$$\eta_{\lambda\text{-conversion}} = \frac{h\nu_2}{h\nu_1} = \frac{\lambda_1}{\lambda_2} \quad (21.3)$$

where λ_1 is the wavelength of the photon absorbed by the phosphor and λ_2 is the wavelength of the photon emitted by the phosphor. Note that wavelength-conversion loss is fundamental in nature. The loss cannot be overcome with conventional λ-converter materials.

However, *quantum-splitting phosphors* allow one to convert one short-wavelength photon into two longer-wavelength photons so that $h\nu_1 = h\nu_2 + h\nu_3$, where $h\nu_1$ is the energy of the photon absorbed by the phosphor and $h\nu_2$ and $h\nu_3$ are the energies of the photons emitted by the phosphors. Several quantum-splitting phosphors have been reported (Justel et al., 1998; Wegh et al., 1999; Srivastava and Ronda, 2003; Srivastava, 2004). The possibility of quantum efficiencies approaching 200% for Eu^{3+}-doped $LiGdF_4$ has been proposed (Wegh et al., 1999). The quantum-splitting phosphor $YF_3:Pr^{3+}$ at room temperature has a quantum efficiency of about 140% for 185 nm excitation (Justel et al., 1998). However, viable quantum phosphors suitable for commercial applications have not yet been demonstrated.

The power-conversion efficiency of a wavelength converter is the product of Eqs. (21.1) and (21.3)

$$\eta_{\lambda\text{-converter}} = \eta_{\lambda\text{-conversion}} \, \eta_{\text{ext}} \, . \tag{21.4}$$

The inherent wavelength-conversion loss is the reason that λ-converter-based white LEDs such as phosphor-based white LEDs have a fundamentally lower efficiency limit than white-light sources based on multiple LEDs.

The wavelength-conversion loss is highest for wavelength conversion from the UV to the red. For example, the conversion from UV (405 nm) to red (625 nm) can have a λ-conversion efficiency of at most 65%. The low λ-conversion efficiency represents a strong driving force to employ red LEDs (rather than red phosphors) in highly efficient lighting systems.

Most white-light emitters use an LED emitting at short wavelength (e.g. blue) and a *wavelength converter*. Some of the light emitted by the blue LED is absorbed in the converter material and then re-emitted as light with a longer wavelength. As a result, the lamp emits at least two different wavelengths. The types and characteristics of wavelength-converter materials will be discussed below.

The possibility that white light can be generated in different ways raises the question as to which is the optimum way to generate white light? There are two parameters that need to be considered: Firstly, the luminous efficiency and, secondly, the color-rendering index. For *signage applications*, the luminous efficiency is of primary importance and the color-rendering index is irrelevant. For *illumination applications*, both the luminous efficiency and the color-rendering index are important.

White-light sources employing two monochromatic complementary colors result in the highest possible luminous efficacy. However, the color-rendering index of such a dichromatic light source is lower than that of broad-band emitters.

The maximum luminous efficacy of radiation, attainable for white light created by two complementary monochromatic colors was calculated by MacAdam (1950). MacAdam showed that luminous efficacies exceeding 400 lm/W can be attained using a dichromatic source for white-light generation. The work of MacAdam (1950) was further refined by Ivey (1963) and Thornton (1971). These authors showed that dichromatic white-light sources have high luminous efficacy but low color-rendering properties, making them perfectly suitable for display applications but unsuitable for daylight illumination applications. In addition, Thornton (1971) showed that trichromatic white-light sources, i.e. sources creating white light by additive mixing of three discrete colors, have a color-rendering index suited for most applications. Thornton reported on an experiment in which 60 observers judged the color rendition of meat, vegetables, flowers, complexions, etc., when illuminated with a trichromatic light source with peak wavelengths at 450, 540, and 610 nm. The color rendition in this experiment was found to be "very good, if not excellent" illustrating the suitability of trichromatic white-light sources as potent daylight illumination sources.

A white-light source duplicating the sun's spectrum would have good color-rendering capability. However, the radiation efficacy of such a light source would be lower than what is possible with other spectral distributions, e.g. a trichromatic distribution. The sun's spectrum has strong emission near the boundaries of the visible spectrum (390 and 720 nm) where the eye sensitivity is very low. Thus exact duplication of the sun's spectrum is not a viable strategy for high-efficiency light sources.

21.2 Wavelength-converter materials

There are several types of converter materials including **phosphors**, **semiconductors**, and **dyes**. Converter materials have several parameters of interest, including the absorption wavelength, the emission wavelength, and the quantum efficiency. A good converter has near 100% quantum efficiency. The overall power-conversion efficiency of a wavelength converter is given by

$$\eta = \eta_{ext}(\lambda_1/\lambda_2) \tag{21.5}$$

where η_{ext} is the external quantum efficiency of the converter, λ_1 is the wavelength of photons absorbed by the phosphor, and λ_2 is the wavelength of photons emitted by the phosphor. Even if the external quantum efficiency is unity, there is always energy loss associated with the wavelength-conversion process, so that the power-conversion efficiency of a wavelength converter is always less than unity.

The most common wavelength-converter materials are **phosphors** and they will be discussed in greater detail in the following section. The optical absorption and emission spectrum of a commercial phosphor is shown in Fig. 21.2. The phosphor displays an absorption band and a lower-energy emission band. The emission band is rather broad, making this particular phosphor suitable for white-light emission. Phosphors are very stable materials and can have quantum efficiencies close to 100%. A common phosphor used for white LEDs is cerium-doped (Ce-doped) YAG phosphor (Nakamura and Fasol, 1997). For Ce-doped phosphors, quantum efficiencies of 75% have been reported (Schlotter et al., 1999).

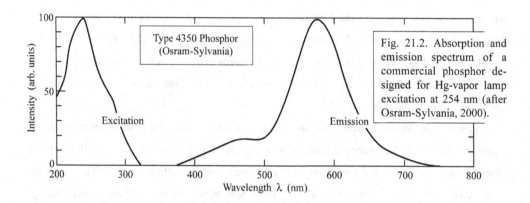

Fig. 21.2. Absorption and emission spectrum of a commercial phosphor designed for Hg-vapor lamp excitation at 254 nm (after Osram-Sylvania, 2000).

Dyes are another type of wavelength converter. Many different dyes are commercially available. An example of a dye optical absorption and emission spectrum is shown in Fig. 21.3. Dyes can have quantum efficiencies close to 100%. However, dyes, as organic molecules, lack the long-term stability afforded by phosphors and semiconductors.

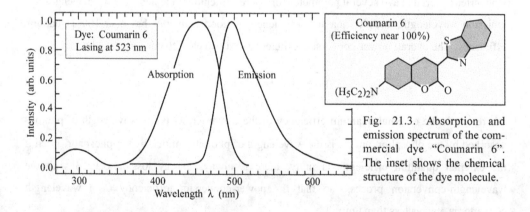

Fig. 21.3. Absorption and emission spectrum of the commercial dye "Coumarin 6". The inset shows the chemical structure of the dye molecule.

Finally, *semiconductors* are another type of wavelength converter. Semiconductors are characterized by narrow emission lines with linewidths of the order of $2kT$. The spectral emission linewidth of semiconductors is narrower than the linewidth of many phosphors and dyes. Thus, semiconductors allow one to tailor the emission spectrum of a semiconductor wavelength converter with good precision.

As for phosphors and dyes, semiconductors can have internal quantum efficiencies near 100%. The light escape problem in semiconductor converters is less severe than it is in LEDs due to the fact that semiconductor converters do not need electrical contacts that could block the light.

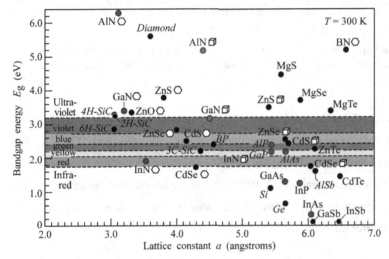

Fig. 21.4. Room-temperature bandgap energy versus lattice constant of common elemental and binary compound semiconductors.

Similar to phosphors and dyes, a great variety of semiconductors is available. Figure 21.4 shows elemental and binary compound semiconductors versus the semiconductor lattice constant. Using ternary or quaternary alloys, wavelength converters operating at virtually any visible wavelength can be fabricated.

21.3 Phosphors

Phosphors consist of an inorganic host material doped with an optically active element. Common hosts are **garnets**, which have the chemical formula $A_3B_5O_{12}$ where A and B are chemical elements and O is oxygen. Among the large group of garnets, yttrium aluminum garnet (YAG), $Y_3Al_5O_{12}$, is a particularly common host material. Phosphors having YAG as a host material are called **YAG phosphors**. The optically active dopant is a rare-earth element, a rare-earth oxide, or another rare-earth compound. Most rare-earth elements are optically active. Rare-earth light-emitting elements include cerium (Ce) used in white-light YAG phosphors, neodymium (Nd) used in lasers (Nd-doped YAG lasers), erbium (Er) used in optical amplifiers, and thorium (Th) oxide used in the mantle of gas lights.

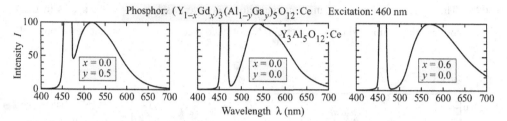

Fig. 21.5. Emssion spectrum of Ce-doped yttrium aluminum garnet (YAG:Ce) phosphor for different chemical compositions. The excitation wavelength is 460 nm (after Nakamura and Fasol, 1997).

The optical characteristics of YAG phosphors can be modified by partially substituting Gd for Y and Ga for Al so that the phosphor host has the composition $(Y_{1-x}Gd_x)_3(Al_{1-y}Ga_y)_5O_{12}$. The emission spectra for a Ce-doped $(Y_{1-x}Gd_x)_3(Al_{1-y}Ga_y)_5O_{12}$ phosphor with different compositions are shown in Fig. 21.5 (Nakamura and Fasol, 1997). The figure reveals that the addition of Gd shifts the emission spectrum to longer wavelengths whereas the addition of Ga shifts the emission spectrum to shorter wavelengths.

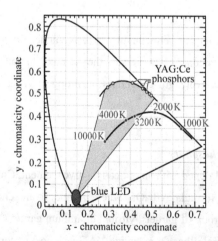

Fig. 21.6. Chromaticity points of YAG:Ce phosphor, and the general area (shaded) accessible to white emitters consisting of a blue LED and YAG:Ce phosphor (adapted from Nakamura and Fasol, 1997). Also shown is the planckian locus with color temperatures.

The chromaticity points of the YAG:Ce phosphors are shown in Fig. 21.6. The shaded region reveals the chromaticities that can be attained by mixing light from a blue LED source with the light of YAG:Ce. The figure reveals that such white emitters can have a very high color temperature.

An alternative to YAG phosphor is **TAG phosphor**, which is based on terbium aluminum garnet, $Tb_3Al_5O_{12}$. Both YAG and TAG crystallize with the garnet structure. Additionally, the Y^{3+} and Tb^{3+} ionic radii are very close (r_{Y3+} = 1.02 Å and r_{Tb3+} = 1.04 Å). Consequently, the X-ray diffraction pattern of YAG does not strongly change as Y in YAG is substituted with Tb, even for Tb mol fractions of 30% (Potdevin et al., 2005). Although TAG phosphors may have slightly lower radiative efficiency than YAG phosphors, TAG phosphors represent a viable alternative to YAG phosphors (Kim, 2005).

21.4 White LEDs based on phosphor converters

A white LED lamp using a phosphor wavelength converter and a blue GaInN/GaN optical excitation LED was first reported by Bando et al. (1996) and reviewed by Nakamura and Fasol (1997). The GaInN/GaN LED used for optical excitation ("optical pumping") was a device reported by Nakamura et al. (1995). The phosphor used as a wavelength converter was Ce-doped YAG with chemical formula $(Y_{1-a}Gd_a)_3 (Al_{1-b}Ga_b)_5 O_{12}$: Ce. The exact chemical composition of the host (YAG) and the dopants (e.g. Ce) is usually proprietary and not publicly available.

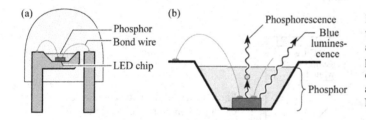

Fig. 21.7. (a) Structure of white LED lamp consisting of a GaInN blue LED chip and a phosphor. (b) Wavelength-converting phosphorescence and blue luminescence (after Nakamura and Fasol, 1997).

The cross-sectional structure of a white LED lamp is shown in Fig. 21.7(a). The figure shows the LED die emitting in the blue and the YAG phosphor surrounding the die. The YAG phosphor can be made as a powder and suspended in epoxy resin. During the manufacturing process, a droplet of the YAG phosphor suspended in the epoxy is deposited on the LED die, so that the resin fills the cup-shaped depression in which the LED die is located, as shown in Fig. 21.7(b). As indicated in the figure, a fraction of the blue light is absorbed by the phosphor and re-emitted as longer-wavelength light.

The emission spectrum of the phosphor-based white lamp thus consists of the blue emission band originating from the semiconductor LED and longer-wavelength phosphorescence, as shown in Fig. 21.8. The thickness of the phosphor-containing epoxy and the concentration of the phosphor suspended in the epoxy determine the relative strengths of the two emission bands. The two bands can thus be adjusted to optimize the luminous efficiency and the color-rendering characteristics of the LED.

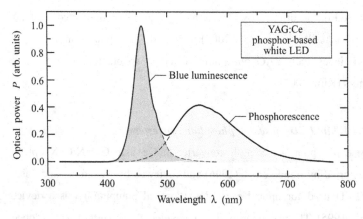

Fig. 21.8. Emission spectrum of a phophor-based white LED manufactured by Nichia Corporation (Anan, Tokushima, Japan).

The location of the white lamp in the chromaticity diagram is shown in Fig. 21.9. The location suggests that the emission color is white with a bluish tint. A bluish white is indeed confirmed when looking at the lamp.

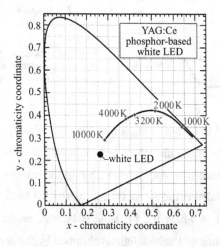

Fig. 21.9. Chromaticity coordinates of a commercial phophor-based white LED manufactured in 2001 by Nichia Corporation (Anan, Tokushima, Japan). Also shown is the planckian locus and associated color temperatures.

First-generation white LEDs from Nichia Corporation were improved in terms of their color rendering capability by adding an additional phosphor that, when excited by 460 nm blue light, has a peak emission wavelength of 655 nm and a full-width at half-maximum of 110 nm (Narukawa, 2004). As a result, the emission can be enhanced in the red range, as shown in Fig. 21.10. Furthermore, by using an optimized phosphor mix, the pronounced notch in the first-generation white LED is reduced. The second-generation white LED lamps from Nichia Corporation (Narukawa, 2004) render red colors better than the first-generation and have a lower color temperature that can range between 2800 K (warm white) and 4700 K depending on the phosphor mix.

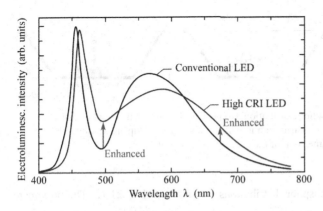

Fig. 21.10. Electroluminescence spectrum of conventional white LED and of high-color-rendering-index white LED. The high CRI results from the broader emission spectrum and the reduction of the notch in the spectrum (after Narukawa, 2004).

Note that the downside of adding red phosphors is a reduced luminous efficiency: The large Stokes shift of red phosphors reduces the efficiency (excitation 460 nm; emission 655 nm) Furthermore, it is well known that red phosphors excitable at 460 nm are comparatively inefficient. Thus, although color rendering capabilities are improved, they are improved at the expense of luminous efficiency.

A concern with white sources is spatial **color uniformity**. The chromaticity of the white source should not depend on the emission direction. Color uniformity can be attained by a phosphor distribution that provides an **equal optical path length** in the phosphor material independent of the emission direction (Reeh et al., 2003).

Spatial uniformity can also be attained by adding **mineral diffusers** to the encapsulant (Reeh et al., 2003). Such mineral diffusers are optically transparent substances, such as TiO_2, CaF_2, SiO_2, $CaCO_3$, and $BaSO_4$, with a refractive index different from the encapsulant. The diffuser will cause light to reflect, refract, and scatter, thereby randomizing the propagation direction and uniformizing the far-field distribution in terms of chromaticity (i.e. spectral composition).

21.5 Spatial phosphor distributions

The **spatial phosphor distribution** in white LED lamps strongly influences the color uniformity and efficiency of the lamp. One can distinguish between **proximate** and **remote phosphor distributions** (Goetz, 2003; Holcomb et al., 2003; Kim et al., 2005; Luo et al., 2005; Narendran et al., 2005). In proximate phosphor distributions, the phosphor is located in close proximity to the semiconductor chip. Proximate phosphor distributions are shown in Fig. 21.11(a) and (b). In remote phosphor distributions, the phosphor is spatially removed from the semiconductor chip. A remote phosphor distribution is shown in Fig. 21.11(c).

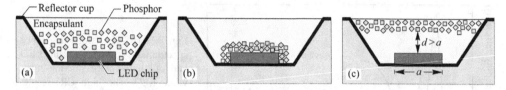

Fig. 21.11. (a) Proximate phosphor distribution, (b) proximate conformal phosphor distribution, and (c) remote phosphor distribution in which phosphor and chip are separated by at least one times the lateral dimension of the chip (after Kim et al., 2005).

Photographs of the different phosphor distributions are shown in Fig. 21.12. The proximate phosphor distribution shown in Fig. 21.12(a) was introduced by Nichia Corporation during the 1990s. The phosphor particles are dissolved in the encapsulation material that is dispensed into the reflector cup. Gravity, buoyancy, and friction generally lead to a distribution of phosphor particles that favors larger phosphor particles to move downward, thereby bringing them closer to the chip surface.

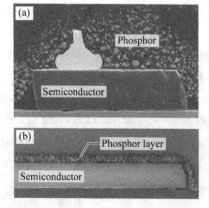
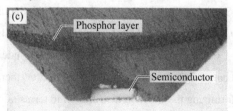

Fig. 21.12. Phosphor distributions in white LEDs: (a) Proximate phosphor distribution. (b) Proximate conformal phosphor distribution. (c) Remote phosphor distribution ((a) and (b) adopted from Goetz, 2003; (c) after Kim et al., 2005).

Another proximate phosphor distribution, called the ***conformal phosphor distribution***, is shown in Fig. 21.12(b). Conformal phosphor distributions are accomplished by wafer-level phosphor dispensation thereby lowering the manufacturing cost as compared with a lamp-level phosphor dispensation. Conformal phosphor distributions provide a small emission area and high luminance, which is particularly relevant for imaging-optics applications. Imaging-optics applications (e.g. automotive headlights) frequently require the use of lenses. Optical design considerations show that point-like sources, i.e. sources with a small emission area, are desirable for these applications.

A general drawback of proximate phosphor distributions is the absorption of phosphorescence by the semiconductor chip. Phosphorescence emitted toward the semiconductor chip can be absorbed by the chip, e.g. by the metal contacts covering the chip. The reflectivity of the semiconductor chip and metal contacts is generally not very high.

This drawback can be avoided by remote phosphor distributions in which the phosphor is spatially distanced from the semiconductor chip (Kim *et al.*, 2005; Luo *et al.*, 2005; Narendran *et al.*, 2005). In such remote phosphor structures, it is less likely that phosphorescence impinges on the low-reflectance semiconductor chip due to the spatial separation between the primary emitter (semiconductor chip) and the secondary emitter (phosphor). The probability that phosphorescence impinges on the semiconductor chip is greatly diminished if the distance between chip and phosphor is equal to or greater than the chip's lateral dimension, i.e. $d > a$, as shown in Fig. 21.11(c). As a result, higher phosphorescence efficiency is enabled. Ray-tracing simulations and experiments using a remote blue phosphor pumped by a GaInN emitter have indeed demonstrated phosphorescence efficiency improvements of 75% and 27%, respectively (Kim *et al.*, 2005; Luo *et al.*, 2005). Narendran *et al.* (2005) reported an average of 61% improvement in light output by using the "scattered photon extraction" (SPE) method. At low currents, the SPE packages exceeded 80 lm/W, compared to 54 lm/W for a typical conventional package.

21.6 UV-pumped phosphor-based white LEDs

White LEDs can also be fabricated with optical excitation of phosphor in the ultraviolet (UV) wavelength range (Karlicek, 1999). Semiconductor sources emitting in the near-UV (320–390 nm) and in the violet, close the edge of the visible spectrum (390–410 nm) are frequently used for such white sources. Semiconductor diodes emitting near 400 nm with remarkably high efficiencies have been reported (Morita *et al.*, 2004).

For deep-UV semiconductor sources (200–320 nm), conventional phosphors, as used in fluorescent lighting, can be used for wavelength conversion. However, the large Stokes shift associated with deep-UV sources is a significant drawback for such sources. Furthermore the development of deep-UV LEDs is challenging due to the low p-type and n-type doping efficiency in AlGaN with high Al content and the difficulties encountered in epitaxially growing high-quality AlGaN with low dislocation and defect densities.

In UV-pumped white LEDs, the entire visible emission originates in the phosphor. Phosphors excited in the deep UV have been used since the 1950s in fluorescent light tubes and since the 1980s in compact fluorescent lamps (CFLs). Phosphors in fluorescent light sources are pumped by the UV emission coming from the low-pressure mercury-vapor discharge occurring inside the tube. The dominant emission of low-pressure mercury-vapor discharge lamps (*Hg lamps*) occurs in the UV at 254 nm. Phosphors with strong absorption in this wavelength range are readily available. The color rendering properties of such phosphors are very suitable for most applications.

A white LED using a UV AlGaInN LED pump source and a tricolor phosphor blend was reported by Kaufmann *et al.* (2001). The LED pump source emitted at 380–400 nm, that is, near the boundary between the visible and UV spectrum. The phosphor blend consisted of three phosphors emitting in the red, green, and blue parts of the spectrum. A color-rendering index of 78 was reported for the lamp.

The color-rendering index (CRI) of UV-excited phosphor mixes ranges between 60 and 100. Excellent CRIs as high as 97 were reported by Radkov *et al.* (2003) for phosphor blends excited near 400 nm. Furthermore, such UV-LED based sources exhibit independence of the phosphor-emission spectrum on the exact UV-LED excitation wavelength, because the visible emission is solely due to the phosphor. Consequently, UV-pumped white lamps are expected to have a highly reproducible optical spectrum so that "binning" will likely not be required. Monte Carlo simulations reported by Radkov *et al.* (2004) indeed showed a very low chromaticity point variation (entirely within the first MacAdam ellipse) for phosphor sources excited with a variety of LEDs coming from chip bins with peak wavelengths ranging between 400 nm and 410 nm. The chromaticity variation was shown to be much lower for UV-LED/phosphor sources than for blue-LED/phosphor sources.

A fundamental drawback of UV-pumped white LEDs is the energy loss (Stokes shift) incurred when converting UV light to white light. The potential luminous efficiency of UV-pumped white LED lamps is therefore markedly lower than that of white sources based on a blue

LED exciting a yellow phosphor.

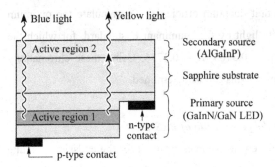

Fig. 21.13. Schematic structure of a photon-recycling semiconductor LED with one current-injected active region (Active region 1) and one optically excited active region (Active region 2) (after Guo et al., 1999).

21.7 White LEDs based on semiconductor converters (PRS-LED)

Light-emitting diodes using semiconductor wavelength converters have been demonstrated by Guo et al. (1999). The schematic structure of the photon-recycling semiconductor LED (PRS-LED) is shown in Fig. 21.13. The figure indicates that a fraction of the light emitted by the blue GaInN LED is absorbed by a AlGaInP secondary active region and re-emitted (or "recycled") as lower-energy photons. In order to obtain white light, the intensity of the two light sources must have a certain ratio that will be calculated below. The schematic power budget of the device is shown in Fig. 21.14. It is assumed that the electrical input power is P_0, and the output powers in the blue and amber spectral range are P_1 and P_2, respectively. The power-conversion efficiency of the blue LED and the photon-recycling semiconductor are assumed to be η_1 and η_2, respectively. The efficiency and luminous efficiency of the device are calculated below.

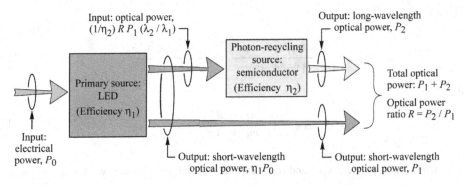

Fig. 21.14. Photon-recycling semiconductor LED power budget with electrical input power P_0 and optical output power P_1 and P_2.

The energy loss occurring in the photon-recycling process must be taken into account when determining the optimum choice of wavelengths for highest efficiency. Note that energy is lost even if the recycling process occurs with unit quantum efficiency. To calculate the optimum wavelength of operation, we represent white light by the Illuminant C standard, for which the chromaticity coordinates are $x_c = 0.3101$, $y_c = 0.3163$, $z_c = 0.3736$. Using these chromaticity coordinates, the pairs of complementary wavelengths can be determined.

21.8 Calculation of the power ratio of PRS-LED

Next, we calculate the light-power ratio between two sources required for the emission of white light and the luminous efficiency of the photon-recycling semiconductor LED. We refer to λ_1 and λ_2 as the primary (short) and secondary (long) wavelength, respectively. For white light, λ_1 and λ_2 are pairs of complementary wavelengths. We define the color masses of the two light sources as

$$m_1 = \bar{x}_1 + \bar{y}_1 + \bar{z}_1 \quad \text{and} \quad m_2 = \bar{x}_2 + \bar{y}_2 + \bar{z}_2 \quad (21.6)$$

where $\bar{x}_1, \bar{y}_1, \bar{z}_1, \bar{x}_2, \bar{y}_2,$ and $\bar{z}_2$ are color-matching functions at the two emission wavelengths λ_1 and λ_2, respectively (Judd, 1951; Vos, 1978; MacAdam, 1950, 1985). We define the power ratio of the two light sources as

$$R = P_2 / P_1 \quad (21.7)$$

where P_1 and P_2 are the optical powers of the short-wavelength source (λ_1) and the long-wavelength source (λ_2), respectively. The chromaticity coordinates of the newly generated color are then given by

$$y_{new} = \frac{P_1 \bar{y}_1 + P_2 \bar{y}_2}{P_1 m_1 + P_2 m_2} = \frac{\bar{y}_1 + R \bar{y}_2}{m_1 + R m_2} \quad (21.8)$$

and

$$x_{new} = \frac{\bar{x}_1 + R \bar{x}_2}{m_1 + R m_2}. \quad (21.9)$$

For a white-light emitter, x_{new} and y_{new} can be chosen to coincide with the chromaticity coordinates of the Illuminant C standard ($x_c = 0.3101$, $y_c = 0.3162$; CIE, 1932; Judd, 1951), i.e. $x_{new} = x_c = 0.3101$ and $y_{new} = y_c = 0.3162$. Solving Eq. (21.9) for the power ratio R yields

$$R = \frac{\bar{y}_1 - y_c m_1}{y_c m_2 - \bar{y}_2}. \quad (21.10)$$

The power ratio as calculated from Eq. (21.10) is shown as a function of wavelength in Fig. 21.15.

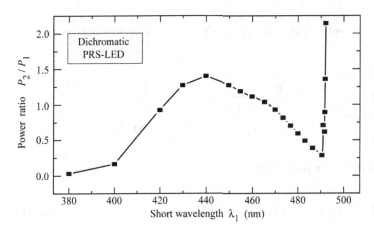

Fig. 21.15. Calculated power ratio between the two optical output powers P_1 and P_2 required to obtain white-light emission (after Guo et al., 1999).

21.9 Calculation of the luminous efficiency of PRS-LED

To produce the optical power P_2 at the wavelength of λ_2 through the recycling of photons from the primary source with wavelength λ_1, the optical power required from the primary source is given by

$$\frac{P_2}{\eta_2} \frac{\lambda_2}{hc} \frac{hc}{\lambda_1} = \frac{P_2 \lambda_2}{\eta_2 \lambda_1} \quad (21.11)$$

where η_2 is the optical-to-optical conversion efficiency of the photon-recycling light source. If P_0 is the electrical input power, then the optical power emitted by the primary LED source is $\eta_1 P_0$, where η_1 is the electrical-to-optical power conversion efficiency of the primary LED. Thus, the optical power emitted by the primary LED is given by

$$P_1 + \frac{P_2 \lambda_2}{\eta_2 \lambda_1} = \eta_1 P_0. \quad (21.12)$$

Solving the equation for the electrical input power and using $P_2 = RP_1$ yields

$$P_0 = P_1\left(\frac{1}{\eta_1} + \frac{R\lambda_2}{\eta_1\eta_2\lambda_1}\right). \tag{21.13}$$

The total optical output power of the PRS-LED is given by

$$P_{out} = P_1 + P_2 = (1+R)P_1 \tag{21.14}$$

so that the overall electrical-to-optical power efficiency of the photon-recycling dichromatic light source is given by

$$\eta = \frac{P_{out}}{P_0} = \frac{P_1(1+R)}{P_1\left(\dfrac{1}{\eta_1} + \dfrac{R}{\eta_1\eta_2}\dfrac{\lambda_2}{\lambda_1}\right)} = \frac{1+R}{\dfrac{1}{\eta_1} + \dfrac{R}{\eta_1\eta_2}\dfrac{\lambda_2}{\lambda_1}}. \tag{21.15}$$

The luminous flux Φ_{lum} of the device is given by

$$\Phi_{lum} = 683\frac{lm}{W}(\bar{y}_1 P_1 + \bar{y}_2 P_2) = 683\frac{lm}{W}(\bar{y}_1 + \bar{y}_2 R)P_1. \tag{21.16}$$

Then the luminous efficacy of radiation (measured in lumens per optical watt) of the photon-recycling semiconductor LED is given by

$$\frac{\Phi_{lum}}{P_{out}} = 683\frac{lm}{W}\frac{\bar{y}_1 + \bar{y}_2 R}{1+R}. \tag{21.17}$$

Thus, the luminous efficiency of the source (measured in lumens per electrical watt) of the PRS-LED is given by

$$\frac{\Phi_{lum}}{P_0} = 683\frac{lm}{W}\eta\,\frac{\bar{y}_1 + \bar{y}_2 R}{1+R}. \tag{21.18}$$

Using this formula, we calculate the luminous efficiency as a function of the primary wavelength. The result of the calculation is shown in Fig. 21.16 for *ideal* sources, i.e. for $\eta_1 = \eta_2 = 100\%$.

The maximum luminous efficiency occurs if the primary source emits at the wavelength 440 nm. A theoretical luminous efficiency of 336 lm/W is obtained for this wavelength. Note that in the calculation we assume that both light sources emit monochromatic light. However, the spontaneous emission from semiconductors has a 1.8 kT spectral width. Taking into account a finite linewidth, the expected luminous efficiency is slightly lower.

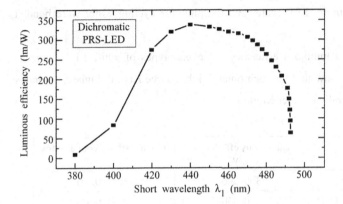

Fig. 21.16. Calculated luminous efficiency of a dichromatic PRS-LED versus its primary emission wavelength (after Guo et al., 1999).

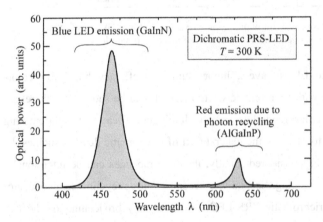

Fig. 21.17. Emission spectrum of dichromatic PRS-LED with current-injected GaInN blue LED primary source and AlGaInP photon recycling wafer (secondary source) emitting in the red (after Guo et al., 2000).

21.10 Spectrum of PRS-LED

PRS-LEDs have been demonstrated using a GaInN/GaN LED emitting in the blue and an electrically passive AlGaInP photon-recycling semiconductor emitting in the red part of the spectrum (Guo et al., 1999). The emission spectrum of the device, depicted in Fig. 21.17, shows the emission line of the primary LED at 470 nm and a second emission line at 630 nm due to absorption of the 470 nm light in the AlGaInP layer and re-emission of light at 630 nm. The recycling semiconductor used in this experiment is an AlGaInP/GaAs double heterostructure. The photon-recycling semiconductor is planar and no surface texturing was performed.

To avoid absorption of light in the GaAs substrate, the GaAs substrate of the AlGaInP epitaxial layer was removed. Firstly, the AlGaInP/GaAs recycling semiconductor was mounted on a glass slide. Subsequently, the GaAs substrate was removed by polishing and selective wet chemical etching. Then the primary LED wafer and the photon-recycling wafer were bonded together.

The approximate theoretical luminous efficiency of several types of white LED lamps is given in Table 21.2. The data show that the dichromatic light source has the highest luminous efficiency as compared to spectrally broader emitters.

Type of LED	Luminous efficiency (lm/W)	Chromaticity coordinates (x, y)
Dichromatic PRS LED	300–360	(0.31, 0.32)
Broadened dichromatic PRS LED	280–350	(0.31, 0.32)
Trichromatic LED	240–340	(0.31, 0.32)
Phosphor-based LED	200–280	(0.31, 0.32)

Table 21.2. Approximate theoretical luminous efficiencies of different types of white LEDs assuming unit quantum efficiency and the absence of resistive power losses.

Generally, dichromatic white LEDs have a higher luminous efficacy but lower color-rendering index (CRI) compared with trichromatic white LEDs. It can be shown that there is a fundamental trade-off between color rendering and the luminous efficacy of light-emitting devices (Walter, 1971). In order to improve the general CRI of dichromatic devices such as the PRS-LED, two possibilities can be considered. Firstly, the emission lines can be intentionally broadened, e.g. by compositional grading. Secondly, a second photon-recycling semiconductor can be added thus creating a trichromatic PRS-LED. However, any broadening of the two emission lines or the addition of an emission line will decrease the luminous efficacy and luminous efficiency of the device.

21.11 White LEDs based on dye converters

White LEDs can also be fabricated using organic dye molecules as wavelength converter materials. The dyes can be incorporated in the epoxy encapsulant (Schlotter *et al.*, 1997). Dyes can also be incorporated in optically transparent polymers.

A drawback of organic dyes is their finite lifetime. Dye molecules "bleach out", i.e. become optically inactive, after a certain number of photon absorption events. Typically a dye molecule is stable for about 10^4–10^6 optical transitions (Jones, 2000). The lack of high molecular stability of dyes is a serious drawback. The lifetime of dyes is considerably shorter than the lifetime of semiconductor or phosphor wavelength converters.

Dyes have a relatively small difference between the absorption and the emission band (Stokes shift). For example, the Stokes shift for the dye *Coumarin 6* is just 50 nm, as discussed earlier in this chapter. This shift is smaller than the Stokes shift required for dichromatic white LEDs that need typical wavelength shifts of 100 nm or more, as inferred from the separation of complementary wavelengths.

References

Bando K., Noguchi Y., Sakano K., and Shimizu Y. (in Japanese) *Tech. Digest, Phosphor Res. Soc.*, 264th Meeting, November 29 (1996)

CIE *Commission Internationale de l'Eclairage Proceedings, 1931* (Cambridge University Press, Cambridge, 1932)

Goetz W. "White lighting (illumination) with LEDs" *Fifth International Conference on Nitride Semiconductors*, Nara, Japan, May 25–30 (2003)

Guo X., Graff J. W., and Schubert E. F. "Photon-recycling semiconductor light-emitting diode" *IEDM Technical Digest*, **IEDM-99**, 600 (1999)

Guo X., Graff J. W., and Schubert E. F. "Photon-recycling for high brightness LEDs" *Compound Semiconductors* **6**, May/June (2000)

Holcomb M. O., Mueller-Mach R., Mueller G. O., Collins D., Fletcher R. M., Steigerwald D. A., Eberle S., Lim Y. K., Martin P. S., and Krames M. "The LED light bulb: Are we there yet? Progress and challenges for solid-state illumination" *Conference on Lasers and Electro-Optics* (*CLEO*), Baltimore, Maryland, June 1–6 (2003)

Ivey H. F. "Color and efficiency of luminescent light sources" *J. Opt. Soc. Am.* **53**, 1185 (1963)

Jones G., personal communication (2000)

Judd D. B. "Report of US secretariat committee on colorimetry and artificial daylight" in *Proceedings of the Twelfth Session of the CIE, Stockholm* Vol. **1**, p. 11 (Bureau central de la CIE, Paris, 1951)

Justel T., Nikol H., and Ronda C. R. "New developments in the field of luminescent materials for lighting and displays" *Angewandte Chemie (International Edition)* **37**, 3084 (1998)

Karlicek Jr. R. F., personal communication (1999)

Kaufmann U., Kunzer M., Köhler K., Obloh H., Pletschen W., Schlotter P., Schmidt R., Wagner J., Ellens A., Rossner W., and Kobusch M. "Ultraviolet pumped tricolor phosphor blend white emitting LEDs" *Phys. Stat. Sol. (a)* **188**, 143 (2001)

Kim J. K., personal communication (2005)

Kim J. K., Luo H., Schubert E. F., Cho J., Sone C., and Park Y. "Strongly enhanced phosphor efficiency in GaInN white light-emitting diodes using remote phosphor configuration and diffuse reflector cup" *Jpn. J. Appl. Phys. – Express Letter* **44**, L 649 (2005)

Luo H., Kim J. K., Schubert E. F., Cho J., Sone C., and Park Y. "Analysis of high-power packages for phosphor-based white-light-emitting diodes" *Appl. Phys. Lett.* **86**, 243505 (2005)

MacAdam D. L. "Maximum attainable luminous efficiency of various chromaticities" *J. Opt. Soc. Am.* **40**, 120 (1950)

MacAdam D. L. *Color Measurement: Theme and Variations* (Springer, New York, 1985)

Morita D., Yamamoto M., Akaishi K., Matoba K., Yasutomo K., Kasai Y., Sano M., Nagahama S.-I., and Mukai T. "Watt-class high-output-power 365 nm ultraviolet light-emitting diodes" *Jpn. J. Appl. Phys.* **43**, 5945 (2004)

Nakamura S., Senoh M., Iwasa N., Nagahama S., Yamada T., and Mukai T. "Superbright green InGaN single-quantum-well-structure light-emitting diodes" *Jpn. J. Appl. Phys. (Lett.)* **34**, L1332 (1995)

Nakamura S. and Fasol G. *The Blue Laser Diode* (Springer, Berlin, 1997)

Narendran N., Gu Y., Freyssinier-Nova J. P., and Zhu Y. "Extracting phosphor-scattered photons to

improve white LED efficiency" *Phys. Stat. Sol. (a)* **202**, R60 (2005)
Narukawa Y. "White light LEDs" *Optics & Photonics News* **15**, No. 4, p. 27 (2004)
Osram-Sylvania Corporation. Data sheet on type 4350 phosphor (2000)
Potdevin A., Chadeyron G., Boyer D., Caillier B., and Mahiou R. "Sol-gel based YAG:Tb^{3+} or Eu^{3+} phosphors for application in lighting sources" *J. Phys. D: Appl. Phys.* **38**, 3251 (2005)
Radkov E., Setlur A., Brown Z., and Reginelli J. "High CRI phosphor blends for near UV LED lamps" *Proc. SPIE* **5530**, 260 (2003)
Radkov E., Bompiedi R., Srivastava A. M., Setlur A. A., and Becker C. "White light with UV LEDs" *Proc. SPIE* **5187**, 171 (2004)
Reeh U., Höhn K., Stath N., Waitl G., Schlotter P., Schneider J., and Schmidt R. "Light-radiating semiconductor component with luminescence conversion element" US Patent 6,576,930 B2 (2003)
Schlotter P., Schmidt R., and Schneider J. "Luminescence conversion of blue light emitting diodes" *Appl. Phys. A* **64**, 417 (1997)
Schlotter P., Baur J., Hielscher C., Kunzer M., Obloh H., Schmidt R., and Schneider J. "Fabrication and characterization of GaN/InGaN/AlGaN double heterostructure LEDs and their application in luminescence conversion LEDs" *Materials Sci. Eng.* **B59**, 390 (1999)
Srivastava A. M. "Phosphors" *Encyclopedia of Physical Science and Technology* 3rd edition **11**, 855 (2004)
Srivastava A. M. and Ronda C. R. "Phosphors" *Interface (The Electrochemical Society)* **12** (2), p. 48 (2003)
Thornton W. A. "Luminosity and color-rendering capability of white light" *J. Opt. Soc. Am.* **61**, 1155 (1971)
Vos, J. J. "Colorimetric and photometric properties of a 2-degree fundamental observer" *Color Res. Appl.* **3**, 125 (1978)
Walter W. "Optimum phosphor blends for fluorescent lamps" *Appl. Opt.* **10**, 1108 (1971)
Wegh R. T., Donker H., Oskam K. D., and Meijerink A. "Visible quantum cutting in $LiGdF_4$:Eu^{3+} through downconversion" *Science* **283**, 664 (1999)

22

Optical communication

LEDs are used in communication systems transmitting low and medium data rates (< 1 Gbit/s) over short and medium distances (< 10 km). These communication systems are based on either *guided light waves* (Keiser, 1999; Neyer et al., 1999; Hecht, 2001; Mynbaev and Scheiner, 2001; Kibler et al., 2004) or *free-space waves* (Carruthers, 2002; Heatley et al., 1998; Kahn and Barry, 2001). In guided-wave communication, individual optical fibers or fiber bundles are used as the transmission medium and LED-based optical communication links are limited to distances of a few kilometers. Optical fiber systems include *silica* and *plastic* optical fibers. Free-space communication is usually limited to a room, even though longer distances are possible. In this chapter we discuss the characteristics of transmission media used for LED communication.

22.1 Types of optical fibers

The cross section of optical fibers consists of a circular core region surrounded by a cladding region. The core region has a higher refractive index than the cladding region. Typically, the core refractive index is about 1% higher than the cladding refractive index. Light propagating in the core is guided inside the core by means of *total internal reflection*. The condition of total internal reflection can be inferred from Snell's law. A light ray is *totally internally reflected* whenever it is incident at the core–cladding boundary. In a ray-optics picture, light rays propagating inside the core follow a *zigzag path*.

There are three types of optical fibers used in communication systems. These types are the (*i*) *step-index multimode* fiber, (*ii*) *graded-index multimode* fiber, and (*iii*) *single-mode* fiber. The three types of fibers are shown in Fig. 22.1 along with the refractive index profiles.

Step-index multimode fibers have a relatively large core diameter. Typical core diameters are 50, 62.5, and 100 μm for silica fibers used in communication systems. Plastic optical fibers have larger core diameters, typically 1 mm. An important advantage of multimode fibers is *easy coupling* of the light source to the fiber. Usually, a ± 5 μm accuracy of alignment is sufficient for multimode fibers with a core diameter of 50 μm. The main disadvantage of multimode fibers is the occurrence of *modal dispersion*.

(a) ***Step-index multimode fiber***
Simple coupling; large modal dispersion

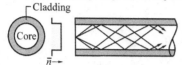

Typical diameters and refractive indices
Core/cladding diameter   62.5/125, 100/140, ... , 1000/1200 μm
Core index   1.45
Index difference   1–2%

(b) ***Parabolically-graded-index multimode fiber***
Simple coupling; difficult fabrication; low or zero modal dispersion

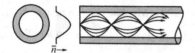

Core/cladding diameter   50/125, 62.5/125, 85/125
Core index at center   1.45
Index difference   1–2% in graded index profile

(c) ***Step-index single-mode fiber***
Difficult coupling; difficult fabrication; no modal dispersion

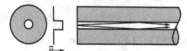

Core/cladding diameter   9/125
Core index   1.45
Index difference   1–2%

Fig. 22.1. (a) Step-index multimode fibers allow for the propagation of several optical modes. (b) Parabolically graded-index multimode fibers allow for the propagation of several modes with similar propagation constant. Graded-index multimode fibers have a lower modal dispersion than step-index multimode fibers. (c) Step-index single-mode fibers have a small core diameter and no modal dispersion.

Because the core diameter in multi-mode fibers is much larger than the operating wavelength, several optical modes can propagate in the waveguide. These optical modes have different propagation constants so that different modes arrive at the end of the fiber at different times, even if they were launched at the same time. This leads to a broadening of the optical pulse and limits the maximum bit rate that can be transmitted over a fiber of a given length.

Modal dispersion is reduced by *graded-index* multimode fibers. Graded-index multimode fibers have a parabolically graded core index leading to a reduction in modal dispersion.

Single-mode fibers have such a small core diameter that only a single optical mode can propagate in the fiber. Typical single-mode core diameters are 5–10 μm. The main advantage of single-mode fibers is the *lack of modal dispersion*. The main disadvantage of single-mode fibers is *difficult coupling* due to the small core diameter. A small core diameter requires light sources with high brightness such as lasers. However, LEDs, in particular edge-emitting LEDs and superluminescent LEDs are also occasionally used with single-mode fibers. Coupling of light into single-mode fibers requires precise alignment with tolerances of a few micrometers.

If the *optical power* to be transmitted over an optical fiber is of prime interest, the core diameter should be as large as possible and the core–cladding index difference should also be as large as possible. Specialty fibers with core diameters of > 1 mm are available. Such fibers are not suitable for communication applications due to the large modal dispersion.

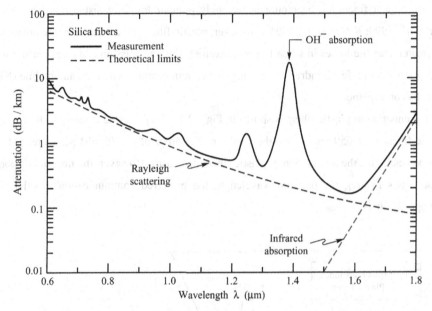

Fig. 22.2. Measured attenuation in silica fibers (solid line) and theoretical limits (dashed lines) given by Rayleigh scattering in the short-wavelength region, and by molecular vibrations (infrared absorption) in the infrared spectral region.

22.2 Attenuation in silica and plastic optical fibers

Silica (SiO_2) has excellent optical properties including great long-term stability. A large variety of glasses and fibers are available. The attenuation of silica fibers is shown in Fig. 22.2. Inspection of Fig. 22.2 reveals that a minimum loss of 0.2 dB occurs at a wavelength of 1.55 µm.

There are several optical "windows" for communication over silica fibers. These communication windows are at 0.85, 1.3, and 1.55 µm. The 0.85 µm communication window is suitable for communication with GaAs-based LEDs and lasers. However, this window is limited to short distances due to the material dispersion and to the high attenuation of silica fibers at that wavelength. The 1.3 µm communication window is also suited for communication with LEDs and lasers. This window has relatively low loss and zero dispersion, allowing for high-bit-rate transmission, in particular in graded-index and single-mode fibers. The 1.55 µm communication window is characterized by the lowest loss of all three windows. Consequently, this window is used for long-distance high-bit-rate communication. To allow for high bit rates, single-mode fibers must be used. Since it is difficult to efficiently couple light emerging from an LED into a single-mode fiber, lasers are preferred over LEDs at 1.55 µm.

Plastic optical fibers are becoming increasingly popular for short-distance communication (Neyer *et al.*, 1999; Kibler *et al.*, 2004). However, plastic fibers have losses that are about 1 000 times greater than the losses in silica fibers. Therefore, the transmission distances are limited to just a few meters to a few hundred meters, e.g. for communication within an automobile (Kibler *et al.*, 2004) or airplane.

The attenuation in plastic fibers is shown in Fig. 22.3. The preferred communication window of plastic fibers is at 650 nm, where the loss is of the order of 0.1–0.2 dB per meter. At even shorter wavelengths, the attenuation in plastic fibers decreases. However, the material dispersion increases, thus making the 650 nm wavelength the preferred communication wavelength in plastic optical fibers.

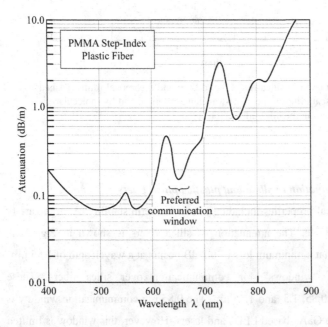

Fig. 22.3. Attenuation of a PMMA step-index plastic optical fiber. At 650 nm, the preferred communication wavelength, the attenuation is about 150 dB/km (after data sheet of Toray Industries Ltd., 2002).

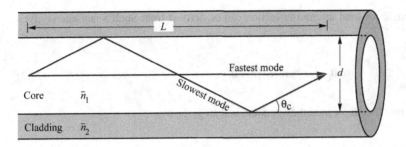

Fig. 22.4. Geometry used for calculation of the modal dispersion in a multimode fiber waveguide.

22.3 Modal dispersion in fibers

Modal dispersion occurs in multimode fibers that have a larger core diameter or a larger index difference between the core and the cladding than single-mode fibers. Typical core diameters range from 50 to 1 000 µm for multimode fibers and 5 to 10 µm for single-mode fibers. In the ray optics model, *different optical modes* correspond to light rays propagating at *different angles* in the core of the waveguide. The derivation of propagation angles in multimode fibers would go beyond the scope of this chapter. Here an approximate calculation will be performed to obtain the modal dispersion.

Consider a fiber waveguide with refractive indices of the core and cladding of $\bar{n}_1$ and $\bar{n}_2$, respectively. Assume that the waveguide supports the propagation of more than one optical mode. Two of these modes are shown schematically in Fig. 22.4. Owing to the difference in optical path length, the mode with the smaller propagation angle θ will arrive earlier at the end of the multimode fiber. The **modal dispersion** is the time delay between the fastest and the slowest optical mode normalized to the length L of the waveguide.

In the calculation, assume that the phase and group velocity are given by $v_{\rm ph} = c / \bar{n}_1 \approx v_{\rm gr}$. The fastest mode has the smallest propagation angle and we approximate the smallest angle by $\theta_{m=0} \approx 0°$. The slowest mode has the largest propagation angle and we approximate the largest angle by $\theta_m \approx \theta_c$, where θ_c is the critical angle of total internal reflection. This approximation can be made without loss of accuracy for multimode fibers which carry many modes.

The propagation times for the fastest and slowest modes per unit length of the fiber are given by

$$\tau_{\rm fast} = \frac{L}{c/\bar{n}_1} \qquad \tau_{\rm slow} = \frac{L/\cos\theta_c}{c/\bar{n}_1} \qquad (22.1)$$

where the critical angle for total internal reflection can be derived from Snell's law and is given by

$$\theta_c = \arccos(\bar{n}_2/\bar{n}_1). \tag{22.2}$$

The time delay per unit length, or modal dispersion, is then given by

$$\frac{\Delta\tau}{L} = \frac{\tau_{slow} - \tau_{fast}}{L} = \frac{n_1}{c}\left(\frac{1}{\cos\theta_c} - 1\right). \tag{22.3}$$

A waveguide supporting *many* modes has a large time delay between the fastest and slowest modes. Thus modal dispersion increases with the number of optical modes supported by the waveguide.

Exercise: *Modal dispersion in waveguides*. Calculate the time delay between the slowest and the fastest modes, and the maximum possible bit rate for a 1 km long multimode fiber waveguide with core refractive index $\bar{n}_1 = 1.45$ and cladding refractive index $\bar{n}_2 = 1.4$.
Solution: Using Snell's law (Eq. 22.2), one obtains $\theta_c \approx 15°$. The time delay calculated from Eq. (22.3) amounts to $\Delta\tau = 170$ ns. The minimum time required to transmit one bit of information is given by $\Delta\tau$. This yields an approximate maximum bit rate of $f_{max} = 1/170$ ns = 5.8 Mbit/s. The calculation shows that modal dispersion can be a significant limitation in optical communication. Graded-index multimode fibers or single-mode fibers are therefore required for high-speed communication systems.

22.4 Material dispersion in fibers

Material dispersion is another mechanism limiting the capacity of optical fibers. Material dispersion is due to the dependence of the refractive index on the wavelength. Figure 22.5 shows, as a function of wavelength, the phase refractive index and the group refractive index of silica. The indices are defined as

$$\bar{n} = \frac{c}{v_{ph}} \quad \text{(phase refractive index)} \tag{22.4}$$

and

$$\bar{n}_{gr} = \frac{c}{v_{gr}} \quad \text{(group refractive index)} \tag{22.5}$$

where v_{ph} and v_{gr} are the phase and group velocity in silica, respectively. The **phase refractive index** and the **group refractive index** are related by

22.4 Material dispersion in fibers

$$\bar{n}_{gr} = \bar{n} - \lambda \frac{d\bar{n}}{d\lambda} = \bar{n} - \lambda_0 \frac{d\bar{n}}{d\lambda_0}. \qquad (22.6)$$

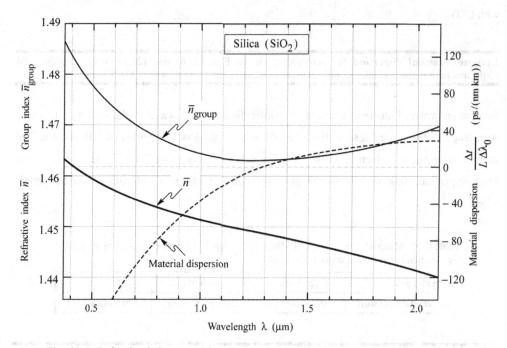

Fig. 22.5. Refractive index, group index, and material dispersion of a silica fibers for an optical signal spectral width $\Delta\lambda_0$ in vacuum. The material dispersion of regular silica fibers is zero at $\lambda = 1.3$ µm.

If the fiber is dispersive, the difference in group velocity between the "slowest color" and the "fastest color" of the optical signal is given by

$$\left|\Delta v_{gr}\right| = \frac{c}{\bar{n}_{gr}^2} \frac{d\bar{n}_{gr}}{d\lambda} \Delta\lambda \qquad (22.7)$$

where $\Delta\lambda$ is the spectral width of the optical signal.

The time delay between the leading edge and the trailing edge of an optical signal after traveling in the fiber for a length L, called the **material dispersion**, is given by

$$\Delta\tau = \frac{L}{v_{gr}^2} \Delta v_{gr} = \frac{L}{c} \frac{d\bar{n}_{gr}}{d\lambda} \Delta\lambda = \frac{L}{c} \frac{d\bar{n}_{gr}}{d\lambda_0} \Delta\lambda_0. \qquad (22.8)$$

22 Optical communication

The material dispersion is measured in ps/(nm km) and it is illustrated for silica fibers in Fig. 22.5. LEDs have a broad emission linewidth. Therefore material dispersion is, along with modal dispersion, the bandwidth-limiting factor in optical fiber communication systems operated with LEDs.

Exercise: *Material dispersion in waveguides.* Derive Eqs. (22.6) and (22.7). Why does material dispersion have a much smaller significance for semiconductor lasers than for LEDs?

Substantial material dispersion exists in plastic fibers at all wavelengths of interest. These wavelengths are the local loss minimum at 650 nm and the low-loss region of 500–600 nm. The material dispersion is given in Table 22.1. The data indicates that 650 nm is the wavelength of least dispersion, making 650 nm the preferred communication wavelength in plastic optical fibers.

Table 22.1. Material dispersion in PMMA plastic optical fibers (courtesy of R. Marcks von Wurtemberg, *Mitel* Corporation, 2000).

Wavelength	525 nm	560 nm	650 nm
Material dispersion	700 ps/(nm km)	500 ps/(nm km)	320 ps/(nm km)

Exercise: *Comparison of material and modal dispersion.* Consider a 62.5 µm core diameter multimode step-index fiber of 3 km length with a core index of $\bar{n}_1 = 1.45$ and a cladding index of $\bar{n}_2 = 1.4$. Assume that the fiber inputs come from either an LED or a laser emitting at 850 nm. Assume that the LED and the laser have a linewidth of 50 and 5 nm, respectively. Calculate the material and the modal dispersion for each case and explain the result.

22.5 Numerical aperture of fibers

Owing to the requirement of total internal reflection, the only light rays that can propagate losslessly in the core of an optical fiber are those that have a propagation angle smaller than the critical angle for total internal reflection. Light rays for which the propagation angle is too large will consequently not couple into the fiber. Here we consider the coupling of light from an LED light source into an optical fiber. We assume that the fiber end has a polished planar surface normal to the optical axis of the fiber, as shown in Fig. 22.6.

As a consequence of the requirement of total internal reflection, only a range of angles will be "accepted" by the fiber for lossless propagation. Outside the *acceptance angle* range, light rays will be refracted into the cladding layer where they will incur losses.

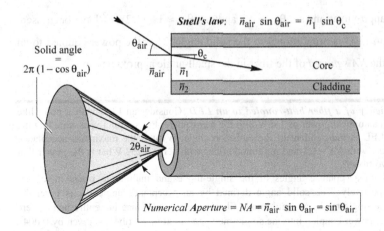

Fig. 22.6. Illustration of the *numerical aperture* (*NA*) of a fiber. For example, the light acceptance angle in air is $\theta_{air} = 11.5°$ for a numerical aperture of $NA = 0.2$.

The range of allowed angles can be inferred from Snell's law. As illustrated in Fig. 22.6, the maximum angle for acceptance in the fiber is given by

$$\bar{n}_{air} \sin \theta_{air} = \bar{n}_1 \sin \theta_c . \tag{22.9}$$

Since the refractive index of air is approximately unity, the maximum acceptance angle in air is given by

$$\theta_{air} = \arcsin(\bar{n}_1 \sin \theta_c). \tag{22.10}$$

The maximum acceptance angle defines a cone of allowed angles, as shown in Fig. 22.6. Light rays incident on the core of the optical fiber with propagation angles within the cone can propagate without loss.

Another way to express the acceptance cone is the **numerical aperture** of the fiber. The numerical aperture (*NA*) is defined as

$$NA = \bar{n}_1 \sin \theta_c = \bar{n}_{air} \sin \theta_{air} = \sin \theta_{air} \approx \theta_{air} \tag{22.11}$$

where the approximation $\sin \theta_{air} \approx \theta_{air}$ is valid for *small* numerical apertures. Typical *NA*s for silica single-mode fibers are 0.1 and typical *NA*s for silica multimode fibers are 0.15–0.25. Plastic optical fibers can have higher *NA*s, typically 0.2–0.4.

The solid angle corresponding to a certain *NA* is given by

22 Optical communication

$$\text{Solid angle} = \Omega = 2\pi(1-\cos\theta_{air}) = 2\pi[1-\cos(\arcsin NA)] \approx \pi NA^2 \qquad (22.12)$$

where the small-angle approximations $\sin\theta_{air} \approx \theta_{air}$ and $\cos\theta_{air} \approx 1 - (1/2)\theta_{air}^2$ have been used. The power emitted by an LED is proportional to the solid angle. Thus the power coupled to an LED is proportional to the *NA squared* of the fiber in the small-angle approximation.

Exercise: *Coupling efficiency of a fiber butt-coupled to an LED.* Consider an LED with a point-like emission region that emits an optical power of 1 mW into the hemisphere. For simplicity, assume that the intensity emitted by the LED is independent of the emission angle. What is the maximum acceptance angle of a single-mode fiber with $NA = 0.1$ and multimode fiber with $NA = 0.25$? What is the power that can be coupled into the two fibers?

Solution: The maximum acceptance angles of the single-mode and multimode fibers in air are $\theta_{air} = 5.7°$ and $14.5°$, respectively. The solid angle defined by an acceptance angle θ_{air} is given by $\Omega = 0.031$ and 0.20 for the single-mode and multimode fiber, respectively. Since the entire hemisphere has a solid angle of 2π, the power coupled into the single-mode and multimode fibers is given by 0.0049 and 0.032 mW, respectively.

22.6 Coupling with lenses

The low coupling efficiency of LEDs to optical fibers can be improved with convex lenses, if the light-emitting region of the LED is smaller than the optical fiber core. In this case, the light-emitting region can be imaged on the fiber core, thereby reducing the angle of incidence. The light source is *adapted* to the NA of the fiber ("*NA-matching*").

A convex lens can produce an image with height I of a light-emitting object with height O. If the image is larger than the object, the angles of the light incident from the lens on the image are less divergent than the light emanating from the object towards the lens. The *smaller divergence* obtained for *magnified* images allows one to increase the coupling efficiency to fibers. The principle of coupling with a convex lens is shown in Fig. 22.7.

The condition for a focused image (minimum image size) is given by the lens equation

$$\frac{1}{d_O} + \frac{1}{d_I} = \frac{1}{f} \qquad (22.13)$$

where d_O and d_I are the distances of the object and the image from the lens, respectively, and f is the focal length of the lens.

The magnification of the image of the LED light source on the core of the fiber is given by

22.6 Coupling with lenses

$$M = \frac{I}{O} = \frac{d_I}{d_O}. \tag{22.14}$$

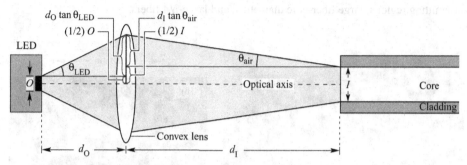

Fig. 22.7. Schematic illustration of coupling with a lens by imaging the light-emitting region of an LED onto the core of an optical fiber. The LED has a circular emission region with diameter O (Object). The emission region is imaged onto the fiber core with diameter I (Image) using a convex lens with focal length f.

As shown in Fig. 22.7, it is

$$\tfrac{1}{2} O + d_O \tan\theta_{LED} = \tfrac{1}{2} I + d_I \tan\theta_{air}. \tag{22.15}$$

If the LED and the core of the optical fiber are much smaller than the diameter of the lens and if the angles are relatively small, then Eq. (22.15) can be approximated by

$$\theta_{LED} = \frac{d_I}{d_O}\theta_{air} = \frac{I}{O}\theta_{air}. \tag{22.16}$$

Since d_I is larger than d_O, the acceptance angle for light emanating from the LED is *larger* than that of the fiber, implying *increased* coupling efficiency. Thus, we can define the numerical aperture of the LED, NA_{LED}, which defines the angle of light emanating from the LED that is accepted by the fiber. Using Eq. (22.14) and the small-angle approximation for NA, NA_{LED} is given by

$$NA_{LED} = \frac{I}{O} NA. \tag{22.17}$$

Since the coupling efficiency is proportional to NA^2 (see Eq. 22.12), the coupling efficiency is increased to

22 Optical communication

$$\text{Coupling efficiency} \propto NA_{LED}^2 = [(I/O)NA]^2. \tag{22.18}$$

The result shows that high coupling efficiencies are obtained for LEDs with small-diameter light-emitting regions, large fiber-core diameters, and large-NA fibers.

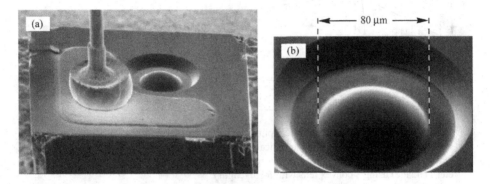

Fig. 22.8. (a) Commercial communication LED chip with integrated lens. (b) Detailed picture of the lens etched by a photochemical process into the GaAs substrate (AT&T ODL product line, 1995).

Lensed LEDs are frequently used in communication applications. A micrograph of an LED with a monolithically integrated lens is shown in Fig. 22.8. The light-emitting region of the LED is 20 µm and the lens shown in Fig. 22.8 has a diameter of about 80 µm.

Exercise: *Coupling efficiency of a fiber coupled to an LED with a lens*. Consider an LED circular emission region with diameter 20 µm coupled to a silica multimode fiber with $NA = 0.2$ and a core diameter of 62.5 µm. The LED emits a power of 1 mW into the hemisphere lying above the planar LED surface. For simplicity, assume that the LED emission intensity is independent of the emission angle. What is the maximum power that can be coupled into the multimode fiber?
Solution: Improved coupling can be obtained by imaging the LED emission region on to the core of the optical fiber. For maximum coupled power, a convex lens with magnification $M = 62.5$ µm / 20 µm = 3.125 can be used. Using the lens, the acceptance angle of the fiber is increased from $\theta_{air} = 11.5°$ to $\theta_{LED} = 35.9°$. The solid angle defined by the LED acceptance angle θ_{LED} is given by $\Omega = 1.19$. Since the LED emits 1 mW into the entire hemisphere (with solid angle $\Omega = 2\pi$), the power coupled into the fiber is given by 0.189 mW.

22.7 Free-space optical communication

Free-space optical communication (Carruthers, 2002; Heatley *et al*., 1998; Kahn and Barry, 2001) is suitable for low to medium bit rates. The most common application of free-space optical communication is the remote control of consumer appliances such as stereos and television sets. Other applications are the remote control of automobile door locks and the cordless interface between computers and peripheral devices such as a mouse, keyboard, and printer.

Free-space optical communication is limited to line-of-sight applications since obstacles such as walls and floors will block the path of light. Furniture may also block the path of light. However, a light beam may be reflected from the ceiling so that communication may still be possible even if there is no direct line of sight connection between the optical transmitter and the receiver.

The wavelength of choice for free-space optical communication is the near infrared. GaAs LEDs emitting with good efficiency are readily available. Infrared light is preferred over visible light sources because the former does not provide a distraction to anyone near the optical transmitter.

Eye safety considerations limit the maximum power of optical transmitters. At a wavelength of 870 nm, the optical power is limited to typically a few mW. Other wavelengths, such as 1 500 nm, allow for higher optical powers. The 1 500 nm wavelength range is termed *"eye safe"*, since the cornea absorbs 1 500 nm light, thus preventing light from reaching the sensitive retina. The wavelength 1 500 nm thus allows for higher optical powers than 870 nm sources.

If we restrict our considerations to small distances, the transmission medium air can be considered to be totally lossless. However, the optical signal strength decreases for uncollimated light beams due to spatial divergence. For *isotropic emitters*, the intensity decreases with the square of the radius, i.e.

$$I = P/(4\pi r^2) \qquad (22.19)$$

where P is the optical power emitted by the source and r is the distance from the source. The decrease in intensity thus has a very different dependence compared with the intensity in fiber communication.

The rapidly decreasing intensity limits the maximum range of optical communication. Collimated light beams can overcome this problem. Transmission distances of several km are possible without significant loss provided that atmospheric conditions are good, i.e. in the absence of fog or precipitation. Semiconductor lasers are used for such collimated transmission systems due to the ability to form collimated beams with very little spatial dispersion.

Multipath distortion or **multipath time delay** severely limits the data rate in free-space optical communication systems. A schematic illustration of multipath distortion is shown in

Fig. 22.9. A light beam emanating from the optical transmitter may take several different paths from the transmitter to the receiver. This is especially true for rooms with high-reflectivity surfaces such as white ceilings, walls, or mirrors. As an approximate rule, the longest path is assumed to be twice as long as the shortest path between the transmitter and the receiver. This approximate rule leads to a multipath distortion time delay of

$$\Delta \tau = L / c \qquad (22.20)$$

where L is the transmitter–receiver distance and c is the velocity of light. The maximum data rate is then limited to

$$f_{max} \approx 1 / \Delta \tau . \qquad (22.21)$$

For a room size of 5 m, the multipath delay is about $\Delta \tau = 17$ ns. Thus the data rate will be limited to about 60 MHz.

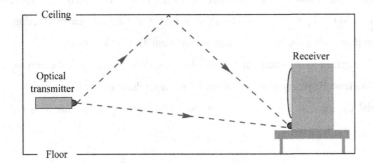

Fig. 22.9. Illustration of multipath distortion of a free-space optical signal, which limits the maximum data rate of the signal.

Another limitation of free-space optical communication is the detector noise. Sunlight and incandescent light sources have strong emission in the infrared. Thus a large DC photocurrent is generated in the detector, especially under direct sunlight conditions. The detector noise can be reduced by limiting the bandwidth of the receiver system. By reducing the bandwidth of the receiver system, and thereby also the system data rate, the detector noise is reduced, since the noise spectrum is much wider than the system bandwidth.

The detector noise due to ambient light sources can also be reduced by using optical band-pass filters, long-wavelength-pass filters, or a combination of both filters. Such filters prevent unwanted ambient light from reaching the detector.

References

Carruthers J. B. "Wireless infrared communications" in *Wiley Encyclopedia of Telecommunications* edited by J. G. Proakis (John Wiley and Sons, New York, 2002)

Heatley D. J. T., Wisely D. R., Neild I., and Cochrane P. "Optical wireless: the story so far" *IEEE Comm. Mag.* **36**, 72 (1998)

Hecht J. *Understanding Fiber Optics* (Prentice Hall, Upper Saddle River, New Jersey, 2001)

Kahn J. M. and Barry J. R. "Wireless infrared communications" *Proc. IEEE* **85**, 265 (2001)

Keiser G. *Optical Fiber Communications* 3rd edition (McGraw-Hill, New York, 1999)

Kibler T., Poferl S., Böck G., Huber H.-P., and Zeeb E. "Optical data buses for automotive applications" *IEEE J. Lightwave Technol.* **22**, 2184 (2004)

Mynbaev D. K. and Scheiner L. L. *Fiber-Optic Communications Technology* (Prentice Hall, Upper Saddle River, New Jersey, 2001)

Neyer A., Wittmann B., and Jöhnck M. "Plastic-optical-fiber-based parallel optical interconnects" *IEEE J. Sel. Top. Quantum Electron.* **5**, 193 (1999)

23

Communication LEDs

LEDs can be used for either free-space communication or for fiber communication applications. *Free-space communication applications* include the remote control of appliances such as television sets and stereos, and data communication between a computer and peripheral devices. LEDs used in optical *fiber communication applications* are suited for distances of a few km and bit rates up to about 1 Gbit/s. Most fibers used with LEDs are multimode (step-index and graded-index) fibers. However, some applications employ single-mode fibers.

23.1 LEDs for free-space communication

Free-space communication LEDs are commonly made with GaAs or GaInAs active regions and are grown on GaAs substrates. The GaInAs layer is pseudomorphic, i.e. sufficiently thin that it is coherently strained, and no dislocations are generated. The emission wavelength of GaAs and coherently strained GaInAs LEDs is limited to wavelengths in the IR ranging from 870 nm (for GaAs active regions) to about 950 nm (for GaInAs active regions).

The wavelength of free-space communication LEDs is in the infrared so that the light emitted is invisible to the human eye and does not distract. Since free-space communication usually involves transmission distances of less than 100 m, the transmission medium (air) can be considered, to a good approximation, to be lossless and dispersionless.

The *total light power* is an important figure of merit in free-space communication LEDs, so that the internal efficiency and the extraction efficiency need to be maximized. The emission pattern (far field) is another important parameter. The emission pattern should be wide to reduce the requirement of aiming the emitter towards the receiver.

23.2 LEDs for fiber-optic communication

Light-emitting diodes are the light source of choice for local area low and medium bit rate optical communication. Owing to the spontaneous emission lifetime of about 1 ns in highly excited semiconductors, the maximum bit rates attainable with LEDs are limited to rates

≤ 1 Gbit/s. Thus, multi-Gbit/s transmission rates are not feasible with LED sources. Transmission rates of several hundred Mbit/s are fully sufficient for many local-area communication applications.

LEDs used for fiber communication applications are very different from LEDs used in lamp applications. In communication LEDs, high coupling efficiency of the light emanating from the LED to the fiber is essential. Only the light emanating from *one* surface, namely the surface of the LED abutting the optical fiber, can be coupled into the fiber. Therefore it is essential to maximize the light emission from *one* surface of the LED. In LEDs used for fiber communication, the *power emitted per unit area* is a useful figure of merit. This is in contrast to free-space communication LEDs where the *total power* emitted by the LED is the appropriate figure of merit.

In order to maximize the LED–fiber coupling efficiency, the light-emitting spot should be smaller than the core diameter of the optical fiber. Typical are circular emission regions with diameters of 20–50 μm for devices used with multimode fibers. Silica multimode fibers have typical core diameters of 50–100 μm.

Plastic optical fibers, on the other hand, can have core diameters as large as 1 mm. Accordingly, LEDs with larger light-emitting areas can be used with plastic fibers.

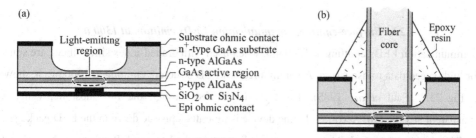

Fig. 23.1. (a) Burrus-type $Al_xGa_{1-x}As$/GaAs DH LED with the opaque GaAs substrate removed above the active region by wet chemical etching. The Burrus-type LED is mounted substrate-side up. (b) Optical fiber coupled to a Burrus-type LED.

23.3 Surface-emitting Burrus-type communication LEDs emitting at 870 nm

One of the first LED structures suitable for optical fiber communication was developed by Charles Burrus of AT&T Bell Laboratories (Burrus and Miller, 1971; Saul *et al.*, 1985). The Burrus-type LED is shown in Fig. 23.1 (a) and consists of a double heterostructure with a GaAs active region grown lattice-matched on a GaAs substrate. The original structure proposed and demonstrated by Burrus was just a homojunction. However, such homojunction LEDs are no longer in use due to unwanted light reabsorption in the layers adjoining the active region.

The Burrus-type structure has several features making it suitable for communication applications. Firstly, the light is generated in an active region of small lateral extent. The lateral size of the active region is determined by the p-type ohmic contact size of the LED. If the p-type confinement layer is sufficiently thin, no current spreading occurs, and the lateral extent of the light-emitting region is the same as the contact size. By design, the lateral extent of the active region is smaller than the core diameter of the optical fiber to maximize coupling efficiency. Secondly, the opaque GaAs substrate is partially removed, as shown in Fig. 23.1, to reduce absorption of light in the substrate. The substrate can be thinned by chemically assisted mechanical polishing to about 150 μm thickness followed by a wet chemical etch. If the GaAs membrane created by etching is too thin, it tends to break easily during fiber coupling. On the other hand, a thick membrane reduces radiative efficiency due to absorption.

Figure 23.1 (b) shows a typical coupling arrangement of the Burrus-type LED to an optical fiber. Epoxy is used to permanently attach the fiber to the LED. Note that the p-type contact also serves as a sink for the heat generated in the active region. The heat sink is particularly efficient if the p-type contact includes a thick layer of electroplated gold.

23.4 Surface-emitting communication LEDs emitting at 1300 nm

Communication LEDs emitting at 1300 nm, when used with graded-index silica fibers, are suited for high-speed data transmission. A communication LED structure emitting at 1300 nm is shown in Fig. 23.2 (Saul *et al.*, 1985). The light is emitted through the InP substrate, which is transparent at 1300 nm. Accordingly, the device is mounted epi-side down in the LED package.

The device has a GaInPAs active region lattice matched to the InP substrate. No current-spreading layer is used so that the light-emitting region is located directly above the contact. At the light-exit point, an optical lens collimates the light to improve the LED-to-fiber coupling efficiency. The lens is etched into the InP substrate by a photo-electrochemical process (Ostermayer *et al.*, 1983).

The emission spectra measured from the surface and from the edge of the device are shown in Fig. 23.3. The two emission spectra are markedly different. The spectrum emitted towards the edge has a smaller spectral width due to self-absorption. During self-absorption, predominantly photons of the high-energy part of the spectrum are reabsorbed.

23.4 Surface-emitting communication LEDs emitting at 1300 nm

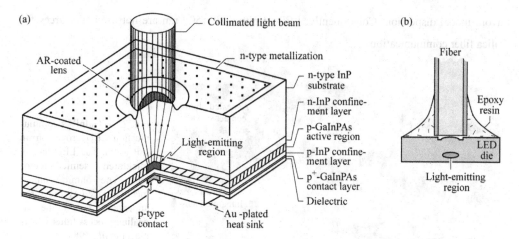

Fig. 23.2. (a) Structure of a communication LED emitting at 1300 nm with a GaInPAs active region lattice-matched to InP. The light generated in the active region is transmitted through the transparent InP substrate. The lateral dimension of the light-emitting region is defined by current injection under the circular ohmic contact with a diameter of 20 µm. An anti-reflection-coated (AR) lens, etched into the substrate, collimates the light beam. (b) Illustration of LED-to-fiber coupling using epoxy resin.

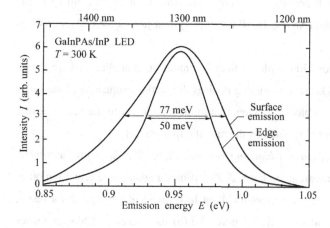

Fig. 23.3. Emission spectra along the edge and surface of a GaInPAs/InP communication LED emitting at 1300 nm. The spectrum emitted along the edge of the LED is narrower due to self-absorption.

A scanning electron micrograph of a GaInPAs/InP LED wafer is shown in Fig. 23.4. The surface of contacts displays roughness due to the annealing process that follows contact deposition. To reduce Fresnel reflection losses at the semiconductor–air boundary, the lens is anti-reflection coated.

Note that the wavelength 1500 nm is of interest for long-distance silica fiber communication. Long-distance communication fibers must have a small core diameter to be single mode and

385

avoid modal dispersion. Consequently, LEDs emitting at 1500 nm are not used as sources for silica fiber communication.

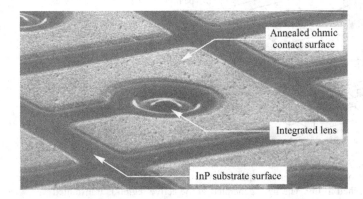

Fig. 23.4. GaInPAs communication LED grown on an InP wafer. The LED has an integrated semiconductor lens. The ohmic contact metal surface has a textured appearance due to the annealing process (after Ostermeyer et al., 1983).

23.5 Communication LEDs emitting at 650 nm

LEDs emitting at 650 nm are useful for plastic optical fiber communication. These fibers have a loss minimum and relatively low dispersion at 650 nm. Communication LEDs emitting in the 600–650 nm range are based on the $(AlGa)_{0.5}In_{0.5}P$ material system just as for 650 nm visible-spectrum LED lamps.

Typical LED structures used for 650 nm plastic fiber communication applications are shown in Figs. 23.5 (a) and (b). Both LEDs are top-emitting devices due to the opaqueness of the GaAs substrate. The LEDs employ current-blocking layers that guide the current to the active region. Light emitted from this region is not obstructed by the metallic ring contact.

The figure shows two types of *current-blocking regions*. Figure 23.5 (a) shows a current-blocking p-n-p structure in the transparent top layer. A Zn-diffused region in the center region of the LED overcompensates donors in the current-blocking layer making it p-type. As a result, current flows only through the Zn-diffused layer. Figure 23.5 (b) shows a current-blocking layer fabricated by epitaxial regrowth. After growth of an n-type blocking layer on top of the p-type confinement layer, the wafer is taken out of the growth system for patterning and etching. Subsequently, the wafer is re-introduced into the growth system for resumption of epitaxial growth. Epitaxial regrowth is a more expensive process and usually the device yield decreases for structures employing regrowth. The transparent window layers can consist of AlGaAs, as shown in Fig. 23.5, GaP, or another transparent semiconductor.

23.5 Communication LEDs emitting at 650 nm

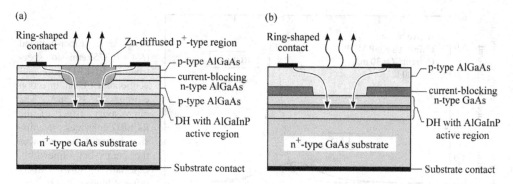

Fig. 23.5. AlGaInP/GaAs LED structures emitting at 650 nm for plastic optical fiber communications. Both LED structures funnel the current to the center of the active region where the emitted light is not obstructed by the top metal contact ring. (a) Structure using an n-type AlGaAs current-blocking layer and a p^+-type diffusion region. (b) Structure fabricated by epitaxial regrowth using an n-type GaAs current-blocking layer.

Resonant-cavity LEDs (RCLEDs) have several advantages over conventional LEDs including high brightness and a narrow spectral width (Schubert *et al.*, 1992, 1994; Schubert and Hunt, 1999). An RCLED emitting at 650 nm is shown in Fig. 23.6 (Streubel and Stevens, 1998; Streubel *et al.*, 1998; Whitaker, 1999; Mitel, 2000). As in the two previous structures, the top contact is ring-shaped and the current is guided to the center opening of the ring. An ion-implanted region is used for current blocking. Hydrogen and oxygen have been employed to render the implanted region insulating. Oxygen implants are more stable than hydrogen implants since small hydrogen atoms tend to easily diffuse out of the semiconductor at moderate annealing temperatures.

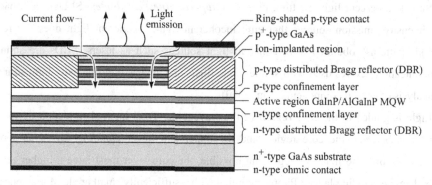

Fig. 23.6. RCLED emitting at 650 nm. Two distributed Bragg reflectors (DBRs) form the cavity. The active region is a GaInP/AlGaInP multiple-quantum well structure (after Whitaker, 1999).

23 Communication LEDs

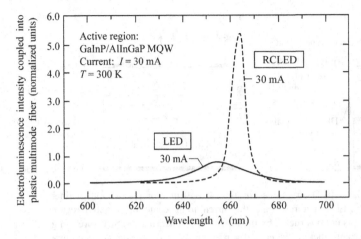

Fig. 23.7. Fiber-coupled ($NA = 0.275$) emission spectrum of RCLED and conventional LED at injection current of 30 mA. The microcavity effect of the RCLED enhances the emission intensity and reduces the emission linewidth, especially for low NA fibers (after Whitaker, 1999).

The emission spectra of an RCLED and a conventional LED are shown in Fig. 23.7. The spectra shown are the spectra of the fiber-coupled light. Inspection of the spectra reveals two features. Firstly, the RCLED exhibits a higher spectral purity thereby reducing chromatic dispersion. Secondly, the fiber-coupled intensity of the RCLED is higher due to the more directed emission pattern. High-speed transmission of 250 Mbit/s over plastic optical fibers has been demonstrated with 650 nm RCLEDs (Streubel and Stevens, 1998).

23.6 Edge-emitting superluminescent diodes (SLDs)

Edge-emitting LEDs are motivated by the need for high-brightness LEDs that allow for high-efficiency coupling to optical fibers. Edge-emitting LEDs comprise an optical waveguide region that guides light emitted along the waveguide by total internal reflection.

Superluminescent light-emitting diodes or *superluminescent diodes* (SLDs) are broad-band high-intensity emission sources that emit incoherent light. Incoherent light does not result in "speckle patterns" obtained from coherent light sources such as lasers. SLDs are suitable as communication devices used with single-mode fibers and also as high-intensity light sources for the analysis of optical components (Liu, 2000).

Light is guided in the *core region* of the waveguide. Total internal reflection occurs at the boundaries between the core region and the upper and lower *cladding layers* as shown in Fig. 23.8. In order to make waveguiding possible, the core layer must have a higher refractive index than the cladding layers. Photons emitted with a sufficiently small angle of incidence at the core–cladding interface will be guided by the waveguide, as indicated in Fig. 23.8.

Since the light is guided by the waveguide, the light intensity emitted by the device linearly increases with the length of the waveguide. Thus, increasing the length of an edge-emitting LED allows one to obtain a higher light output intensity. However, the electrical current required to drive the LED also increases with the stripe length.

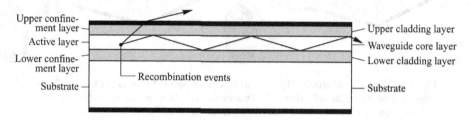

Fig. 23.8. Waveguide geometry showing guided light rays in the core layer with low angles of incidence.

Superluminescent diodes are edge-emitting LEDs that are pumped at such high current levels that *stimulated emission* occurs. In the stimulated emission process, one photon stimulates the recombination of an electron–hole pair and the emission of another photon. The photon created by the stimulated emission process has the same propagation direction, phase, and wavelength as the stimulating photon. Thus SLDs have greater coherence compared with LEDs. In the stimulated emission regime, spontaneous emission towards the top surface of the LED is reduced and emission into waveguide modes is enhanced.

SLDs are quite similar to semiconductor laser structures with one important difference: SLDs lack the optical feedback provided by the reflectors of a semiconductor laser. Two typical SLD structures are shown in Fig. 23.9. The SLD structure shown in Fig. 23.9 (a) has a reflective backside reflector facet; the front-side facet, however, is coated with an anti-reflection (AR) coating. To prevent lasing, the front-side facet must have a reflectivity of $\leq 10^{-6}$ (Liu, 2000; Saul *et al.*, 1985). Exceeding the required reflectivity results in unwanted lasing of the device. Owing to the high-quality AR coating requirement, SLDs with an AR coating are expensive to manufacture.

A lower-cost alternative SLD structure is shown in Fig. 23.9 (b). This structure uses a *lossy region* near the back-side facet of the diode. The lossy region is not covered by the top metal contact and thus is not pumped by the injection current. Practically no feedback occurs if the length of the lossy region is much longer than the absorption length of the core region, i.e.

$$\text{length of lossy region} \gg \alpha^{-1}, \quad (23.1)$$

23 Communication LEDs

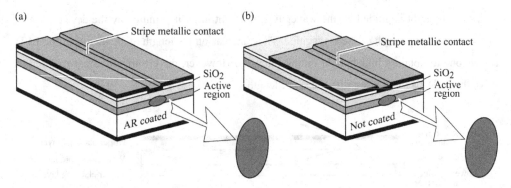

Fig. 23.9. Common structures of superluminescent diodes (SLDs). (a) SLD with cleaved facets coated with anti-reflection (AR) coatings. (b) SLD with cleaved, reflecting facets and stripe contact injecting current over the partial length of the device.

where α is the absorption coefficient of the core region. The absorption coefficient in III–V semiconductors near the band edge is $\alpha \approx 10^4$ cm^{-1}. Thus for lossy-region lengths exceeding several tens of micrometers, the optical feedback from the back-side facet is negligibly small.

In addition to absorption losses, diffraction losses occur in the region not pumped by the electrical current. Gain guiding occurs in the region injected by the electrical current but not in the lossy region. Thus both absorption and diffraction losses prevent this type of SLD from lasing.

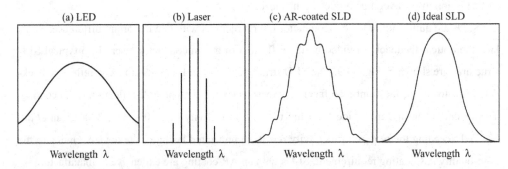

Fig. 23.10. Spectrum of (a) regular LED with a theoretical linewidth of $1.8\,kT$, (b) multi-mode semiconductor laser, (c) superluminescent diode (SLD) fabricated by AR coating of a multimode laser, (d) ideal SLD with linewidth less than kT (after Liu, 2000).

Emission spectra of an LED, SLDs and a laser are shown in Fig. 23.10. The LED has a broad spontaneous emission spectrum. The spectrum of an SLD with a residual small facet reflectivity exhibits periodic oscillations in the emission spectrum due the Fabry–Perot cavity enhancement. An ideal SLD has a smooth spectrum and does not exhibit any oscillations. The spectral width of SLDs is narrower than that of LEDs due to increased coherence caused by stimulated emission. Also shown is the spectrum of a Fabry–Perot laser with several laser modes.

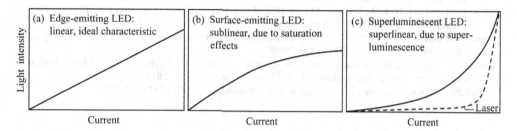

Fig. 23.11. Light-versus-current (*L–I*) characteristic of different LEDs. (a) Edge-emitting LED with little or no saturation effects. (b) Surface-emitting LED with small active area exhibiting saturation effects due to carrier overflow. (c) Superluminescent LED. Also shown is the *L–I* characteristic of a laser that exhibits a distinct threshold current.

A comparison of the *L–I* curves of an LED, an SLD, and a laser is shown in Fig. 23.11. Surface-emitting LEDs with a small light-emitting region diameter, tend to have **sublinear** *L–I* characteristics. At high injection current densities, the small active volume of surface-emitting LEDs is swamped with carriers leading to saturation. Edge-emitting LEDs operating in the spontaneous emission regime have **linear** *L–I* characteristics, as expected for ideal LEDs. SLDs have a **superlinear** *L–I* characteristic due to stimulated emission. In the stimulated emission regime, an increasing number of photons are guided by the waveguide. The number of photons emitted into waveguide modes increases with injection current as stimulated emission becomes dominant. As for SLDs, semiconductor lasers have superlinear emission characteristics. However, the *L–I* curve of lasers exhibits a more distinct threshold than that of SLDs.

References

Burrus C. A. and Miller B. I. "Small-area double heterostructure AlGaAs electroluminescent diode sources for optical fiber transmission lines" *Opt. Commun.* **4**, 307 (1971)

Liu Y. "Passive components tested by superluminescent diodes" February issue of *WDM Solutions* p. 41 (2000)

Mitel Corporation. RCLEDs were first manufactured by the Mitel Corporation. RCLED emitting at 650 nm has Part # 1A466 (2000). Large-scale production of RCLEDs was started in 2001 by Osram Optosemiconductors Corporation, *see* Wirth R., Huber W., Karnutsch C., and Streubel K. "High-efficiency resonant-cavity LEDs emitting at 650 nm" *Compound Semiconductors* **8**, 49 (2002)

Ostermayer Jr. F. W., Kohl P. A., and Burton R. H. "Photoelectrochemical etching of integral lenses on GaInPAs/InP light-emitting diodes" *Appl. Phys. Lett.* **43**, 642 (1983)

Saul R. H., Lee T. P., and Burrus C. A. "Light-emitting-diode device design" in *Lightwave Communications Technology* edited by W. T. Tsang, *Semiconductors and Semimetals* **22** Part C, (Academic Press, San Diego, 1985)

Schubert E. F. and Hunt N. E. J. "Enhancement of spontaneous emission in microcavities" in *Vertical Cavity Surface Emitting Lasers* edited by C. Wilmsen, H. Temkin, and L. A. Coldren (Cambridge University Press, Cambridge, UK, 1999)

Schubert E. F., Wang Y.-H., Cho A. Y., Tu L.-W., and Zydzik G. J. "Resonant-cavity light-emitting diode" *Appl. Phys. Lett.* **60**, 921 (1992)

Schubert E. F., Hunt N. E. J., Micovic M., Malik R. J., Sivco D. L., Cho A. Y., and Zydzik G. J. "Highly efficient light-emitting diodes with microcavities" *Science* **265**, 943 (1994)

Streubel K. and Stevens R. "250 Mbit/s plastic fibre transmission using 660 nm resonant cavity light emitting diode" *Electron. Lett.* **34**, 1862 (1998)

Streubel K., Helin U., Oskarsson V., Backlin E., and Johansson A. "High brightness visible (660 nm) resonant-cavity light-emitting diode" *IEEE Photonics Technol. Lett.* **10**, 1685 (1998)

Whitaker T. "Resonant cavity LEDs" *Compound Semiconductors* **5** (4), 32 (May 1999)

24

LED modulation characteristics

LEDs are the most commonly used light source for local-area communication systems operating from very short (< 1 m) to medium distances (5 km). Typical bit rates are tens of Mbit/s up to about 1 Gbit/s.

LEDs are non-linear devices and as such the series resistance, shunt resistance, and capacitance depend strongly on the applied voltage. A thorough analysis must take into account these non-linearities. However, much can be learned by considering the LED as a linear device, even though some important aspects of the LED modulation behavior cannot be inferred from the linear model. In this chapter, we first analyze LEDs as linear devices and subsequently discuss modulation characteristics, including some non-linear modulation characteristics.

24.1 Rise and fall times, 3 dB frequency, and bandwidth in linear circuit theory

A simple *RC* circuit is shown in Fig. 24.1 (a). When subjected to a step-function input pulse, the output voltage increases according to

$$V_{out}(t) = V_0 [1 - \exp(-t/\tau_1)] \tag{24.1}$$

where $\tau_1 = RC$ is the time constant of the *RC* circuit. When the input voltage returns to zero, the output voltage decreases according to

$$V_{out}(t) = V_0 \exp(-t/\tau_2) . \tag{24.2}$$

For an *RC* circuit, it is $\tau_1 = \tau_2$. The rise and fall times are defined as the time difference between the 10% and 90% points of the voltage, as shown in Fig. 24.1 (b). The rise and fall times of the signal are related to the time constants τ_1 and τ_2 by

$$\tau_r = (\ln 9)\tau_1 \approx 2.2\tau_1 \quad \text{and} \quad \tau_f = (\ln 9)\tau_2 \approx 2.2\tau_2 . \tag{24.3}$$

24 LED modulation characteristics

The *voltage* transfer function $H(\omega)$ is given by

$$H(\omega) = (1 + i\omega\tau)^{-1}. \tag{24.4}$$

The **bandwidth** of the system, Δf, corresponds to the frequency at which the power transmitted through the system is reduced to half of its low-frequency value. This condition can be written as $|H(2\pi f)|^2 = 1/2$. Thus the bandwidth of the RC circuit is given by

$$\Delta f = f_{3dB} = \frac{1}{2\pi\tau} = \frac{\ln 9}{2\pi\tau_r} = \frac{\ln 9}{\pi(\tau_r + \tau_f)} \approx \frac{0.70}{(\tau_r + \tau_f)}. \tag{24.5}$$

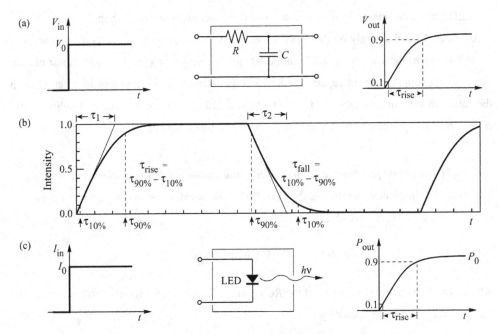

Fig. 24.1. (a) Illustration of the system response of a linear RC system with rise time τ_r. (b) Rise and fall time of a signal with an exponential time dependence and the time constants τ_1 and τ_2. (c) Illustration of the light output power as a function of time for an LED with a rise time of τ_r.

The bandwidth is also called the **3 dB frequency**, since the power transmitted at this frequency is reduced by 3 dB compared with its low-frequency value.

Next, consider an LED with a rise time τ_r as illustrated in Fig. 24.1 (c). As a step-function input current is applied to the LED, the optical output power increases according to

24.2 Rise and fall time in the limit of large diode capacitance

$$P_{\text{out}}(t) = P_0 [1 - \exp(-t/\tau_r)]. \tag{24.6}$$

In analogy to Eqs. (24.1) and (24.4), the *power* transfer function is given by

$$H_{\text{LED}}^2(\omega) = (1 + i\omega\tau)^{-1}. \tag{24.7}$$

The absolute value of the power transfer function is reduced to half at the 3 dB frequency of the LED. Thus the 3 dB frequency of an LED is given by

$$\Delta f = f_{\text{3dB}} = \frac{\sqrt{3}}{2\pi\tau} = \frac{\sqrt{3}\ \ln 9}{2\pi\tau_r} = \frac{\sqrt{3}\ \ln 9}{\pi(\tau_r + \tau_f)} \approx \frac{1.2}{(\tau_r + \tau_f)}. \tag{24.8}$$

Exercise: ***Derivation of equations***. Derive Eqs. (24.3), (24.4), (24.5) and (24.8).

Exercise: ***Rise and fall time and 3 dB frequency***. Consider an LED with a rise time of 1.75 ns. Assume that the fall time of the LED is identical to the rise time. What is the 3 dB frequency of the device? Give the physical reasons as to why Eq. (24.8) gives only an *approximate* value of the 3 dB frequency.
Solution: A 3 dB frequency of 343 MHz is expected on the basis of Eq. (24.8). In practice the 3 dB frequency can be lower or higher than the calculated value since the rise and fall are frequently not exponential. As a practical rule, the numerical factor 1.2 in the numerator of Eq. (24.8) can vary between 1.0 and 1.5.

24.2 Rise and fall time in the limit of large diode capacitance

In diodes used for solid-state lamp applications, the current-injected p-n junction area is large, sometimes as large as the entire LED die. Such diodes have a *large* capacitance. Denoting the diode capacitance as C and the overall series resistance of the drive circuit and the diode as R, the rise and the fall time of the diode are equal and these times are given by the RC time constant.

In communication LEDs, the current-injected active region is much smaller, so that the spontaneous lifetime rather than the diode capacitance limits the maximum modulation frequency. As a result, communication LEDs can be modulated. Since LEDs do not exhibit strictly exponential changes in power, as predicted by Eq. (24.6), Eq. (24.8) is only an approximation.

Consider an LED in which the p-n junction region extends over the entire area of the die. The LED has a small contact area that determines the size of the light-emitting spot. Such LEDs are used for communication applications. At zero bias, the capacitance of the LED is given by the *depletion* capacitance (space-charge capacitance) of the diode. Since the area of the diode is large (e.g. 250 μm × 250 μm), the capacitance is large and can amount to 200–300 pF.

As the diode is turned on, the current *crowds* in the area below the contact. When the p-n junction is forward biased, the capacitance of the LED is given by the *diffusion* capacitance. The relevant area is, however, not the entire LED die but just the region injected with the diode current.

The reduction of the diode capacitance increases the LED modulation bandwidth. The *depletion* capacitance can be reduced by mesa etching. However, the mesa should be larger than the contact size in order to avoid surface recombination effects.

There is no viable way to reduce the *diffusion* capacitance. The diffusion capacitance can be reduced by purposely introducing defects that act as luminescence killers. Such defects reduce the minority carrier lifetime and thereby the diffusion capacitance. Such LEDs can be modulated at several GHz. However, the light output intensity decreases as well so that the overall benefit of such lifetime killers is questionable.

24.3 Rise and fall time in the limit of small diode capacitance

Next, we discuss the rise and fall time in the limit of small diode capacitance. This consideration applies, for example, to surface-emitting communication LEDs that have a small-area active region. Consider an LED driven by a constant current that is switched on at $t = 0$. Electrons are injected into the active region and the carrier concentration builds up. At the same time, the optical output intensity of the LED increases. In the case of the monomolecular recombination model, the light output intensity is directly proportional to the injected minority carrier concentration.

The monomolecular rate equation is given by

$$\frac{I}{eAd} = \frac{dn_a}{dt} = \frac{n_a}{\tau} \tag{24.9}$$

where n_a is the carrier concentration in the active region, A is the current-injected area of the active region, and d is the thickness of the active region. The steady-state current flow of magnitude I causes a steady-state minority carrier concentration $n_a = I\tau/(eAd)$. The mean lifetime of the carriers is the spontaneous recombination lifetime τ.

Next, consider that the diode is initially in the "off" state and, starting at $t = 0$, the diode is injected with a constant current I. When the diode is in the "off" state, the minority carrier concentration in the active region is very low and we approximate the concentration with $n_a \approx 0$.

Solving the differential Eq. (24.9) for the initial condition $n_a \approx 0$ at $t = 0$ for a constant injection current yields that the carrier concentration in the active region increases according to

$$n_a(t) = n_a \left(1 - e^{-t/\tau}\right) = \frac{I\tau}{eAd}\left(1 - e^{-t/\tau}\right). \tag{24.10}$$

The equation reveals that it takes the spontaneous recombination time τ to fill the active region with the steady-state carrier concentration. The light output intensity follows the minority carrier concentration directly. Thus, the rise time is given by the spontaneous recombination time.

A similar consideration applies to the fall time of the diode. Once the diode has been switched off, the decay constant for emission is, of course, the spontaneous recombination lifetime. Thus, the fall time of an LED is given by the spontaneous recombination lifetime.

In the case of an undoped active region, the monomolecular recombination equation no longer applies and the bimolecular recombination equation must be used to describe the carrier dynamics. Also, the carrier lifetime is no longer a constant. In this case, the *shortest* lifetime, i.e. the lifetime that applies when the carrier concentration is at the *highest* level, can be used.

It should be noted that there are methods to *reduce* both the rise and the fall times *below* the limit of the spontaneous recombination lifetime. The rise time can be reduced by *current shaping*. The fall time can be reduced by *carrier sweep-out*. Both methods will be discussed below.

24.4 Voltage dependence of the rise and fall times

The measurement of the rise time and fall time of an LED is shown in Fig. 24.2 (Schubert et al., 1996). As indicated in the figure, the rise and fall times are measured from the 10–90% values of the optical signal. The photocurrent of a p-n junction photodetector is used in the measurement. It must be ensured in the measurement that the rise and fall time of both the pulse generator and the photodiode are much faster than the LED rise and fall time. The measured rise and fall times include the time constant of the pulse generator, the LED, the detector, detector amplifier circuit, and the oscilloscope. However, the time constant of the LED is the longest and hence the dominant time constant. The time constants shown in Fig. 24.2 are *upper limits* to the true time constants of the LED.

24 LED modulation characteristics

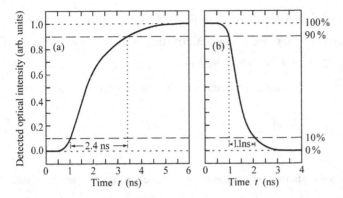

Fig. 24.2. Measured (a) rise time (τ_{rise} = 2.4 ns) and (b) fall time (τ_{fall} = 1.1 ns) of an RCLED. The photodetector used in the measurement is much faster than the LED, so that the measured times are essentially the rise and fall times of the LED.

RCLED T = 300 K
V_{on} = 1.4 V V_{off} = 0.0 V
f_{3dB} = 490 MHz

Inspection of Fig. 24.2 reveals that the rise time is much longer than the fall time. The large difference between the rise and the fall times displayed in Fig. 24.2 is not expected based on the theoretical model discussed above.

To gain a better understanding of the difference between the rise time and the fall time, the times have been measured as a function of the diode bias conditions. In the "on" state, the diode is biased with a voltage of 1.4 V. However, a range of voltages can be chosen for the "off" state, since a p-n junction diode does not emit light for voltages even slightly below the turn-on voltage.

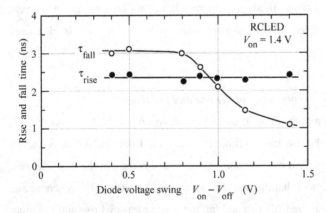

Fig. 24.3. Rise time and fall time as a function of the voltage swing. The fall time of the diode decreases with increasing voltage swing due to sweep-out of carriers out of the active region.

The experimental results are shown in Fig. 24.3. Whereas the "on" voltage is kept constant at V_{on} = 1.4 V, the "off" voltage is varied from 0 to 1.0 V. Inspection of Fig. 24.3 reveals a strong voltage dependence of the fall time. The voltage dependence is caused by carrier sweep-out of the active region. In contrast, the rise time is practically independent of voltage.

24.5 Carrier sweep-out of the active region

The voltage dependence of the fall time shown in Fig. 24.3 can be explained by voltage-dependent carrier sweep-out of the active region. Figure 24.4 shows the active region band diagram in the "off" state for large (a) and small (b) voltage swings. For the case of a small voltage swing, carriers essentially remain in the active region until they recombine. As a result, it will take the spontaneous lifetime for the carriers to recombine and the light intensity to decay.

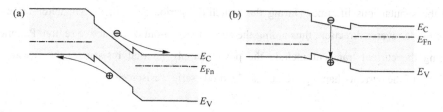

Fig. 24.4. Illustration of two mechanisms determining the fall time. (a) If the diode is, in its off state, at zero or reverse bias, carrier sweep-out of the active region results. The sweep-out time can be very short, << 1 ns. (b) If the modulating voltage amplitude is smaller, that is, if the diode is, in its off state, forward-biased below the threshold voltage, carriers are *not* swept out of the active region, so that the intensity decay is determined by the spontaneous recombination lifetime.

The situation is quite different for large voltage swings. At zero bias, the band diagram of the active region is highly sloped due to the built-in electric field of the p-n junction. As a result, free carriers are swept out of the active region into the neutral n- and p-type confinement regions of the semiconductor. The carrier sweep-out is most efficient for large voltage swings, i.e. when a high electric field is created in the space-charge region of the p-n junction. The sweep-out time can be much shorter than the spontaneous lifetime. Thus, the fall time is determined *not* by the spontaneous recombination lifetime but by the shorter sweep-out time. Considering the magnitude of the built-in electric field and the carrier mobility, the sweep-out time is estimated to be in the picosecond range.

Exercise: *Calculation of carrier sweep-out time*. Calculate the carrier sweep-out time for typical values of the electric field in the p-n junction depletion region, typical carrier velocity, and an active region thickness of 0.1–1 μm.

Solution: The carrier sweep-out time can be very short. For typical diode parameters, the carrier sweep-out time is about 1–100 ps, i.e. much shorter than the spontaneous recombination time. As an example, let us assume that a carrier drifts with the drift-saturation velocity, which is about 10^7 cm/s, across a reverse-biased active region. The time needed to drift across a 1.0 μm thick active region is given by 1.0 μm / 10^7 cm/s = 10 ps.

24.6 Current shaping

A common method to reduce the rise and the fall time is *current shaping* (Lee, 1975; Zucker, 1978; Saul *et al.*, 1985). The diagram of a current-shaping circuit is shown in Fig. 24.5 (a). The current-shaping circuit is essentially a capacitor and a resistor in series with the LED. The capacitor creates a current transient when the LED is switched on or off, as shown in Fig. 24.5 (b). During the switch-on transient, the excess current flowing through the capacitor helps in reaching the steady-state carrier concentration in the active region within a time shorter than the spontaneous lifetime. During the switch-off period ($V = 0$), the capacitor biases the diode in the reverse direction, thus aiding the current sweep-out of the active region. Parameters entering the current transient include the power supply internal resistance, the resistor and capacitor of the current-shaping circuit, and the diode series resistance.

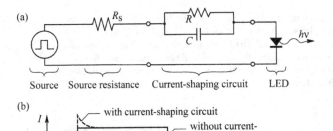

Fig. 24.5. (a) Illustration of an RC "current-shaping circuit" used to decrease the rise time of LEDs. (b) Diode current pulse versus time in the limit of small diode capacitance (solid line) and effect of current-shaping circuit on diode current (dashed line).

A reasonable design criterion for the current-shaping circuit is that the RC time constant of the circuit introduced by the RC current-shaping circuit should be equal to the spontaneous recombination lifetime and that the initial current at the beginning of the voltage pulse should be twice the magnitude of the steady-state current. In this case, the value of the resistor R of the current-shaping circuit is chosen to equal the differential resistance of the diode. The capacitance of the circuit can be chosen so that the RC time constant of the entire circuit coincides with the LED rise time (Schubert *et al.* 1996). As a result of the current-shaping circuit, the 3 dB bandwidth increases.

Note that the diode series resistance is strongly voltage-dependent, and therefore linear circuit theory can only provide *estimates* rather than accurate values. Therefore, experimental or numerical methods are required to optimize the RC current-shaping circuit.

The current-shaping circuit requires an increased operating voltage and this reduces the overall efficiency of the drive circuit. However, the power efficiency is usually of little relevance due to the low overall power consumption of communication LEDs.

24.7 3 dB frequency

The 3 dB frequency of an LED can be determined by measuring the frequency at which the detector signal decreases to one-half of the low-frequency value. The frequency response of the detector needs to be taken into account in the measurement. A 3 dB frequency of about 500 MHz was determined for the LED for which the rise and fall times are shown in Fig. 24.2. Comparison of the 3 dB frequencies for different LED structures revealed that the highest 3 dB frequencies are attained with devices having low parasitic resistances and capacitances. This can be achieved by a small p-type contact area, a thick SiO_2 isolation layer on the p-type side of the device, a small area of the bonding pad on the p-type side of the device, and a mesa etch limiting the p-n junction depletion capacitance at zero bias.

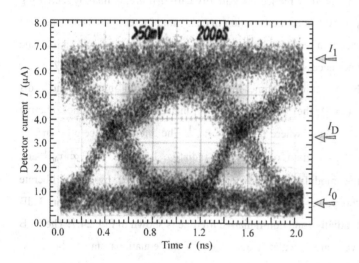

Fig. 24.6. "Eye diagram" of the received optical signal of an RCLED. The optical signal is measured as the photocurrent of a Si photodiode measured with a sampling oscilloscope. The data rate of the RCLED is 622 Mbit/s. Also indicated are the three current levels for the "0" state (I_0), "1" state (I_1), and decision current (I_D).

24.8 Eye diagram

The *eye diagram* allows one to estimate the overall performance of an optical communication system. The eye diagram is the receiver signal of a randomly generated digital signal. An eye diagram of an LED operating at 622 Mbit/s is shown in Fig. 24.6. A data rate of 622 Mbit/s is used in the well-known "synchronous optical network" (SONET) standard. The figure indicates the level of the "1" state and of the "0" state and of the *decision level*, i.e. the boundary between what is interpreted by the receiver as a "0" and "1". The figure also reveals the "eye". An "open eye" such as the one shown in the figure, allows for a low bit-error rate. The "eye" shown in Fig. 24.6 is wide open, indicating that low bit-error-rate data transmission is possible at that frequency using LEDs. The "on" and "off" pulse-generator voltages of the diode were 1.4 and 1.1 V, respectively. A pulse-shaping *RC* circuit with $R = 20\ \Omega$ and $C = 100$ pF was used for the measurement. Minimizing parasitic elements (e.g. bond pad capacitance) and employment of an *RC* pulse shaping circuit should make transmission rates of 1 Gbit/s possible.

As the data rate is increased, the eye will close, i.e. the photocurrent of the "0" and "1" level cannot be clearly distinguished. This results in an increase in the bit-error rate.

24.9 Carrier lifetime and 3 dB frequency

Shortening the minority carrier lifetime through either very high doping of the active region or deliberate introduction of deep traps will increase the maximum modulation frequency. Deep traps have a two-fold effect. Firstly, they reduce the minority carrier lifetime, thereby increasing the 3 dB frequency. Secondly, they reduce the emission intensity and increase the heat generated inside the LED. Very high concentrations of a shallow dopant, on the other hand, will shorten the carrier lifetime and often, but not necessarily, hurt the device efficiency (Ikeda *et al.*, 1977).

We therefore analyze the effect of lifetime reduction on emission intensity and 3 dB frequency. According to Eq. (24.8), the 3 dB frequency of an LED depends on the radiative lifetime according to $f_{3\ \text{dB}} = 3^{1/2}/(2\pi\tau)$, where $\tau^{-1} = \tau_r^{-1} + \tau_{nr}^{-1}$. The internal device efficiency is given by $\eta_{\text{int}} = \tau_{nr}/(\tau_r + \tau_{nr})$. In the limit of small non-radiative lifetimes, $f_{3\ \text{dB}} \propto \tau_{nr}^{-1}$ and $\eta_{\text{int}} \propto \tau_{nr}$. Thus, although the modulation bandwidth *can* be increased by the deliberate introduction of deep traps, the power-bandwidth product *cannot*. The relation between 3 dB frequency, output power, and radiative and non-radiative lifetime is shown in Fig. 24.7. The 3 dB frequencies and intensity levels are calculated according to the equations stated above. At frequencies much higher than the 3 dB frequency, a linear decrease of the optical intensity on the log–log scale can be assumed (Wood, 1994).

If, however, one were to succeed in decreasing the minority carrier lifetime without affecting the efficiency, the modulation speed of an LED could be increased. Although modulation rates as high as 1.7 GHz have been demonstrated for devices having a highly doped active region, the increase in bandwidth has been accompanied by a decrease in device efficiency (Chen *et al.*, 1999).

Chen *et al.* (1999) proposed very high Be doping ($N_A = 2 \times 10^{19}$ to 7×10^{19} cm^{-3}) of a GaAs active region to shorten the lifetime without significantly degrading the internal quantum efficiency. For a test device that was doped at 2×10^{19} cm^{-3}, the authors found a cutoff frequency of 440 MHz and an internal quantum efficiency of 25–30%. The authors found a cutoff frequency of 1.7 GHz and an internal quantum efficiency of 10% for a device that was doped at 7×10^{19} cm^{-3}.

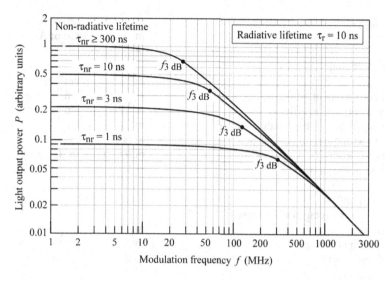

Fig. 24.7. Calculated LED output power versus modulation frequency for different values of the non-radiative lifetime. A radiative lifetime of 10 ns is assumed.

References

Chen C. H., Hargis M., Woodall J. M., Melloch M. R., Reynolds J. S., Yablonovitch E., and Wang W. "GHz bandwidth GaAs light-emitting diodes" *Appl. Phys. Lett.* **74**, 3140 (1999)

Ikeda K., Horiuchi S., Tanaka T., and Susaki W. "Design parameters of frequency response of GaAs-AlGaAs DH LEDs for optical communications" *IEEE Trans. Electron Dev.* **ED-24**, 1001 (1977)

Lee T. P. "Effect of junction capacitance on the rise time of LEDs and on the turn-on delay of injection lasers" *Bell Syst. Tech. J.* **54**, 53 (1975)

Saul R. H., Lee T. P., and Burrus C. A. "Light-emitting-diode device design" in *Lightwave Communications Technology* edited by W. T. Tsang, *Semiconductor and Semimetals*, **22** Part C (Academic Press, San Diego, 1985)

Schubert E. F., Hunt N. E. J., Malik R. J., Micovic M., and Miller D. L. "Temperature and modulation characteristics of resonant-cavity light-emitting diodes" *IEEE J. Lightwave Technol.* **14**, 1721 (1996)

Wood D. *Optoelectronic Semiconductor Devices* p. 98 (Prentice Hall, New York, 1994)

Zucker J. "Closed-form calculation of the transient behavior of (AlGa)As double heterojunction LEDs" *J. Appl. Phys.* **49**, 2543 (1978)

Appendix 1

Frequently used symbols

Symbol	Explanation	SI unit	Common unit
a_0	lattice constant	m	nm or Å
a_B	Bohr radius	m	nm or Å
A	area	m^2	cm^2
$\mathcal{A}$	Einstein $\mathcal{A}$ coefficient	s^{-1}	s^{-1}
α	absorption coefficient	m^{-1}	cm^{-1}
α_0	absorption coefficient at $h\nu = 2E_g$	m^{-1}	cm^{-1}
α_{fc}	free-carrier absorption coefficient	m^{-1}	cm^{-1}
α_g	absorption coefficient at the bandgap energy	m^{-1}	cm^{-1}
B	bimolecular recombination coefficient	m^3/s	cm^3/s
$\mathcal{B}$	Einstein $\mathcal{B}$ coefficient	m^3/(J s^2)	m^3/(J s^2)
c	speed of light in vacuum	m/s	m/s
C	capacitance	F	F
CRI	color-rendering index	–	–
C–V	capacitance-versus-voltage characteristic	–	–
d	layer thickness	m	μm
D	diffusion constant of dopants	m^2/s	cm^2/s
$\mathcal{D}$	dielectric displacement	C/m^2	C/cm^2
DOS	density of states, see N_c, N_v	–	–
D_n, D_p	diffusion constant of electrons, holes	m^2/s	cm^2/s
$\Delta n, \Delta p$	change in electron concentration, hole concentration	m^{-3}	cm^{-3}
$\Delta E_C, \Delta E_V$	band discontinuity of conduction band, valance band	J	eV
ΔE^*	color difference	–	–
e	elementary charge	C	C
$\mathcal{E}$	electric field	V/m	V/cm
E	energy	J	eV
E_0	energy of lowest state of a quantum well	J	eV
E_C	energy of conduction band edge	J	eV
E_F	Fermi energy	J	eV
E_{Fi}	Fermi energy in intrinsic semiconductor	J	eV
E_{Fn}	quasi-Fermi energy in n-type region	J	eV
E_{Fp}	quasi-Fermi energy in p-type region	J	eV
E_g	energy of semiconductor bandgap	J	eV
E_{Ryd}	Rydberg energy	J	eV
E_T	energy of trap or deep level	J	eV
E_V	energy of valence band edge	J	eV
$\varepsilon = \varepsilon_r \varepsilon_0$	dielectric permittivity	A s/(V m)	A s/(V m)
ε_0	absolute dielectric constant	A s/(V m)	A s/(V m)
ε_r	relative dielectric constant	–	–
η	efficiency	%	%
η_{ext}	external quantum efficiency	%	%
$\eta_{extraction}$	extraction efficiency	%	%

Symbol	Description	SI Unit	Common Unit
η_{int}	internal quantum efficiency	%	%
η_{power}	power efficiency	%	%
f	frequency	Hz	Hz
F	finesse of cavity	–	–
ϕ	angle	°	°
ϕ_c	critical angle for total internal reflection	°	°
Φ	electrostatic potential	V	V
Φ	angle	°	°
Φ_{lum}	luminous flux	lm	lm
G	generation rate	m^{-3}/s	cm^{-3}/s
G_0	generation rate in equilibrium	m^{-3}/s	cm^{-3}/s
G_{excess}	excess generation rate	m^{-3}/s	cm^{-3}/s
h	Planck constant	J s	J s or eV s
$\hbar$	Planck constant divided by 2π	J s	J s or eV s
$h\nu$	photon energy	J	eV
I	optical intensity	W/m^2	W/cm^2
I	current	A	A
I_s	reverse saturation current	A	A
I–V	current-versus-voltage characteristic	–	–
J	current density	A/m^2	A/cm^2
J_s	saturation current density	A/m^2	A/cm^2
k	wave number ($2\pi/\lambda$)	m^{-1}	cm^{-1}
k	Boltzmann constant	J/K	J/K or eV/K
L	length	m	m
L_{cav}	cavity length	m	µm
L_n, L_p	diffusion length of electrons, holes	m	µm
L–I	light-output-power-versus-current characteristic	–	–
L^*	CIE uniform color space coordinate of 1986	–	–
λ	wavelength	m	nm
λ_{Bragg}	Bragg reflection wavelength	m	nm
m	mass	kg	kg
m_e	free-electron mass	kg	kg
m^*	effective mass	kg	kg
m_e^*, m_h^*	effective mass of electron, hole	kg	kg
m_{lh}^*, m_{hh}^*	effective mass of light hole, heavy hole	kg	kg
$\mu = \mu_r \mu_0$	magnetic permeability	V s/(A m)	V s/(A m)
μ_0	absolute magnetic constant	V s/(A m)	V s/(A m)
μ_r	relative magnetic constant	–	–
μ_n, μ_p	mobility of electrons, holes	m^2/(V s)	cm^2/(V s)
n	n-type semiconductor material	–	–
n$^-$	lightly doped n-type material	–	–
n$^+$	heavily doped n-type material	–	–
n	electron concentration	m^{-3}	cm^{-3}
n_0	equilibrium electron concentration	m^{-3}	cm^{-3}
n^{2D}	two-dimensional electron concentration	m^{-2}	cm^{-2}
n_{ideal}	diode ideality factor	–	–
n_p	electron concentration in p-type material	m^{-3}	cm^{-3}
$\bar{n}$	refractive index	–	–
$\bar{n}_{air}$	refractive index of air	–	–

$\bar{n}_{gr}$	group index	–	–
$\bar{n}_s$	refractive index of semiconductor	–	–
N_A, N_D	acceptor, donor concentration	m^{-3}	cm^{-3}
N_c	effective density of states at conduction band edge	m^{-3}	cm^{-3}
N_T	concentration of trap or deep level	m^{-3}	cm^{-3}
N_v	effective density of states at valence band edge	m^{-3}	cm^{-3}
ν	frequency of optical radiation	Hz	Hz
p	p-type semiconductor material	–	–
p$^-$	lightly doped p-type material	–	–
p$^+$	heavily doped p-type material	–	–
p	momentum	kg m/s	kg m/s
p	hole concentration	m^{-3}	cm^{-3}
p_0	equilibrium hole concentration	m^{-3}	cm^{-3}
p^{2D}	two-dimensional hole concentration	m^{-2}	cm^{-2}
p_n	hole concentration in n-type material	m^{-3}	cm^{-3}
P	power	W	W
$P(\lambda)$	spectral power density	W/m	W/nm
Q	electrical charge	C	C
Q	cavity quality factor	–	–
r	optical field reflection coefficient	–	–
R	optical power reflection coefficient	–	–
R	resistance	Ω	Ω
R	recombination rate	m^{-3}/s	cm^{-3}/s
R_0	recombination rate in equilibrium	m^{-3}/s	cm^{-3}/s
R_{excess}	excess recombination rate	m^{-3}/s	cm^{-3}/s
R_s	series resistance	Ω	Ω
ρ	charge density	C/m^3	C/cm^3
ρ	resistivity	Ω m	Ω cm
ρ_c	specific contact resistance	Ω m^2	Ω cm^2
ρ_{DOS}^{3D}	three-dimensional density of states	m^{-3}/J	cm^{-3}/eV
ρ_{DOS}^{2D}	two-dimensional density of states	m^{-2}/J	cm^{-2}/eV
S	surface recombination velocity	m/s	cm/s
σ	conductivity	(Ω m)$^{-1}$	(Ω cm)$^{-1}$
t	time	s	s
t	layer thickness	m	μm
T	temperature	K	K or °C
T	optical power transmission coefficient	–	–
T_c	carrier temperature	K	K or °C
T_j	junction temperature	K	K or °C
τ	recombination lifetime	s	s
τ_{cav}	recombination lifetime in a cavity	s	s
τ_n, τ_p	recombination lifetimes of electrons, holes	s	s
u, u'	CIE uniform chromaticity coordinate of 1960, 1976	–	–
u^*	CIE uniform color space coordinate of 1986	–	–
v	velocity	m/s	m/s
v_{gr}	group velocity	m/s	m/s
v_{ph}	phase velocity	m/s	m/s
v, v'	CIE uniform chromaticity coordinate of 1960, 1976	–	–
v^*	CIE uniform color space coordinate of 1986	–	–

V	voltage	V	V
V_D	diffusion voltage	V	V
V_f	forward voltage	V	V
V_{f1}, V_{f2}, V_{f3}	forward voltage measured at a specified current	V	V
V_j	junction voltage	V	V
V_{th}	threshold voltage	V	V
$V(\lambda)$	eye sensitivity function	–	–
W	width	m	m
W_D	depletion layer width	m	µm
W_{DH}	double heterostructure width	m	µm
x	chromaticity coordinate	–	–
$\bar{x}$	CIE color-matching function (red)	–	–
X	CIE tristimulus value (red)	–	–
y	chromaticity coordinate	–	–
$\bar{y}$	CIE color-matching function (green)	–	–
Y	CIE tristimulus value (green)	–	–
$\bar{z}$	CIE color-matching function (blue)	–	–
Z	CIE tristimulus value (blue)	–	–

Note: This list does not contain some symbols that are used only in the section where they are defined.

Appendix 2

Physical constants

a_B	=	0.5292 Å	Bohr radius	($a_B = 0.5292 \times 10^{-10}$ m)
ε_0	=	8.8542×10^{-12} A s/(V m)	absolute dielectric constant	
e	=	1.6022×10^{-19} C	elementary charge	
c	=	2.9979×10^{8} m/s	speed of light in vacuum	
E_{Ryd}	=	13.606 eV	Rydberg energy	
g	=	9.8067 m/s^2	acceleration on earth at sea level due to gravity	
G	=	6.6873×10^{-11} m^3/(kg s^2)	gravitational constant	($F = G M m / r^2$)
h	=	6.6261×10^{-34} J s	Planck constant	($h = 4.1356 \times 10^{-15}$ eV s)
$\hbar$	=	1.0546×10^{-34} J s	$\hbar = h/(2\pi)$	($\hbar = 6.5821 \times 10^{-16}$ eV s)
k	=	1.3807×10^{-23} J/K	Boltzmann constant	($k = 8.6175 \times 10^{-5}$ eV/K)
μ_0	=	1.2566×10^{-6} V s/(A m)	absolute magnetic constant	
m_e	=	9.1094×10^{-31} kg	free electron mass	
N_{Avo}	=	6.0221×10^{23} mol^{-1}	Avogadro number	
$R = k N_{Avo}$	=	8.3145 J K^{-1} mol^{-1}	ideal gas constant	

Note: The *dielectric permittivity* of a material is given by $\varepsilon = \varepsilon_r \varepsilon_0$ where ε_r and ε_0 are the *relative* and *absolute* dielectric constant, respectively. The *magnetic permeability* of a material is given by $\mu = \mu_r \mu_0$ where μ_r and μ_0 are the *relative* and *absolute* magnetic constant, respectively.

Useful conversions

$$1 \text{ eV} = 1.6022 \times 10^{-19} \text{ C V} = 1.6022 \times 10^{-19} \text{ J}$$
$$E = h\nu = hc/\lambda = 1239.8 \text{ eV} / (\lambda / \text{nm})$$
$$kT = 25.86 \text{ meV} \quad (\text{at } T = 300 \text{ K})$$
$$kT = 25.25 \text{ meV} \quad (\text{at } T = 20 \text{ °C} = 293.15 \text{ K})$$

Appendix 3

Room temperature properties of semiconductors: III–V arsenides

Quantity	Symbol	AlAs	GaAs	InAs	(Unit)
Crystal structure		Z	Z	Z	–
Gap: Direct (D) / Indirect (I)		I	D	D	–
Lattice constant	$a_0 =$	5.6611	5.6533	6.0584	Å
Bandgap energy	$E_g =$	2.168	1.42	0.354	eV
Intrinsic carrier concentration	$n_i =$	10	2×10^6	7.8×10^{14}	cm^{-3}
Effective DOS at CB edge	$N_c =$	1.5×10^{19}	4.4×10^{17}	8.3×10^{16}	cm^{-3}
Effective DOS at VB edge	$N_v =$	1.7×10^{19}	7.7×10^{18}	6.4×10^{18}	cm^{-3}
Electron mobility	$\mu_n =$	200	8500	33 000	cm^2/Vs
Hole mobility	$\mu_p =$	100	400	450	cm^2/Vs
Electron diffusion constant	$D_n =$	5.2	220	858	cm^2/s
Hole diffusion constant	$D_p =$	2.6	10	12	cm^2/s
Electron affinity	$\chi =$	3.50	4.07	4.9	V
Minority carrier lifetime	$\tau =$	10^{-7}	10^{-8}	10^{-8}	s
Electron effective mass	$m_e^* =$	$0.146\, m_e$	$0.067\, m_e$	$0.022\, m_e$	–
Heavy hole effective mass	$m_{hh}^* =$	$0.76\, m_e$	$0.45\, m_e$	$0.40\, m_e$	–
Relative dielectric constant	$\varepsilon_r =$	10.1	13.1	15.1	–
Refractive index near E_g	$\bar{n} =$	3.2	3.4	3.5	–
Absorption coefficient near E_g	$\alpha =$	10^3	10^4	10^4	cm^{-1}

- D = Diamond. Z = Zincblende. W = Wurtzite. DOS = Density of states. CB = Conduction band. VB = Valence band.
- The Einstein relation relates the diffusion constant and mobility in a non-degenerately doped semiconductor: $D = \mu\,(kT/e)$
- Minority carrier diffusion lengths are given by $L_n = (D_n \tau_n)^{1/2}$ and $L_p = (D_p \tau_p)^{1/2}$
- The mobilities and diffusion constants apply to low doping concentrations ($\approx 10^{15}\ cm^{-3}$). As the doping concentration increases, mobilities and diffusion constants decrease.
- The minority carrier lifetime τ applies to doping concentrations of $10^{18}\ cm^{-3}$. For other doping concentrations, the lifetime is given by $\tau = B^{-1}(n+p)^{-1}$, where $B_{GaAs} = 10^{-10}\ cm^3/s$.

Appendix 4

Room temperature properties of semiconductors: III–V nitrides

Quantity	Symbol	AlN	GaN	InN	(Unit)
Crystal structure		W	W	W	–
Gap: Direct (D) / Indirect (I)		D	D	D	–
Lattice constant	$a_0 =$	3.112	3.191	3.545	Å
	$c_0 =$	4.982	5.185	5.703	Å
Bandgap energy	$E_g =$	6.28	3.425	0.77	eV
Intrinsic carrier concentration	$n_i =$	9.4×10^{-34}	1.9×10^{-10}	920	cm^{-3}
Effective DOS at CB edge	$N_c =$	6.2×10^{18}	2.3×10^{18}	9.0×10^{17}	cm^{-3}
Effective DOS at VB edge	$N_v =$	4.9×10^{20}	1.8×10^{19}	5.3×10^{19}	cm^{-3}
Electron mobility	$\mu_n =$	300	1500	3200	cm^2/Vs
Hole mobility	$\mu_p =$	14	30	–	cm^2/Vs
Electron diffusion constant	$D_n =$	7	39	80	cm^2/s
Hole diffusion constant	$D_p =$	0.3	0.75	–	cm^2/s
Electron affinity	$\chi =$	1.9	4.1	–	V
Minority carrier lifetime	$\tau =$	–	10^{-8}	–	s
Electron effective mass	$m_e^* =$	$0.40\, m_e$	$0.20\, m_e$	$0.11\, m_e$	–
Heavy hole effective mass	$m_{hh}^* =$	$3.53\, m_e$	$0.80\, m_e$	$1.63\, m_e$	–
Relative dielectric constant	$\varepsilon_r =$	8.5	8.9	15.3	–
Refractive index near E_g	$\bar{n} =$	2.15	2.5	2.9	–
Absorption coefficient near E_g	$\alpha =$	3×10^5	10^5	6×10^4	cm^{-1}

- D = Diamond. Z = Zincblende. W = Wurtzite. DOS = Density of states. CB = Conduction band. VB = Valence band.
- The Einstein relation relates the diffusion constant and mobility in a non-degenerately doped semiconductor: $D = \mu\,(kT/e)$
- Minority carrier diffusion lengths are given by $L_n = (D_n \tau_n)^{1/2}$ and $L_p = (D_p \tau_p)^{1/2}$
- The mobilities and diffusion constants apply to low doping concentrations ($\approx 10^{15}$ cm^{-3}). As the doping concentration increases, mobilities and diffusion constants decrease.
- The minority carrier lifetime τ applies to doping concentrations of 10^{18} cm^{-3}. For other doping concentrations, the lifetime is given by $\tau = B^{-1}(n+p)^{-1}$, where $B_{GaN} \approx 10^{-10}$ cm^3/s.

Appendix 5

Room temperature properties of semiconductors: III–V phosphides

Quantity	Symbol	AlP	GaP	InP	(Unit)
Crystal structure		Z	Z	Z	–
Gap: Direct (D) / Indirect (I)		I	I	D	–
Lattice constant	$a_0 =$	5.4635	5.4512	5.8686	Å
Bandgap energy	$E_g =$	2.45	2.26	1.35	eV
Intrinsic carrier concentration	$n_i =$	0.044	1.6×10^0	1×10^7	cm^{-3}
Effective DOS at CB edge	$N_c =$	2.0×10^{19}	1.9×10^{19}	5.2×10^{17}	cm^{-3}
Effective DOS at VB edge	$N_v =$	1.5×10^{19}	1.2×10^{19}	1.1×10^{19}	cm^{-3}
Electron mobility	$\mu_n =$	60	110	4600	cm^2/Vs
Hole mobility	$\mu_p =$	450	75	150	cm^2/Vs
Electron diffusion constant	$D_n =$	1.6	2.8	120	cm^2/s
Hole diffusion constant	$D_p =$	11.6	1.9	3.9	cm^2/s
Electron affinity	$\chi =$	3.98	3.8	4.5	V
Minority carrier lifetime	$\tau =$	10^{-6}	10^{-6}	10^{-8}	s
Electron effective mass	$m_e^* =$	0.83 m_e	0.82 m_e	0.08 m_e	–
Heavy hole effective mass	$m_{hh}^* =$	0.70 m_e	0.60 m_e	0.56 m_e	–
Relative dielectric constant	$\varepsilon_r =$	9.8	11.1	12.4	–
Refractive index near E_g	$\bar{n} =$	3.0	3.0	3.4	–
Absorption coefficient near E_g	$\alpha =$	10^3	10^3	10^4	cm^{-1}

- D = Diamond. Z = Zincblende. W = Wurtzite. DOS = Density of states. CB = Conduction band. VB = Valence band.
- The Einstein relation relates the diffusion constant and mobility in a non-degenerately doped semiconductor: $D = \mu (kT/e)$.
- Minority carrier diffusion lengths are given by $L_n = (D_n \tau_n)^{1/2}$ and $L_p = (D_p \tau_p)^{1/2}$.
- The mobilities and diffusion constants apply to low doping concentrations ($\approx 10^{15}$ cm^{-3}). As the doping concentration increases, mobilities and diffusion constants decrease.
- The minority carrier lifetime τ applies to doping concentrations of 10^{18} cm^{-3}. For other doping concentrations, the lifetime is given by $\tau = B^{-1} (n+p)^{-1}$, where $B_{GaP} \approx 10^{-13}$ cm^3/s.

Appendix 6

Room temperature properties of semiconductors: Si and Ge

Quantity	Symbol	Si	Ge	(Unit)
Crystal structure		D	D	–
Gap: Direct (D) / Indirect (I)		I	I	–
Lattice constant	$a_0 =$	5.43095	5.64613	Å
Bandgap energy	$E_g =$	1.12	0.66	eV
Intrinsic carrier concentration	$n_i =$	1.0×10^{10}	2.0×10^{13}	cm^{-3}
Effective DOS at CB edge	$N_c =$	2.8×10^{19}	1.0×10^{19}	cm^{-3}
Effective DOS at VB edge	$N_v =$	1.0×10^{19}	6.0×10^{18}	cm^{-3}
Electron mobility	$\mu_n =$	1500	3900	cm^2/Vs
Hole mobility	$\mu_p =$	450	1900	cm^2/Vs
Electron diffusion constant	$D_n =$	39	101	cm^2/s
Hole diffusion constant	$D_p =$	12	49	cm^2/s
Electron affinity	$\chi =$	4.05	4.0	V
Minority carrier lifetime	$\tau =$	10^{-6}	10^{-6}	s
Electron effective mass	$m_e^* =$	$0.98\, m_e$	$1.64\, m_e$	–
Heavy hole effective mass	$m_{hh}^* =$	$0.49\, m_e$	$0.28\, m_e$	–
Relative dielectric constant	$\varepsilon_r =$	11.9	16.0	–
Refractive index near E_g	$\bar{n} =$	3.3	4.0	–
Absorption coefficient near E_g	$\alpha =$	10^3	10^3	cm^{-1}

- D = Diamond. Z = Zincblende. W = Wurtzite. DOS = Density of states. CB = Conduction band. VB = Valence band.
- The Einstein relation relates the diffusion constant and mobility in a non-degenerately doped semiconductor: $D = \mu\,(kT/e)$.
- Minority carrier diffusion lengths are given by $L_n = (D_n \tau_n)^{1/2}$ and $L_p = (D_p \tau_p)^{1/2}$.
- The mobilities and diffusion constants apply to low doping concentrations ($\approx 10^{15}$ cm^{-3}). As the doping concentration increases, mobilities and diffusion constants decrease.
- The minority carrier lifetime τ applies to doping concentrations of 10^{18} cm^{-3}. For other doping concentrations, the lifetime is given by $\tau = B^{-1}(n+p)^{-1}$, where $B_{Si} \approx 5 \times 10^{-14}$ cm^3/s, $B_{Ge} \approx 5 \times 10^{-13}$ cm^3/s.

Appendix 7

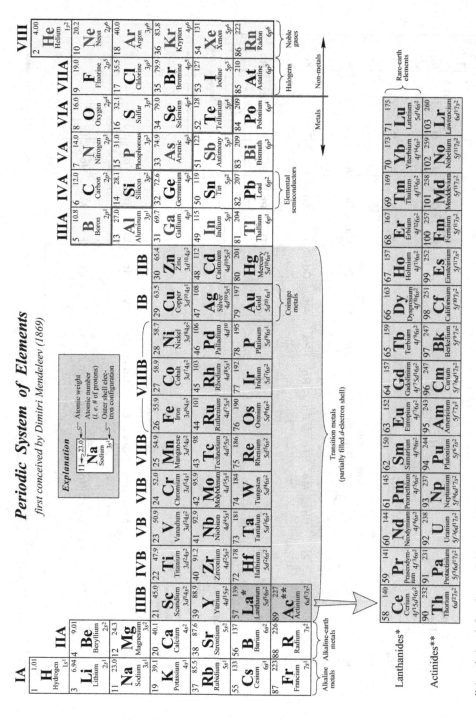

Appendix 8

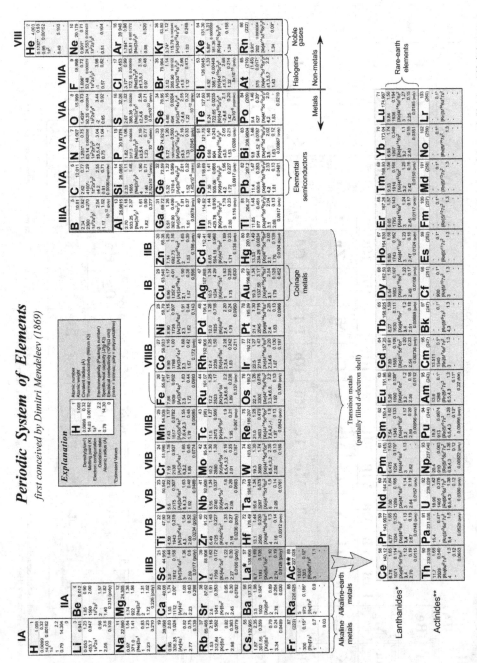

Index

absorption, 50–53, 145–149, 259
    coefficient, 53, 146, 148
    free carrier, 147–149
absorption constant, *see* absorption, coefficient
acceptance angle, *see* numerical aperture
active region, 71, 75–84, 149, 232
active region thickness, 115
adaptive color shift, *see* color vision
adaptive color vision, *see* color vision
additive color mixing, *see* color mixing
adiabatic transport, 75, 83
AlAs, 130
AlGaAs, 4–8, 206–209
AlGaInN, 222–236
AlGaInP, 19–21, 209–211, 265–268
AlGaN, 222–236
AlN, 107, 223
Al_2O_3 and Al_xO_y, 160, 180
alumina, *see* Al_2O_3
anneal, post-growth, 17
anti-reflection coating, 159, 160, 385
attenuation, *see* fiber loss
Auger recombination, *see* recombination
avalanche multiplication, 3, 17

bandgap bowing, 73, 223, 224
bandgap, direct, *see* direct–indirect transition
bandgap discontinuity, 113
bandgap energy, 107, 223
    temperature coefficient, 105–108
    temperature dependence, 105–108
bandgap, indirect, *see* direct–indirect transition
bandwidth, 393–395
bimolecular recombination, *see* recombination
binning, 358
black-body radiation, *see* planckian source
blocking layer, *see* electron-blocking layer, *see also* current-blocking layer
Boltzmann statistics, 61
bowing, *see* bandgap bowing
Bragg wavelength, 173, 174, 179
brightness, 286, 287, 302
buffer layer, 9
built-in voltage, *see* diffusion voltage
Burrus-type LED, *see* LED

CaF_2, 160
candela, definition, 277, 280
candle, 277
candlepower, definition, 277

capacitance, 395–397
carborundum, 1, 2
carrier leakage, 75–83, 121, 122
carrier lifetime, *see* lifetime
carrier loss, *see* carrier leakage
carrier overflow, *see* carrier leakage
carrier sweep-out, 399
carrier temperature, *see* temperature
cathodo-luminescence, 124
cavity quality factor, 243
CdS, 4
characteristic temperature, *see* temperature
chip shaping, *see* die shaping
chromatic adaptation, 324
chromaticity coordinates, 294, 297, 298, 299
    uniform, 297–299
chromaticity diagram, 292–302, 308, 309, 314, 320
    and LEDs, 301, 302, 315
CIE, 280–283
CIE illuminant, 308
CIE reference object, *see* test-color sample
circadian cycle, *see* circadian rhythm
circadian rhythm, 287, 288
circular contact, *see* contact
cladding layer, *see also* confinement layer
    of fiber, 367, 368, 371
    of planar waveguide, 388, 389
cluster, 230
coherence, 389
coherent strain, *see* strain
color, 302
color and wavelength of LED, *see* LED
color difference, 299
color, dominant, *see* wavelength, dominant
color gamut, 314, 315
color-matching function, 292–300, 304, 305
color mixing, 313–315
color, primary, 315
color purity, 300, 301
color rendering, 21, 313–326
    index, 315–326, 340, 341, 355
        general, 315–326
        special, 315–326
color saturation, *see* color purity
color space, uniform, 321
color temperature, 306–311, 343, 352
    correlated, 309–311
color vision, 292–311
    adaptive, 324

415

Index

colorimetry, 292–302
communication, *see* optical communication
communication LED, *see* LED
compensation, 5, 59
complementary wavelength, 333–335
composition grading, 71–75, 82, 83
concentration, critical, 119
cone cells, *see* eye
confinement layer, 75–83, 113–115, 119–122, 149
conformal phosphor distribution, *see* phosphor distribution
conservation of radiance theorem, 189
contact,
    circular, 135, 136
    geometry, 156
    interdigitated, 140, 234
    reflectivity, 164–168
    stripe, 133–135
    transparent, 167, 168
core of fiber, 367, 368, 371
core region, 388
correlated color temperature, *see* color temperature
coupling, *see* fiber coupling
cracking, 235, 236
CRI, *see* color-rendering index
critical angle for DBR reflectance, 178–180
critical angle for total internal reflection, 91, 168
    see also total internal reflection
critical concentration, *see* concentration
critical thickness, *see* thickness
cross-hatch pattern, 124
current-blocking layer, 142, 143, 386, 387
current crowding, 135–140
current densities, typical, 63
current flow, 127–143
current shaping, 400, 401
current spreading
    layer, 5, 20, 127–136, 157
    length, 135, 138–140
current–voltage characteristic, 59–69, 83, 101–111

dangling bonds, 41, 124,
data rate, *see* maximum data rate
DBR, *see* distributed Bragg reflector
decay of luminescence, *see* luminescence
deep levels, 35–39, 44, 45, 117, 227–231
defect, 35–39, 44, 45, 117, 227–231
depletion region, 59, 60, 65, 73
dichromatic source, 333–337
die shaping, 150–153
diffuse reflector, *see* reflector

diffuse surface, *see* surface
diffuser, 98, 198, 355
diffusion,
    capacitance, 396
    length, 69
    of impurities, 4, 118, 119, 234
    voltage, 60–62
diode equation, *see* Shockley equation
diode ideality factor, *see* ideality factor
diode voltage, *see* forward voltage
direct gap, *see* direct–indirect transition
direct–indirect transition, 8–11, 20, 38, 130, 207, 210
dislocation, *see* misfit dislocation
disorder parameter, 34
dispersion,
    material, 372–374
    modal, 371, 372, 374
distributed Bragg reflector, *see* reflector
dominant color, *see* color
dominant wavelength, *see* wavelength
donor–acceptor pair recombination, *see* recombination
doping activation, 17, 18, 226, 227
doping, GaN p-type, 17, 18, 226, 227
doping of active region, 116–119
doping of confinement layer, 119–122
doping, using superlattice, 227
double heterostructure, *see* heterostructure
drive circuit, 110, 111
dye, 349–351, 364, 365

edge-emitting LED, *see* LED
efficacy, luminous, 283–286
efficiency,
    coupling, *see* fiber coupling
    external, quantum, 87, 145, 217
    internal, quantum, 45, 46, 86, 87, 113–126, 145
    light extraction, 86, 87, 97, 98, 145–160,
    luminous, 214–216, 284–286, 361, 362, 364
    power, 4, 87
    quantum, *see* efficiency, internal or external
    radiative, 56
    wall-plug, *see* efficiency, power
wavelength converter, 347–349
Einstein A coefficient, 57
Einstein B coefficient, 57
Einstein model, 56, 57
Einstein relation, 69
electroluminescence, 1, 16
electron-beam irradiation, 17
electron-blocking layer, 81–83
electron–hole recombination, *see* recombination

electrostatic discharge, 193–195
emission
    by avalanche multiplication, 2, 3
    by minority carrier injection, 2, 3
    energy, 69, 90, 103
    enhancement, 248–253
    linewidth, 90
    pattern, 93–97
    pattern, lambertian, 94–97
    spectrum, 87–91, 217, 263, 354, 363, 384, 385, 390
    spectrum, gaussian, 249, 334, 335
    temperature dependence, 54–56, 98–100, 111, 218
    wavelength, *see* emission energy
encapsulant, 97, 98, 196–199
    epoxy resin, 97, 196–198
    graded-index, 198
    PMMA, 198
    silicone, 198
energy gap, *see* bandgap energy
epoxy resin dome, *see* encapsulant
epoxy resin encapsulant, *see* encapsulant
equal-energy point, 295
escape cone, *see* light escape cone
ESD, *see* electrostatic discharge
excitation,
    low-level, 28–32
    high-level, 32, 33
exponential decay, 29–31
extended defect, *see* defect
external quantum efficiency, *see* efficiency
extraction, *see* light extraction
extraction efficiency, *see* efficiency
eye
    diagram, 401, 402
    ganglion cell, 287–288
    human, 275–288
    retina, 275
    retinal cone cell, 275
    retinal ganglion cell, 287
    retinal rod cell, 275
    sensitivity, 275–278
        function, 280–282, 290, 291

Fabry–Perot resonators, 155, 241–248
fall time, 393–401
far-field pattern, *see* emission pattern
feedback, *see* optical feedback
Fermi's Golden Rule, 48
fiber
    coupling, 376–378
    graded-index multimode, 367, 368
    loss, 369, 370
    plastic, 369, 370
    PMMA, 370
    silica, 369, 370
    single-mode, 367, 368
    step-index multimode, 367, 368
finesse of cavity, 243
flip-chip packaging, 160
forward voltage, 59–69, 83, 84, 104–111, 158, 219, 236, *see also* threshold voltage
    temperature coefficient, 107–110
    temperature dependence, 104–111, 219, 220
    one, V_{f1}, 67
    two, V_{f2}, 67
    three, V_{f3}, 67
free-carrier absorption, *see* absorption
free-space communication, *see* communication
free spectral range, 243
Fresnel reflection, *see* reflection

GaAs, 4–8, 53, 106, 107, 148, 206–209
GaAsP, 5–15, 201–206
GaAsP:N, 10–15, 201–206
gain, *see* optical gain
GaInAs, 260–265
GaInN, 15–19, 211–213, 231–233
GaInPAs, 118
gamut, *see* color gamut
GaN, 15–19, 53, 106, 148, 223
ganglion cell, *see* eye
GaP, 10–15, 18, 20, 53, 106, 107, 128, 148, 157, 201–206
GaP:N, 10–15, 18, 201–206
GaSb, 107
gaussian emission spectrum, *see* emission
Ge, 53, 106, 107
general illumination, 332
generation current, 65
graded-index encapsulant, *see* encapsulant
graded-index fiber, *see* fiber
grading, *see* composition grading
group refractive index, *see* refractive index

heatsink, 195, 196, 236
Heisenberg uncertainty principle, *see* uncertainty principle
hemispherical LED, *see* LED
heterostructure, 70–81, 113–126, 149, 207, 208, 234
high-level excitation, *see* excitation
history of LEDs, *see* LED
homojunction, 69, 70, 114
hue, 302
human eye, *see* eye
human vision,

linearity, 286–287
mesopic, 276
photopic, 276, 281, 282
scotopic, 276, 281

ideal isotropic emitter, *see* isotropic emitter
ideality factor, 63, 64, 105, 133
illuminance, 278
illuminant, *see* CIE illuminant
impact ionization, *see* avalanche multiplication
implantation, 266, 387
impurity
    diffusion, *see* diffusion
    isoelectronic, 10–12, 14, 201–206
    optically active, 10–12, 14, 256, 351, 352
    rare-earth, 351, 352
InAs, 107
incandescence, 1, 3, 311
indirect gap, *see* direct–indirect transition
infrared communication LEDs, *see* LED
InGaN, *see* GaInN
injection laser, *see* laser
InN, 107, 213, 223, 224
InP, 53, 106, 107, 148
InSb, 107
interdigitated contact, *see* contact
internal quantum efficiency, *see* efficiency
isoelectronic impurity, *see* impurity
isotropic emitter, 95, 96
    ideal, 264

joint density of states, 89
joint dispersion relation, 89
junction displacement, 118, 119
junction temperature, *see* temperature

Kohlrausch decay, 34, 35

Lambert's cosine law, 185
lambertian emission pattern, *see* emission pattern
laser
    blue, 18
    injection, 4, 5, 8, 18, 20
    microdisk, 170, 273
    vertical-cavity surface-emitting, 143, 180, 259, 260
    zero-threshold, 240, 270
lateral injection, 140–142
lattice match and mismatch, 75, 123–126, 129, 202
leadframe, 192
leakage, *see* carrier leakage
LED

Burrus-type, 383, 384
color and wavelength of, 284
communication, 382–391
edge-emitting, 388–391
hemispherical, 96, 150
history of, 1–23
infrared communication, 382–391
monochromatic nature of, 90
organic, 93, 98, 256, 272
parabolic-shaped, 96
pedestal-shaped, 152
photon-recycling, 268–270
photon-recycling semiconductor, 359–364
photonic crystal, 272, 273
planar, 96
polymer, *see* LED, organic
resonant-cavity, 7, 240, 255–273, 387, 388
single-mode, 240
superluminescent, 388–391
surface-emitting communication, 383–388
ultraviolet, 231–236, 231–236
visible-spectrum, 6–22, 201–220
lens, 376–378
lifetime, *see also* reliability
    carrier, 29–46, 57, 60, 61, 70, 90, 117, 402–403
light-emitting diode, *see* LED
light escape cone, 91–93, 150-152
light extraction, 86, 87, 235, *see also* extraction
light, trapped, 86, 150
lightness, 302
light-versus-current characteristic, 391
linewidth, *see* emission linewidth
lorentzian lineshape, 250
loss, *see* absorption, carrier loss, carrier leakage, fiber loss
low-level excitation, *see* excitation
lumen, definition, 278
luminance, 185, 186, 278, 279, 280
luminescence decay, 33–35
luminescence killer, *see* defect
luminous efficacy, *see* efficacy
luminous efficiency, *see* efficiency
luminous flux, 216, 278, 280, 284
luminous intensity, 277, 280
lux, definition, 278

MacAdam ellipses, 297–300
material dispersion, *see* dispersion
Matthews–Blakeslee law, *see* critical thickness
maximum data rate, 380
maximum modulation frequency, 395, 402
mesa, 122
mesopic vision, *see* human vision

metal mirror, *see* reflector
metal reflector, *see* reflector
metal–insulator–semiconductor emitter, 16
metal–semiconductor diode, *see* Schottky diode
metal–semiconductor emitter, *see* Schottky diode
microcavity, *see* resonant cavity
minimum flicker method, 282
minority carrier lifetime, *see* lifetime
mirror, *see* reflector
mirror loss, 166
MIS, *see* metal–insulator–semiconductor
misfit dislocation, 75, 123–126, 129, 202, 212, 227–231
mixing of light, *see* color mixing
modal dispersion, *see* dispersion
modes, *see* optical modes
modulation characteristics, 393–403
momentum of carrier and photon, 88, 89
monochromatic nature of LEDs, *see* LED
monomolecular recombination, *see* recombination
MQW, *see* multiple quantum well
multimode fiber, *see* fiber
multipath delay, *see* multipath distortion
multipath distortion, 379, 380
multiple quantum well, *see* quantum well
Munsell color system, 317
Munsell test-color sample, *see* test-color sample

native defect, *see* defect
noise, 380
non-adiabatic injection, 83, 84
non-exponential decay, 32–35
    see also Kohlrausch decay
non-radiative recombination, *see* recombination
numerical aperture, 374–376

ODR, *see* omnidirectional reflector
omni-directional reflector, *see* reflector
optical communication, 265, 367–380
    fiber, 367–377, 382, 383
    free space, 378–380, 382
    window, 369, 370
optical density, 169
optical feedback, 389, 390
optical fiber, *see* fiber
optical gain, 390
optical mode, 241–249, 258
optical mode density, 244–248, 258
optical power, 277
optically active impurity, *see* impurity

packaging, 191–199

parabolic LED, *see* LED
parallel resistance, *see* resistance
parallelepiped, 151, 152
parasitic diode, 67
parasitic resistance, *see* resistance
PBG, *see* photonic bandgap
PC, *see* photonic crystal
pedestal-shaped LED, *see* LED
penetration depth, 175
pentachromatic, 344
phase refractive index, *see* refractive index
phonon emission, 83
phosphor, 349–359
    conformal distribution, 356, 357
    proximate distribution, 356, 357
    quantum splitting, 347, 348
    remote distribution, 356, 357
    TAG, 353
    YAG, 351–353
phosphorescence, 350, 353, 354
photocurrent, 65, 66
photometric units, 277–280
photonic bandgap, *see* photonic crystal
photonic crystal, 272, 273
photon recycling, 268–270
photon-recycling semiconductor LED, *see* LED
photopic vision, *see* human vision
planar LED, *see* LED
Planck black-body radiation, *see* planckian source
planckian locus, 308–310
planckian source, 51, 52, 306–311
plastic optical fiber, *see* fiber
PMMA encapsulant, *see* encapsulant
PMMA fiber, *see* fiber
p-n junction, 59–63
p-n junction displacement, *see* junction displacement
point defect, *see* defect
polarization effect, 224–226
polymer encapsulant, *see* encapsulant
polymer light-emitting diode, *see* LED
post-growth anneal, *see* anneal
power efficiency, *see* efficiency
power ratio of dichromatic source, 334
premature turn-on, *see* sub-threshold turn on
primary color, *see* color
proximate phosphor, *see* phosphor distribution
p-type doping of GaN, 16–19

quality factor, *see* cavity quality factor
quantum dot, 216
quantum efficiency, *see* efficiency

quantum well, 20, 33, 79–84, 114, 115, 225, 226, 232, 233, 336

radiance, 184–186
radiant flux, *see* optical power
radiation pattern, *see* emission pattern
radiative recombination, *see* recombination
radiometric units, 277–280
rare-earth impurity, *see* impurity
RC time constant, 400
RC time circuit, 400
RCLED, *see* resonant-cavity light-emitting diode
recombination
    Auger, 40, 44
    bimolecular, 28–35, 53–56, 71,
    bimolecular, coefficient, 31, 32, 53–56
    current in diode, 65
    donor–acceptor pair, 336
    doping dependence, 54–56,
    electron-hole, 27–46, 51, 65, 71, 87
    lifetime, *see* lifetime
    monomolecular, 30–32
    non-radiative, 27–46, 65, 75, 122, 229–231
    radiative, 27–46, 48–57, 71, 229–231
    rate 48–57
    Shockley–Read, 35–39, 65, 77
    spontaneous, 27–46, 48–57, 239–253, 239–253, 244, 245
    stimulated, 48, 56, 57, 239, 389
    surface, 41–44, 122, 123, 131
    surface, velocity, 42, 43
    temperature dependence, 54–56
reduced mass, 89
reference illuminant, *see* reference source
reference light source, *see* reference source
reference object, *see* test-color sample
reference source, 316, 321
reflection,
    Fresnel, 97, 159, 164, 165, 385
    metal, 166
    resonant cavity, 262
    total internal, 86, 91, 97, 98
reflector, 163–189
    cup, 192
    diffuse, 184–189
    distributed Bragg (DBR), 20, 170–181
    lambertian, *see* reflector, diffuse
    loss, *see* mirror loss
    metal, 164–168, 260, 261
    mixed diffuse-specular, 187, 188
    omnidirectional (ODR), 181–184
    specular, 184–189
    total internal, 168–170

refractive index, 51–53, 165
    effective, 173, 174
    group, 372, 373
    phase, 372, 373
regrowth, 142, 143
reliability, 209
remote phosphor distribution, *see* phosphor distribution
resistance
    contact, 138, 234
    DBR, 180, 181
    ohmic, 63–69, 83
    parallel, 63–69
    series, 63–69, 71–75
    series, evaluation, 67–69
    thermal, 195, 196
resonant-cavity LED, *see* LED
resonant-cavity structure, 239–253, 255–273
resonant cavity, tunable, 272
retina, *see* eye
reverse saturation current, *see* saturation current
rise time, 393–398
rod cells, *see* eye
roughened surface, *see* surface texturing

saturation current, 61, 67
saturation of color, *see* color purity
Schottky diode, 2, 3, 15–17, 195
scotopic vision, *see* human vision
self-absorption, *see* absorption
series resistance, *see* resistance
shaping of die, *see* die shaping
Shockley equation, 60, 61, 64, 65
Shockley–Read recombination, *see* recombination
shunt, *see* resistance
Si, 53, 106, 107
SiC, 1–4
silica, *see* SiO_2
silica fiber, *see* fiber
silicon carbide, *see* SiC
silicone encapsulant, *see* encapsulant
silver mirror, *see* metal mirror
Si_3N_4, 160
single-mode fiber, *see* fiber
single-mode LED, *see* LED
SiO_2, 160, 177
Snell's law, 91, 94, 145, 150, 168, 375
solar blind range, 222
solar spectrum, 306–307
solder-bump bonding, 160,
solid-state lighting, 332–365
solubility of nitrogen, 205
sorting, *see* binning

spectral purity, 255, 264, 388
spectrum, *see* emission spectrum
specular reflector, *see* reflector
spontaneous emission, *see* recombination
spontaneous emission enhancement, *see* emission enhancement
spontaneous emission lifetime, *see* carrier lifetime
spontaneous recombination, *see* recombination
step-index fiber, *see* fiber
stimulated emission, *see* recombination
stop band, 170, 172, 173, 177
strain, 125
stretched exponential decay, 34, 34
stripe contact, *see* contact
sub-threshold turn on, 64, 67
sublinear light-vs.-current characteristic, 391
superlinear light-vs.-current characteristic, 391
superluminescent LED, *see* LED
surface, diffuse, 154–156
surface-emitting communication LED, *see* LED
surface recombination, *see* recombination
surface recombination velocity, *see* recombination
surface reconstruction, 41
surface roughening, *see* surface texturing
surface states, 42
surface texturing, 154–156, 186–189
surfactant, 117
synchronous optical network, 401

TAG phosphor, *see* phosphor
temperature,
    carrier, 101–111
    characteristic, 99, 111
    dependence of bandgap energy, *see* bandgap energy
    dependence of emission, *see* emission
    dependence of forward voltage, *see* forward voltage
    junction, 101–111
terbium aluminum garnet, *see* TAG
test-color sample, 317–319, 328–331
test illuminant, *see* test source
test light source, *see* test source
test source, 316, 321
tetrachromatic, 344
textured surface, *see* surface texturing
thermal resistance, *see* resistance
thickness, critical, 125, 236
three dB frequency, *see* 3 dB frequency
threshold voltage, 61–63
thresholdless laser, *see* zero-threshold laser
TiO_2, 160

titania, *see* TiO_2
total internal reflection, *see* reflection
    see also critical angle for total internal reflection
transfer function, 394
transparent contact, *see* contact
transparent substrate technology, 20, 156–159
trap, *see* defect
trapped light, *see* light
trichromacy, 293
trichromatic source, 338–344
tristimulus value, 294
truncated inverted pyramid, 152, 153
tunable resonant cavity, *see* resonant cavity
tunnel junction, 132
turn-on voltage, *see* threshold voltage

ultraviolet, *see* UV
uncertainty principle, 10, 11
uniform chromaticity coordinates, *see* chromaticity coordinates
uniform color space, *see* color space
Urbach energy, *see* Urbach tail
Urbach tail, 130, 146
UV, 222, 223
UV-A radiation, 222
UV-B radiation, 222
UV-C radiation, 222
UV, deep, 222, 223
UV, extreme, 222, 223
UV LED, *see* LED
UV, near, 222, 223
UV, vacuum, 222, 223

van Roosbroeck–Shockley equation, 52
van Roosbroeck–Shockley model, 50–54
Varshni formula, 104, 106
    see also temperature dependence of bandgap energy
vertical-cavity surface-emitting laser, *see* laser
vertical transition, 89
visible-spectrum LED, *see* LED
vision, *see* human vision

wall-plug efficiency, *see* efficiency
wake-sleep rhythm, *see* circadian rhythm
waveguide, *see* fiber
waveguided modes, 186, 371, 389
wavelength converter, 346–365
wavelength, dominant, 300, 301
wafer bonding, 157, 158
white LED lamp, 332–344, 346–365
    based on semiconductors, 19, 332–344

based on wavelength converters, 19, 346–365
Wien's law, 307, 308
window layer, *see* current-spreading layer

YAG phosphor, *see* phosphor
yellow luminescence, 39
yttrium aluminum garnet, *see* YAG

Zener diode, 194, 195
Zn-O, 11–13
ZnS, 4, 160
zirconia, *see* ZrO_2
ZrO_2, 260

0.01 rule, 343
3 dB frequency, 393–395, 401–403

Printed in the United States
By Bookmasters